Pleistocene Mammals of Europe

Pleistocene Mammals of Europe

Björn Kurtén

LONDON AND NEW YORK

Pleistocene Mammals of Europe

Björn Kurtén

LONDON AND NEW YORK

First published 1968 by Transaction Publishers

Published 2017 by Routledge
2 Park Square, Milton Park, Abingdon, Oxon OX14 4RN
711 Third Avenue, New York, NY 10017, USA

Routledge is an imprint of the Taylor & Francis Group, an informa business

Library of Congress Catalog Number: 2007020579

Library of Congress Cataloging-in-Publication Data

Pleistocene mammals of Europe.
p. cm.
Orginially published: London : Weidenfeld & Nicolson ; Chicago : Aldine Pub. Co., 1968..
Includes bibliographical references and index.
ISBN 978-0-202-30953-8 (pbk.: acid-free paper)—
ISBN 978-0-202-30946-0 (pbk.: acid-free paper)
1. Mammals, Fossil—Europe. 2. Paleontology—Pleistocene.
3. Paleontology—Europe. I. Title.

QE881.K8 2007
569.094--dc22 2007020579

ISBN 13: 978-0-202-30953-8 (pbk)

Contents

Foreword vii

Part One Faunal Sequence in Europe

1 SETTING THE STAGE 3
2 THE VILLAFRANCHIAN, PRELUDE TO THE ICE AGE 8
3 CHRONOLOGY OF THE ICE AGE 18
4 THE AGE OF INTERGLACIALS 22
5 THE AGE OF GLACIATIONS 28

Part Two Pleistocene Mammal Species

6 INTRODUCING THE PLEISTOCENE MAMMALS 39
7 ORDER INSECTIVORA 42
8 ORDER CHIROPTERA 51
9 ORDER PRIMATES 58
10 ORDER CARNIVORA 62
11 ORDER PROBOSCIDEA 130
12 ORDER PERISSODACTYLA 139
13 ORDER ARTIODACTYLA 153
14 ORDER RODENTIA 191
15 ORDER LAGOMORPHA 226

Part Three The Changing Fauna

16 THE SPECIES PROBLEM IN THE QUATERNARY 237
17 SIZE AND NUMBERS 244
18 ORIGINATION OF SPECIES 253
19 FAUNAL TURNOVER 260
20 ANIMAL GEOGRAPHY 265
21 MAN AND THE FAUNA 270

Appendix Stratigraphic Range of Species 275

References 286

Index 304

Foreword

THE mammals of the Pleistocene are but now being discovered by the evolutionists. Here we get the sought-for tie-in between zoology and palaeontology: many Pleistocene mammals are still in existence and may be studied as living beings against the background of their own fossil history, extending back for millennia through geological time.

But the Pleistocene of Europe and its mammalian fauna is of interest to many others beside the zoologist, the evolutionist, and the palaeontologist. It is of exceptional importance to the student of human origins and of Stone Age archaeology. Among others who may profit from a study of this topic may be mentioned the geologist, the palaeoclimatologist and the palaeogeographer, and indeed any one who takes an interest in natural history and wishes to re-enter a colourful past.

The first version of this book was a semipopular paperback in the Swedish Aldus series. The present edition is completely rewritten and greatly expanded, but I have retained the non-technical approach to make the story accessible to readers with varying backgrounds. The first part of the book is an outline of the Pleistocene history of Europe, with its climatic changes and succession of mammalian faunas. In the second part are listed all the species of Mammalia known from the Pleistocene and Postglacial of Europe, with the evolution, range in time and space, and mode of life set down for each species, as far as known. The final part is an evaluation of the story in terms of evolution and palaeogeography.

The species is the pivot of the present treatment. The species is, of course, the most important taxonomic category in the study of evolution [195], and to write a comprehensive zoology of all the species of an entire geological epoch struck me as an exciting and challenging project. In most palaeontological texts the genus is the basic category, not the species. With increasing precision of study the species is gradually taking over this role, and the greatest progress is being made in the case of the Quaternary, for which we have a more detailed record than for the more distant past.

Originally the study had been planned to cover only the Carnivora, but it was soon found necessary to introduce material on the other mammalian orders; in the end equal emphasis was given to all of them. It should be noted, however, that much of what is said about carnivores here is based on original research, while this is not the case as regards the other orders of mammals.

Thanks to a three-year travel and research grant from the University of Helsinki I have had opportunity to study most of the larger collections of European Pleistocene mammals on the spot. Much of the work was done during two long sojourns in England in 1961 and 1962, during which Dr K. A. Joysey (Cambridge) and Dr A. J. Sutcliffe (London) provided most valuable assistance and criticism. I have also profited greatly from stimulating discussions with Dr K. Kowalski (Kraków). Dr M. Crusafont Pairó (Barcelona) kindly checked the material on Spanish fossils and provided important new information. Many other colleagues have contributed by showing me the collections in their care, and giving generously of their time and advice. Among these I wish particularly to mention Drs K. D. Adam (Stuttgart), C. Arambourg (Paris), A. Azzaroli (Florence), H. Bohlken (Kiel), M. Degerbøl (Copenhagen), K. Ehrenberg (Vienna), E. W. Guenther (Kiel), D. A. Hooijer (Leiden), J. Hürzeler (Basle), V. Jaanusson (Uppsala), H. D. Kahlke (Weimar), F. E. Koby (Basle), J. P. Lehman (Paris), K. P. Oakley (London), H. E. P. Spencer (Ipswich), D. Starck (Frankfurt), L. Thaler (Montpellier), E. Thenius (Vienna), H. Tobien (Mainz), J. F. de Villalta Comella (Barcelona), R. West (Cambridge), P. Woldstedt (Bonn) and H. Zapfe (Vienna). Mrs Sonia Cole read a preliminary draft of the text and made valuable suggestions for improving it. To all these persons and institutions I wish to express my sincere gratitude. Finally, I wish to dedicate this book to three inspiring teachers: Pontus Palmgren (Helsingfors), Birger Bohlin (Uppsala) and George Gaylord Simpson (Cambridge, Massachusetts).

Helsingfors BJÖRN KURTÉN

Part One
Faunal Sequence in Europe

Chapter 1

Setting the Stage

A HUNDRED MILLION years ago, during the reign of the dinosaurs, the earth was a moist, warm planet under a tropical sun. Much of Europe was flooded by the Chalk Sea, teeming with reptilian life, while ponderous monsters moved slowly over the endless beaches and plains of the low-lying land. This was the crest of one of the great heat waves in geological time. It lasted more than 200 million years.

To find the trough preceding it we have to move back into Permian times, 250 million years ago and more; there once again we find the presence of continental glaciers, a sight which was to become so familiar during the period with which we are concerned, the Pleistocene. After Permian times the temperature curve ascended gradually and the face of the globe became more uniform. Mountains were eroded and as they were planed down the landscape became more and more monotonous. The sea level rose and the continental margins were flooded. The pace of geological processes had become imperceptible, as though in a world left over to entropy.

After this peak the temperature curve began to trend downward again. The sea, which had spread halfway across the continents, very gradually, and with many halts and reversals, began to recede. The coastal lagoons dried up and new land was laid bare. Once more the inner forces of the earth ground into gear and new mountains were built up. The poles cooled off more rapidly than the tropics and once again the earth was girded by distinctive climatic belts. The giant reptiles became scarce and finally none were left. The mammals, which had led a timid life in the background, developed larger and more varied forms. Thus the Cretaceous passed into the Tertiary Period, 65 million years ago.

Still the temperature fell. This is the basic theme throughout the Tertiary, the background against which we must view the long, complicated evolutionary history of the mammals: over millions of years the temperature fell, the sea retreated and the mountain chains rose. These tendencies fluctuated in strength and direction, but they were always there. Out of the great Mediterranean sea that extended from Europe through Asia in the beginning of the Tertiary, known as the Tethys Sea, there arose a mountainous archipelago which became consolidated and grew higher. Eventually the first snow glittered on its highest peaks and the Alps, the Himalayas and the American Cordillera were born.

In Europe the climate was subtropical and as late as 15 million years ago lush jungles covered most of the continent. But as the temperature continued to fall, about 10 million years ago in the early Pliocene there came a sharply marked shift. The climate became drier and great grasslands, savannas and steppes spread over the continent.

To us who observe it in the foreshortened perspective of the geological time scale, the shift appears dramatically sudden; but it must have gone on for many thousands of years and to the living beings of the time it would not have been noticeable. It is easy enough for us to observe the descent of the temperature curve – by ingenious methods it can actually be directly measured [74]. Yet the change in the face of the earth at that time was slow in comparison with what was to follow during the Ice Age: this was like an explosion, a total revolution in the tempo of geological events. The world we now live in is the world of the Ice Age. Even if we disregard the influence of man it is a world which is dynamically changing at an abnormal rate; it is in a state of flux unequalled since the Permian Ice Age more than 250 million years ago.

Measured by our everyday standards even this tempo may seem slow. Yet we are now able within a single generation to observe such sub-phases of the climatic evolution as the evident amelioration during the first half of this century. This is typical of the Ice Age. Temperature fluctuations of a kind requiring millions of years in the Tertiary now take place within thousands of years. The rate of change has been intensified by a factor of a thousand or more.

What actually did take place at the end of the Tertiary? We might perhaps say that a 'short-wave' climatic oscillation with a wave-length of perhaps some 50,000 years was superimposed on the 'long-wave' curve with a wave-length of 250 million years. It has been suggested that the short-wave factor is always present but that its influence becomes effective only at sufficiently low basic temperatures [318]. There would be a threshold effect: when the long-range trend approaches a critical minimum, the dips in the short-wave curve become exaggerated, a threshold is reached and inland ice is formed. The crests of the short-wave curve correspond with the interglacial phases when the ice melts.

If this is correct, the series of glaciations of the Ice Age should be preceded by a series of merely cool fluctuations. Actually something of this kind may be observed during the long prelude to the Ice Age called the Villafranchian.

The causes of the Ice Age are still under debate. There are many theories, some of them extremely ingenious and plausible, but none has so far won general acceptance. That the Ice Age is connected with the general processes of geological evolution seems probable, because continental ice sheets seem only to form at times of intense mountain building [295]. During such periods the sea level tends to recede (the geological term is regression, while a rise of the sea level is called a transgression). This may be due to sinking of parts of the ocean floor. As a result the continental blocks emerge out of the water and since dry land does not retain heat as efficiently as water, the heat loss of the earth will increase. Mountains are pushed up to meet the snow line and perhaps to affect the atmospheric circulation [220]. All these factors, as well as many others, may contribute to the making of an Ice Age. For recent discussion of these problems see [43; 75; 77].

We divide geological time into Eras; the Era we live in is called the Cenozoic. It is, in turn, divided into two Periods, the Tertiary and the Quaternary, and these into Epochs. The Tertiary Epochs are the Paleocene, Eocene, Oligocene, Miocene and Pliocene; the Quaternary Epochs are the Pleistocene and the Holocene. The Pleistocene coincides with the Ice Age and the Holocene is the Recent Epoch of geology, usually taken to represent the last ten thousand years. The Holocene is also often called the Post-glacial; in old literature the terms Diluvium (for the Pleistocene) and Alluvium (for the Holocene) may be found.

An Epoch may be further subdivided into Ages. There are three Pleistocene Ages: the Early (Lower), the Middle, and the Late (Upper) Pleistocene. The Early Pleistocene is often called the Villafranchian (after the town Villafranca d'Asti in Piedmont, Italy, where rocks formed in this phase were first characterized). Villafranchian fauna is mainly of Tertiary type, but at this time we also find the first evidence of climatic oscillations of a type suggestive of the Ice Age [68; 175; 191]. Towards the end of the Villafranchian, and especially in the earliest part of the Middle Pleistocene, the warm oscillations have a pattern typical of real interglacials; on the other hand it is only in the later half of the Middle Pleistocene that the cold oscillations assume fully glacial proportions. In the late Middle Pleistocene and the Late Pleistocene there was a regular alternation between glaciations and interglacials.

A shorter break in a glaciation, when the ice sheets melted only partially, is termed an interstadial. But the warm oscillations came in all sizes and there is a complete gradation between an interstadial and an interglacial.

The geological sediments give important information on the climatic conditions under which they were formed, both by their own nature and that of the fossil organic material that they may contain [46]. The principle is self-evident in the case of fossils of plants or animals which are still in existence and the climatic requirements of which we are familiar with. A particularly useful method is the study of the microscopic pollen flora contained in many sediments. Such studies have revealed, for instance, that each interglacial had its own individual climatic and vegetational history, so that an interglacial may be identified on its pollen profile [307]. In general the lowermost sediments in an interglacial deposit will contain a vegetation of cool type dating from the phase during and immediately after the melting of the glaciers; later on, warmth-loving plants and animals immigrate during the climatic optimum of the interglacial; and at the end the cold flora and fauna return with the deterioration of climate heralding the next glaciation.

Many mammals are important as climatic indicators. In the interglacials southern forms like hippopotami and monkeys may invade Europe, but mainly our interglacials are characterized by a temperate-type fauna with deer, boar, elephants and rhinoceroses of woodland type. The glacial deposits, in contrast, carry a fauna of tundra or taiga type with woolly mammoth, woolly rhinoceros, reindeer, lemming and arctic fox.

Even sediments devoid of fossils may be indicative of past climates [46]. Morainic deposits of various kinds are formed directly by glaciers. Eskers, which are terminal moraines and were formed at the ice margin, are particularly conspicuous features in a landscape recently moulded by the inland ice like that of Scandinavia today. Of greater geological importance, because of the vast areas that they cover, are tills; these great sheets of glacial material are much more persistent than terminal moraines. Alternation between tills and 'warm' deposits gives direct evidence of climatic fluctuation.

Other 'cold' deposits are loess, a wind-blown dust resulting from frost weathering, then trapped by steppe vegetation; and solifluction deposits which form at the foot of slopes by the effect of spring floods on partly frozen earth. Sediments of this type are characteristic of the area immediately surrounding the land ice, the so-called periglacial zone. 'Warm' deposits, on the other hand, include fossil soils, formed by chemical agencies during warm intervals.

In glacial times a great volume of water was bound up in the inland ice. Since the water originally came from the sea through evaporation, the sea level must have gone down considerably, probably 400 ft. or more [110]. The effect on river deposits is interesting. During regressions the rivers cut down and formed deep valleys. Many flooded valleys of this type, dating from the last glaciation, are easily spotted on a map. During transgressions, on the other hand, rivers built up their beds by delta sedimentation. Since the oscillation of the sea level is superimposed on a long-range, continuous regression, these shifts in the history of the rivers were enacted at successively lower levels for each new interglacial; the net result is a series of interglacial river terraces, each of which may be correlated with its corresponding interglacial shore line (or 'raised beach').

Higher up the river the history becomes somewhat different. During glaciations the streams are overloaded with gravels resulting from frost weathering; unable to transport such masses of rocks, the rivers build up their beds. In interglacial times, on the other hand, the water supply is rich and frost weathering negligible, so that the rivers will cut down, the glacial bed remaining as a terrace. Here too the fluctuations appear at a successively sinking level, because the erosion gradually dissects the landscape, scours out the river valleys, and causes the ground water level to sink. Thus the river terraces are glacial near the sources of the rivers and interglacial near their mouths. This is of great importance to the palaeomammalogist, since fluviatile deposits may contain great numbers of fossil bones.

Chapter 2

The Villafranchian, Prelude to the Ice Age

THERE is no definite agreement at present regarding the boundary between the Pliocene (the last of the Tertiary epochs) and the Pleistocene. Originally, only what we now call the Middle and Late Pleistocene were included in the Pleistocene, while the Villafranchian was regarded as the Late Pliocene. However, evidence soon accumulated to indicate that the climate had been fairly cold in the Villafranchian, or at least part of it, so that it would seem more natural to include it in the Pleistocene [92; 202]. In addition, the mammalian fauna of the Villafranchian contains a number of modern forms, which are absent in the Pliocene proper [111]. Among the most important are the one-toed horses (genus *Equus*), the true elephants and mammoths, and the cattle-like bovids (genus *Bos* and related genera like *Bison* and *Leptobos*). The true horses were migrants from North America, and entered the Old World across the Bering Strait, while the elephants and cattle evolved in the Old World.

Deposits containing mammalian fossils of Villafranchian age have been found at many sites in Europe. Some of the best known lie in central and southern France and northern Italy and include both fluviatile and volcanic deposits. In addition, Villafranchian cave deposits are common in a belt further to the east, especially in Hungary and neighbouring areas. The correlation between the eastern and western localities is sometimes difficult, for the cave sediments mostly contain the bones of small mammals, whereas larger forms are predominant at the open-air sites.

Comparison between the Villafranchian faunas of various sites often reveals considerable differences, partly due to local factors but partly also because they may differ in age. It is now thought that the Villafranchian lasted a very long time, perhaps as much as two million years, and that many local assemblages represent but brief stages in this history. Such local assemblages may be arranged in a chronological sequence, in which there is a gradual modernization of the fauna as new species evolve or immigrate and old ones die out. The earliest Villafranchian faunas strongly resemble those of the immediately preceding stage, the Late Pliocene or Astian; while the latest Villafranchian faunas are transitional to those of the Middle Pleistocene [58; 64; 84; 175; 245; 302].

When a historical event can be analysed in detail, it is often found that a seemingly dramatic and sudden change may in fact consist of a series of separate events. This is also true of the transition from the Astian to the Villafranchian – or, in other words, from the Pliocene to the Pleistocene. Paleontologists often talk about 'faunal waves' to describe the rejuvenation of the animal stocks in the transition from one stage to another and one might easily picture this as a sort of breaker, crashing into the scene. In actual fact the wave is built up gradually as new immigrants are added, while some local forms die out and others continue their evolution.

The Astian faunas in Europe, known from various sites in northern Italy, from Roussillon (Perpignan) and Montpellier in France, from Malusteni in Roumania, and various sites in Hungary, were typical forest faunas with mastodonts, deer and other woodland animals; the flora was subtropical [226]. Once more the climate had reverted to a more oceanic type after the drier interval of the Middle Pliocene.

If we proceed now to the earliest Villafranchian faunas, as found at Villafranca d'Asti in Piedmont [175], at Vialette, Le Puy [278] and the Etouaires Ravine at Mt Perrier near Issoire (Puy-de-Dôme) [241; 243; 302], we again find mostly forest animals. Many are of Astian type, for instance the deer (many species), mastodonts, a rhinoceros, a tapir. But on the other hand various new forms appeared, such as the cheetah, the hunting hyena and various antelopes. Out of the three Pleistocene guide fossils mentioned above, only one was present at this stage: the large bovids, represented by the genus *Leptobos*. True horses and true elephants had not yet arrived.

As our knowledge of the Astian fauna increases, the essential continuity between it and the early Villafranchian fauna may stand out even more clearly. For instance, the raccoon-dog and the Perrier Hyena, hitherto regarded as new Villafranchian elements, are now known to have immediate ancestors in the Astian of Europe.

It cannot be decided as yet whether these early Villafranchian forest faunas are climatically different from those of the Astian. Was this a temperate forest of Pleistocene type, or a subtropical forest as in the

Astian? In the marine deposits from Italy, which may be correlated with the Villafranchian, there suddenly appear shells of a type nowadays found in the North Atlantic, but not in the Mediterranean; this must point to a great deterioration in the climate [86]. Fossil floras of Villafranchian date in the neighbourhood of Milan are north-temperate in type with alder, pine, spruce, chestnut, heather and ferns [206]. But it is not yet certain whether the climatic change occurred as early as Etouaires times.

In southern regions, not directly affected by the formation of ice sheets, glaciations may take the form of pluvials or times of increased precipitation. The normal reaction to this is a spread of forests, while grasslands tend to increase in area during the warm, dry interpluvial phases. This is the sort of oscillation suggested in the faunal history of the Villafranchian; so that it is possible that the initial forest episode recorded at Etouaires, Vialette and Villafranca d'Asti might be a pluvial, the first of its kind, foreboding the great continental glaciations that were to follow more than a million years later.

At higher levels of the Mt Perrier (Roccaneyra and Pardines; see figure 1) a steppe fauna has been found [175] and a Spanish assemblage at Villaroya in Logrono [50] seems to be of the same age. Almost all the Astian deer of Etouaires times have vanished and instead there are antelopes, gazelles, and other steppe animals. The horses are particularly interesting. At the beginning of the steppe episode (Roccaneyra, Villaroya) only the ancient, three-toed *Hipparion* is present, a Pliocene relict. At the Loubières de Pardines level, the first one-toed horses migrated into Europe: a great caballine-type horse, *Equus bressanus*. This was the second of the Pleistocene index fossils to invade Europe. The flora and fauna of Villaroya [51] clearly indicate a warmer climate than at Saint-Vallier (see below).

Now follows another forest episode, presumably representing a cool oscillation; it is recorded at Saint-Vallier near Lyon [302]. The same phase is probably also represented at Chagny, Le Puy [61], but here are found also some of the earliest Villafranchian species, so that more than one phase is probably recorded at Chagny (figure 2). The sediment at Saint-Vallier (figure 3) is of loess type, which could suggest periglacial conditions, but it does not seem likely that the climate was really very cold. The fauna indicates a varied environment with woods, some grassland and a rich supply of water (both beaver and otter are present). The flora, with cedar, pine, oak, etc., suggests a temperate climate.

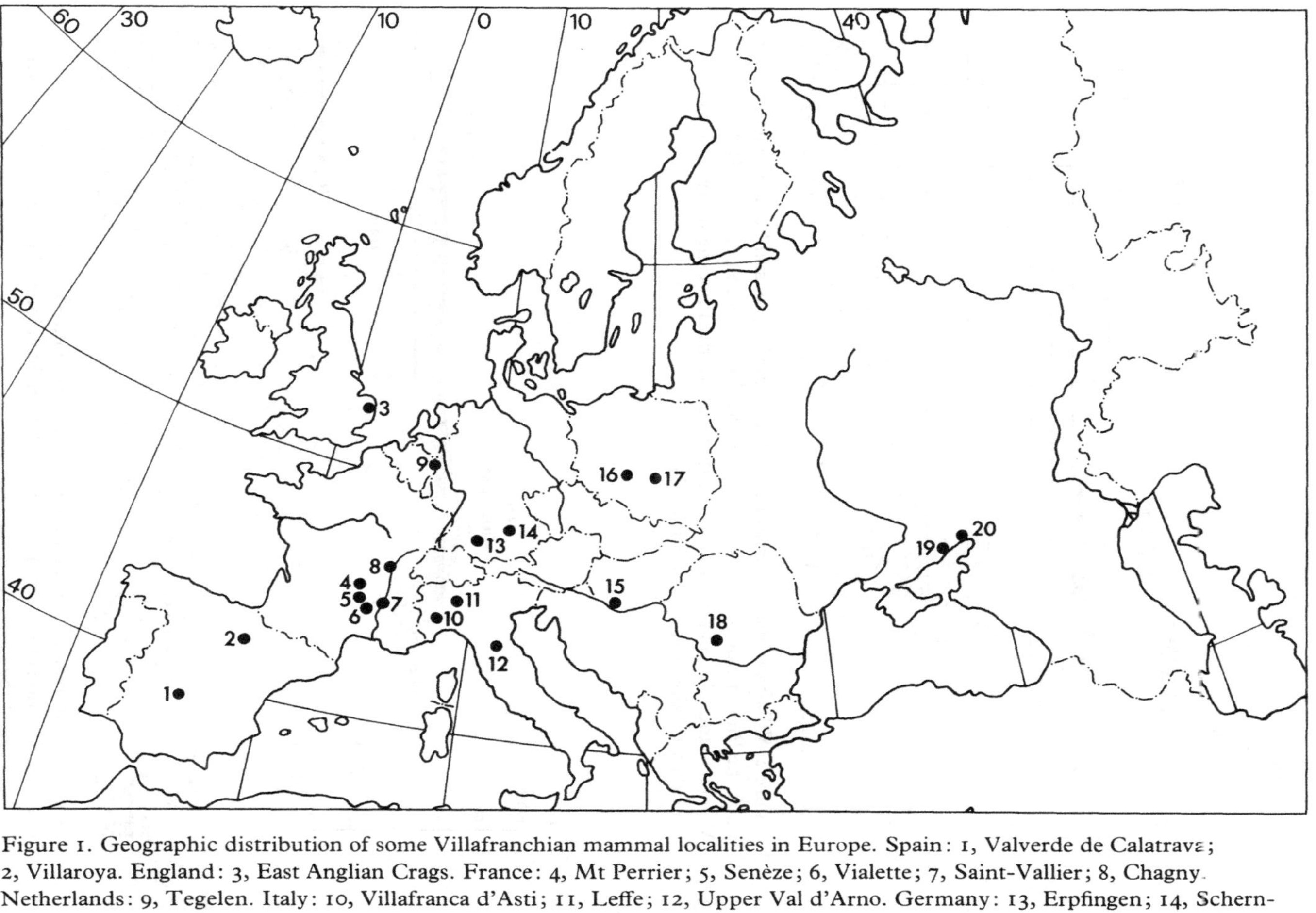

Figure 1. Geographic distribution of some Villafranchian mammal localities in Europe. Spain: 1, Valverde de Calatrava; 2, Villaroya. England: 3, East Anglian Crags. France: 4, Mt Perrier; 5, Senèze; 6, Vialette; 7, Saint-Vallier; 8, Chagny. Netherlands: 9, Tegelen. Italy: 10, Villafranca d'Asti; 11, Leffe; 12, Upper Val d'Arno. Germany: 13, Erpfingen; 14, Schernfeld. Hungary: 15, Villány. Poland: 16, Rebielice; 17, Kadzielnia. Roumania: 18, Oltet. USSR: 19, Mariupol; 20, Taganrog.

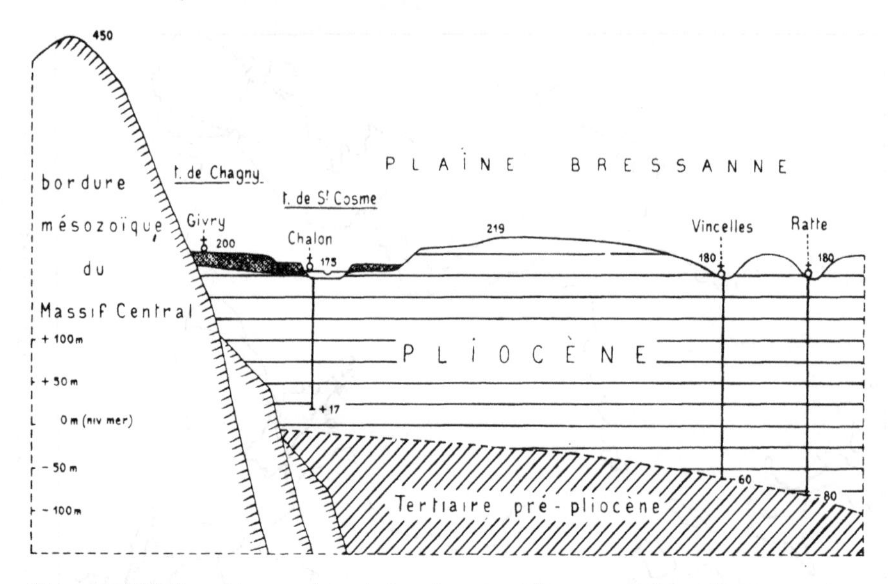

Figure 2. Villafranchian sands and clays (grey) at Chagny and Saint-Cosme, resting upon the Pliocene deposits forming the Bresse Plain. Deep borings at Ratte, Vincelles and Chalon reveal the great thickness of the Pliocene strata. Distance Vincelles-Chalon 40 km. After Bourdier.

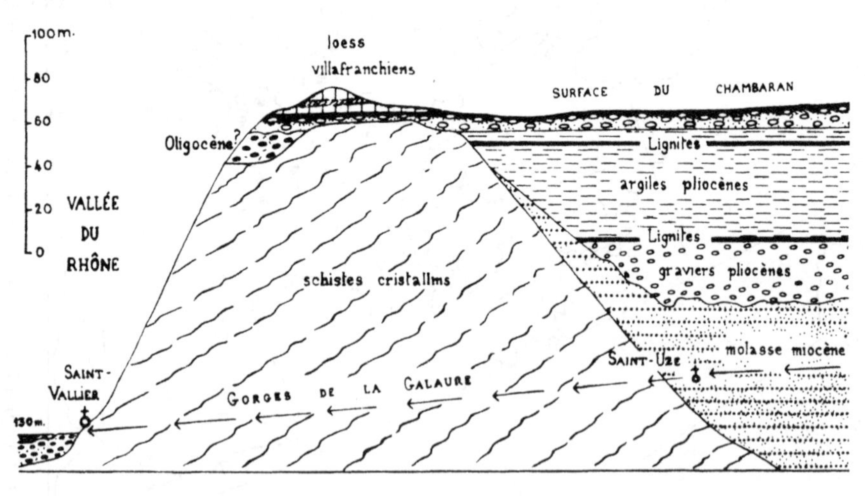

Figure 3. Left flank of the Rhone Valley between Saint-Vallier and Saint-Uze, showing situation of mammal-bearing Villafranchian loess. The section is parallel to the gorge of the Galaure. After Bourdier.

At Saint-Vallier and Chagny the third index fossil of the Pleistocene, the elephant, enters the European scene. Though true elephants appeared in Europe thousands of years later than cattle and horse, in India they actually appeared earlier than *Bos* and *Equus*. This early Indian species was the first known true elephant, *Archidiskodon planifrons*, which sooner or later invaded almost all of the Old World, evolving into local races or species as it did so.

With the rich fauna of Senèze [242] we apparently enter a new steppe phase with a dry, hot climate. Senèze is a village in the department of Haute-Loire southeast of the town Brioude and lies in a basin, the slopes of which are covered with Villafranchian deposits. The volcanic tuffs on the western side are richly fossiliferous, many of the remains being complete articulated skeletons, perhaps of animals killed by gases and ash falls during eruptions. The mighty volcanoes of Auvergne are extinct today, but their peaks may still be identified by the traveller. The fauna of nearby Chilhac [242] is essentially an impoverished version of Senèze and evidently of the same date.

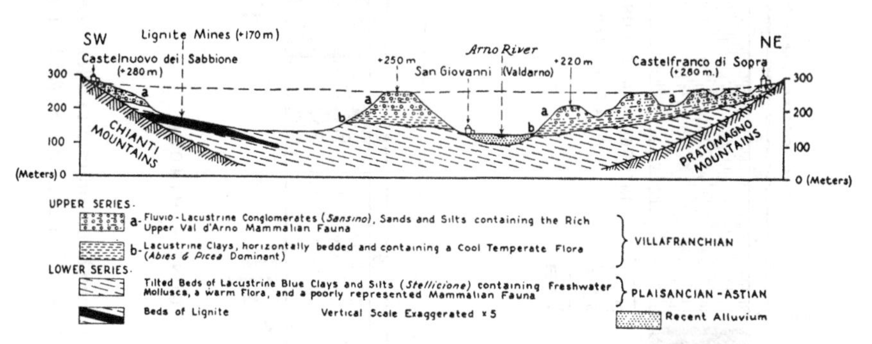

Figure 4. Section through the basin of the upper Val d'Arno. After Movius.

The uppermost Villafranchian is represented at the famous sites of the Magra and Arno river valleys in Tuscany (figure 4). The upper Val d'Arno (upstream from Florence) and Olivola (Val di Magra) are especially rich in fossils of this age, suggesting forested land and abundant water [11; 175; 211; 242]. At another late Villafranchian site, Leffe in Bergamo on the southern slopes of the Alps [191; 264], a direct connection with the Alpine glaciations is suggested. The long stratigraphic sequence at this locality shows a series of cold–warm fluctuations, revealed by the pollen analytical work of Professor Lona, and is capped by deposits from the first great continental glaciation,

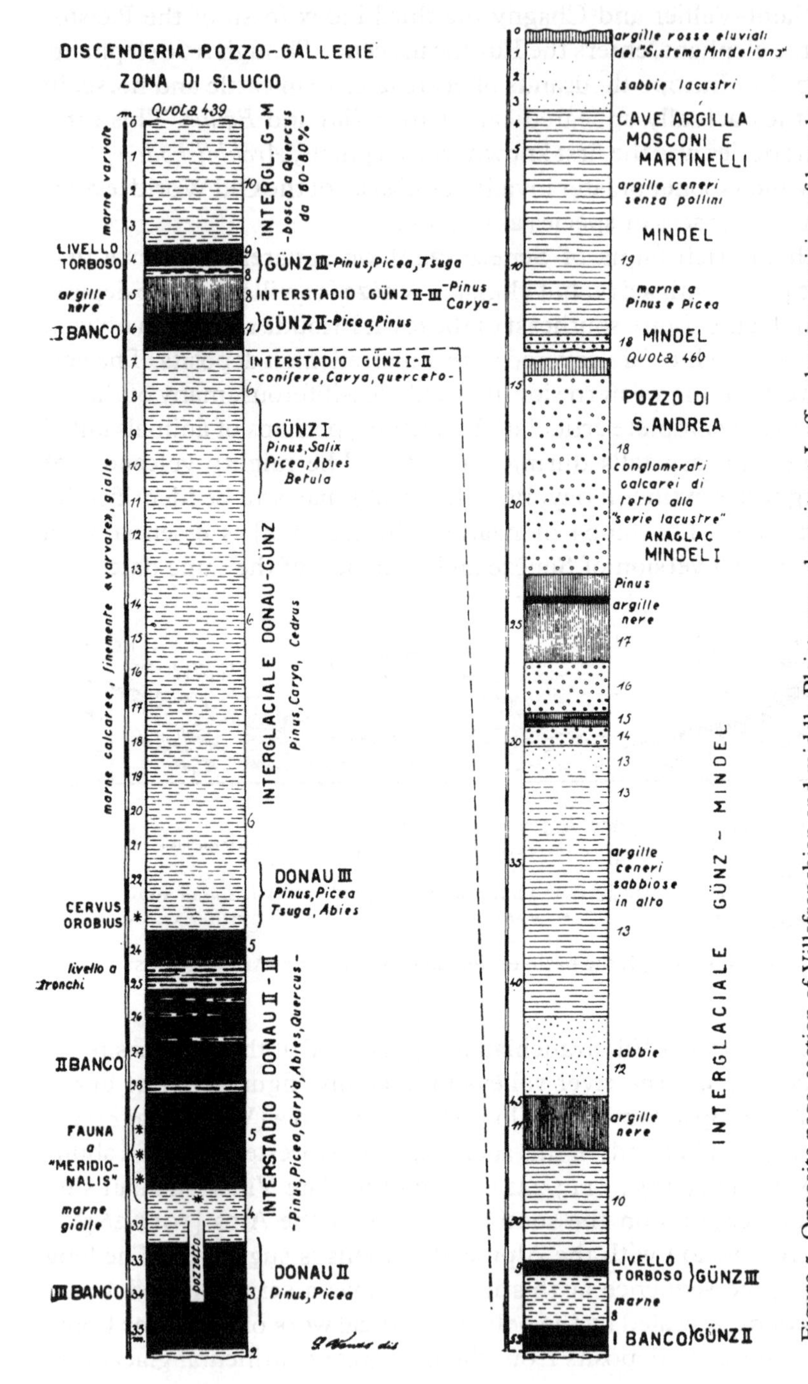

Figure 5. Opposite page, section of Villafranchian and middle Pleistocene deposits at Leffe, showing position of bore holes at S. Lucio and S. Andrea. Above, profiles at S. Lucio and S. Andrea with fauna, pollen flora, and suggested correlations. After Lona.

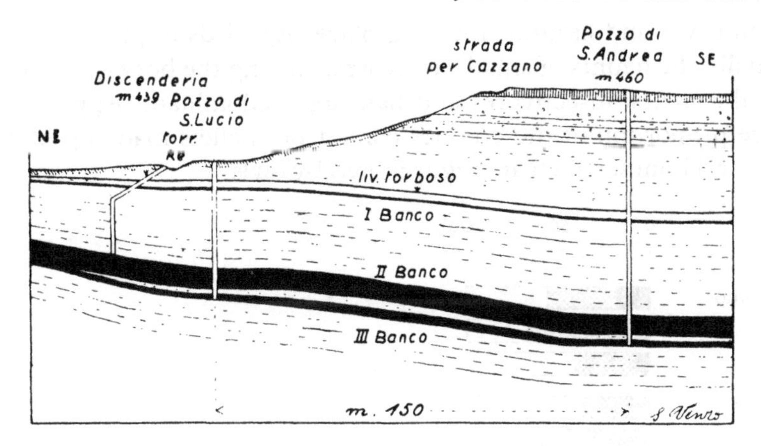

the 2-Mindel (figure 5). The mammalian fauna of Leffe comes from one of the earliest warm oscillations in this sequence. This means that there were several, probably three, interglacial phases after the Leffe and Val d'Arno stage (the two faunas are practically indistinguishable) and before the temperature curve made its first great plunge at the beginning of the 2-Mindel glaciation.

It is possible that the first of these post-Leffe interglacials may be identical with the one represented at Tegelen in Limburg, in the Netherlands. The interglacial clays of Tegelen carry a rich mammalian fauna, which appears slightly more advanced than that of Val d'Arno, and is transitional to the Middle Pleistocene [247; 297]. This is the type locality of the so-called Tiglian interglacial. The clays at Durfort in southeastern France contain a fauna and flora which has been correlated with Tegelen [39].

In the Villafranchian, then, there was an alternation between forest and steppe faunas in southern and southwestern Europe [39; 175] (see figure 6). From the Alps we know of glacial terraces antedating those of the Middle Pleistocene and they are generally regarded as evidence of local glaciation during cold phases in the Villafranchian; they form the so-called Donau (Danube) complex [68].

Certain cave deposits from southern Germany can be correlated with the Late Villafranchian, because they contain remains of large mammals. The Villafranchian cave of Erpfingen [189] has a fauna closely resembling that of Senèze (70 per cent of the Erpfingen species are also found at Senèze, 57 per cent at Val d'Arno). The rich fossil sequence in the cave deposits of eastern Europe is more difficult to compare with the one that has been described here. Most sites were small caves and fissures, roosting places for birds of prey and occasionally the haunts of small carnivores. Among the bones, remains of small rodents, insectivores and bats are predominant. Apparently a large proportion of these bones come from pellets made up of hair, skin and bones, which are regurgitated by owls.

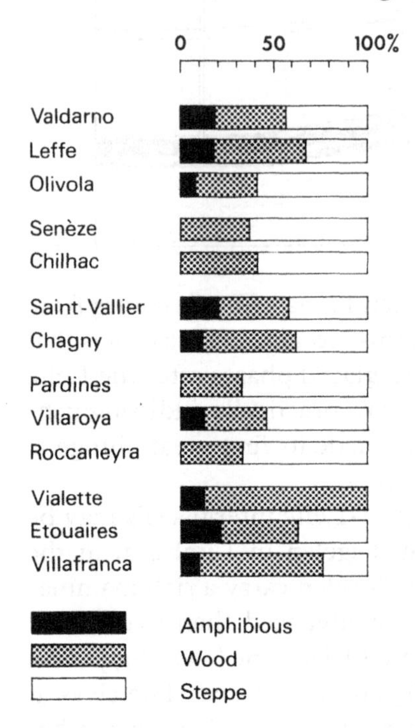

Figure 6. Variation in relative numbers of amphibious, woodland and steppe-living species of mammals at different Villafranchian localities, to illustrate climatic fluctuation. After Kurtén.

It is interesting to see that the caves from various periods in Hungary may often be identified simply on the basis of the orientation of their long axis, suggesting that the strike of the karstic fissuring was affected by tectonic stresses [163]. The early Villafranchian and/or late Astian caves extend in a NNE–SSW direction, the later Villafranchian strike E–W and the early Middle Pleistocene N–S. A great number of ancient fissure fillings at Villány, Csarnóta and Nagyharsányhegy have furnished the basis for a detailed stratigraphical sequence, forming a most important standard of comparison [145; 162; 163]. The local so-called Villanyian faunal stage is approximately, though not exactly, a correlative of the Villafranchian as delimited here.

Late Astian and/or earliest Villafranchian faunas have also come from the Ivanovce fissure near Trencin in Czechoslovakia [82] and the Weze fissure near Dzialoszyn in Poland [152; 261; 270]. At Hajnácka near Filákovo in Czechoslovakia, lake sediments have yielded a sequence of two fossiliferous horizons from the Lower and Middle Villafranchian respectively [83]. The older fauna occurs in limnic sands, while the younger fauna, representing mass killing by volcanic eruption, is preserved in basaltic tuffs. A fissure filling at Podlesice in the Kraków–Wieluń Highlands in southern Poland [149] also contains an early fauna, originally thought to be early Middle Pleistocene in date but later regarded as Villafranchian; recently, however, Kowalski [157] has assigned a Pliocene age to it. A somewhat more recent fauna, perhaps early Villafranchian, comes from the Rebielice fissure in the northern part of the same region [153]. Finally, karst pits at Kadzielnia in the town of Kielce, Poland, contain a fauna that may be ascribed to the Tiglian [150].

Chapter 3

Chronology of the Ice Age

IN the Middle Pleistocene our stage shifts to the north. It is particularly southern Germany and eastern England that will be the focus of our interest, together with the French and Hungarian areas. But before going into the Middle Pleistocene faunal successions, a general survey of the climatic history of the Middle and Late Pleistocene will be useful.

There were at least four different glaciations. They were originally characterized in the Alps by means of moraines, tills, and river terraces, by the geologists Penck and Brückner [215]. To make it easier to distinguish them, the name of each glaciation will be preceded by a number. The glaciations are then as follows, in chronological order: 1-Günz, 2-Mindel, 3-Riss and 4-Würm.

It was also noticed that the glaciations were subdivided into two or more cold phases or stadials, separated by interstadials. Stadials are denoted by Roman numerals; 4-Würm II, for instance, is the second cold phase of the Würm glaciation.

During the glacial phases most of the Alpine mountain arc was covered by a continuous ice sheet, out of which the highest peaks rose as nunataks. The size of the Alpine ice sheet was small in comparison with the tremendous Scandinavian land ice, which extended southwest, south and east from its centre in the Scandinavian mountains; it reached into southern England, central Germany and southern Poland. The history of the Scandinavian ice sheets has been compiled through study of moraines, tills, loess deposits and the alternating 'warm' sediments. It has been possible to distinguish three major glaciations, apparently corresponding to Alpine Nos. 2–4; no certain correlative of the 1-Günz has been found. The Alpine 1-Günz is known to have been less extensive than the later Alpine glaciations; if this was true for the Scandinavian ice sheet also, it is possible that the traces of the first Scandinavian glaciation may have been obliterated by later ice sheets. The later Alpine glaciations have Scandinavian correlatives as follows: 2-Mindel and 2-Elster; 3-Riss and 3-Saale; 4-Würm and 4-Warthe-Weichsel (stadials of the last glaciation). Although this correlation is of long standing, the evidence for it remains insecure and many authors suggest that the Alpine nomenclature should not be used except in the Alps [318]. To most readers, however, the use of the Alpine names is probably less confusing.

In North America there is a sequence of four different glaciations, perhaps corresponding to those in the Alps.

The glaciations alternate with warm phases, the interglacials. Obviously there were at least three, but actually the number of interglacials is greater. There is a late Villafranchian interglacial, the Tiglian, which apparently antedates the 1-Günz. In addition there seem to be at least two separate warm oscillations between the Tiglian and the 2-Mindel, suggesting that 1-Günz is divisible into two well-separated stadials. Finally, there seems to be at least one 'extra' interglacial towards the end of the Middle Pleistocene. If we include the interglacial in which we now live, the total will be seven interglacials. They will be numbered alphabetically from A to G in analogy with the glaciations, as follows:

A-Tegelen or the Tiglian interglacial, antedating 1-Günz [297].

B-Waalian, perhaps an important interstadial separating 1-Günz I and 1-Günz II [39].

C-Cromerian, the interglacial between 1-Günz and 2-Mindel [317].

D-Holsteinian, the interglacial between 2-Mindel and 3-Riss [171].

E-Ilford. This is the 'extra' interglacial [271]. It may perhaps be regarded as an important interstadial subdividing the 3-Riss complex.

F-Eemian, the interglacial between 3-Riss and 4-Würm.

G, the Holocene interglacial.

It will be evident that the climatic history of the Pleistocene is highly complicated and it would of course be naïve to think that we now have a definitive chronology; future research will add much to the detail of this picture.

So far nothing has been said about the time scale of this series of events. In the case of the Pleistocene, the problem has been especially difficult. Dating of the geological history is based on radioactive substances which disintegrate and form new elements. In principle the method is to measure the amount of the radioactive element (uranium, for instance) as well as its fission products (in this case, uranium-lead and helium) present in the rock to be dated; when the rate of atomic disintegration is known, the age since the formation of the rock can be calculated. Unfortunately the uranium–lead method is impracticable for the short time span since the beginning of the Pleistocene. On the other hand the well-known radiocarbon method [176] gives excellent dates for the Holocene, but it can only be used for dates going back about 50,000 years so that it does not even cover all of 4-Würm. The main part of the Pleistocene is too young for the uranium method and too old for the radiocarbon method. In this situation resort has been made to extrapolation of sedimentation rates based on a short terminal radiocarbon dated part [75; 77]. Another method of dating uses the radiation curve of Milankovitch [197]. Both methods are open to serious objections.

With the discovery of various new dating techniques, of which the potassium–argon (or K–Ar) method is the most important at the

Table 1

Epoch or Age	*Climatic Phase*	*Date BP*
Holocene	G Postglacial	0–10,000
Late Pleistocene	4-Würm II stadial	10,000–30,000
	4-Würm I-II interstadial	30,000–40,000
	4-Würm I stadial	40,000–70,000?
	F-Eem interglacial	100,000?
Middle Pleistocene	3-Riss II stadial	
	E-Ilford interglacial?	
	3-Riss I stadial	
	D-Holstein interglacial	230,000
	2-Mindel glacial	>400,000
	C-Cromer interglacial	
	1-Günz II stadial	
	B-Waalian interglacial	
	1-Günz I stadial	~1,000,000?
Villafranchian	A-Tegelen interglacial	
	Val d'Arno, forest	
	Senèze, steppe	
	Saint-Vallier, forest	
	Pardines, steppe	
	Etouaires, forest	3·3 m.y.
Astian	Late Pliocene forest faunas	4 m.y.?

moment, it has finally become possible to date the phases of the Pleistocene [208]. At present relatively few dates have been processed, but at least a preliminary time scale is beginning to emerge. Recent work [79] indicates that the Villafranchian and its North American correlative the Blancan may have begun about 3 (in Europe) or 3·5 (America) million years ago.[1] Mid-Villafranchian deposits at Olduvai in East Africa date back some 1·8 million years, while an early glaciation in North America, perhaps a correlative of one of the Donau or Günz stages, has an age of about 1 million years. Italian material from Torre in Pietra, perhaps dating from the end of the 2-Mindel, is dated at about 430,000 years, while a date for the D-Holstein Interglacial gives an age of 230,000 years. The accompanying table gives a summary of the time scale and the main climatic phases during the Quaternary Period. In addition to the K–Ar method, the study of protactinium/ionium ratios (the Pa^{231}/Th^{230} method) is now assuming importance, especially in the dating of deep-sea sediments [228].

[1] In a recent symposium on the origin of man, figures of 3·3 million years for Etouaires and 2·6 million years for Roccaneyra were cited by Dr Curtis from the geochronological laboratory in Berkeley.

Chapter 4

The Age of Interglacials

FROM King's Lynn in the north to Clacton-on-Sea in the south, the coast of East Anglia forms a semicircle. To the north lies The Wash, a shallow and difficult body of water where the tides rush across miles of gravel banks. The geological processes at work here are the same as those that formed the Villafranchian gravel sheets, the so-called Crags, which cover much of the peninsula. Unfortunately the distribution of the Crags is patchy and direct stratigraphic superposition is rarely to be seen, so that the evidence for the relative ages of the various types of Crag is somewhat circumstantial.

The Coralline Crag, probably the oldest deposit, contains corals of the warm Astian seas, animals that today would face certain death in the cold waters of the North Sea. In later deposits the corals vanish, to be replaced by shells of molluscs now inhabiting the northern seas. Boswell's sequence [36] as quoted by Zeuner [318] is:

Weybourne Crag (cold, with 21 per cent Arctic species)
Chillesford Crag }
Norwich Crag } (moderate, 10–11 per cent Arctic)
Newer Red Crag }
Older Red Crag (warm, 2 per cent Arctic)
Coralline Crag (no Arctic forms)

The Crags contain the bones and teeth of land animals which have been washed into the offshore deposits. Unfortunately the impression of this land fauna is somewhat confused because fossils out of older deposits, for instance the Eocene London Clay, may also have been washed out and redeposited with the Villafranchian gravels. The same sort of thing is still going on along the coast of East Anglia as fossils weather out of the Pleistocene cliff and get into the present-day shore gravels together with the bones of modern animals. Even the latter may contain oddities. Mr H. E. P. Spencer, of the Ipswich Museum Association, tells me of an animal jawbone found on the shore just like a Pleistocene bone fallen from the cliff. It turned out to belong to a camel. Perhaps it came from a circus animal that died on a transport and was thrown in the sea.

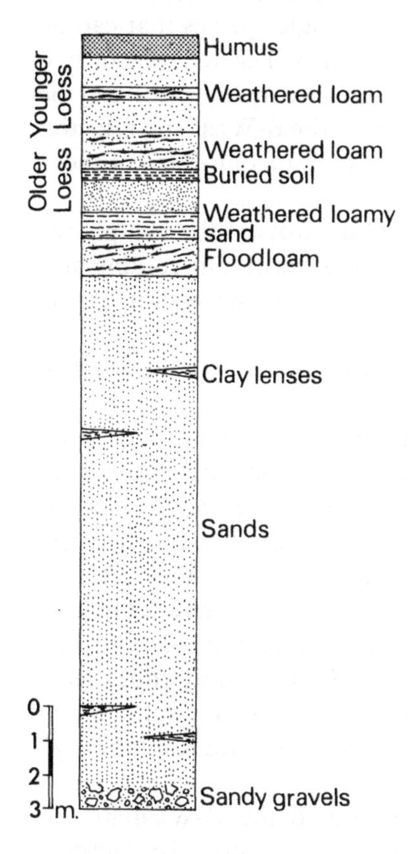

Figure 7. Profile of the Grafenrain pit at Mauer near Heidelberg. The fossiliferous sands are overlain by (1) fluviatile deposits and (2) a sequence of loesses with intercalated weathering horizons. After Müller-Beck & Howell.

The Crags are overlain by the Forest Bed Series, a sequence of interglacial deposits with a very rich fossil fauna and flora [317]. It was suspected that the sequence is heterogeneous and recently West [306; 307; 310] has been able to demonstrate this by means of pollen analytical studies. There are two distinct interglacials with individual floral sequences, separated by a cold oscillation. The fauna of the older interglacial is particularly well represented in the neighbourhood of Bacton on the eastern coast, while the younger fauna may be studied at Cromer further to the north. In both cases the faunas are of a temperate forest type. There are some subtropical elements such as macaque and hippopotamus.

Compared to the Villafranchian faunas, the Forest Bed assemblage is markedly rejuvenated: now, for the first time, modern species in appreciable numbers turn up. Almost all the Villafranchian mammals belong to species that have since become extinct: even the late Villafranchian fauna of Senèze does not have a single species that can be identified with one still living. At the stage of A-Tegelen, one or two species (only a few per cent of the total number in any case) may be identical with species now in existence, but in the B-Waalian of Bacton and the C-Cromer the percentage rose to 30 or 40, one-third or more of the total.

Above the Forest Bed, the thick glacial deposits of the two phases of the 2-Mindel form the high cliff along the coast. The folding and other disturbances of these layers give a vivid impression of the immense forces that were unleashed here.

Faunas dating from B-Waalian and C-Cromerian times are also found in central Europe. The gravel pit at Mauer (figure 7), where the famous jaw of Heidelberg Man was found, has a rich fauna, quite similar to that of C-Cromer [229]. It is a temperate fauna with *Hippopotamus*. On the other hand, the faunas in the gravels of Mosbach near Wiesbaden reflect a colder environment. The upper layers (Mosbach 2–3) are now thought to date from the beginning of the 2-Mindel; the 'main fauna' (Mosbach 2) represents a dry-temperate phase postdating Mauer but antedating the Mindel proper [2]. The lowermost gravels (Mosbach 1) are older than Mauer and would seem to date from 1-Günz II. Thus the actual C-Cromer level, though bracketed by the Mosbach gravels, is missing here. (See Kahlke [123], for Mosbach faunal list.)

Older continental faunas, corresponding to the B-Waalian, are known slightly further to the south. They are characterized by the persistence of the primitive vole *Mimomys*, in which the cheek teeth had closed roots. In modern voles of the genus *Arvicola*, a derivative of *Mimomys*, the cheek teeth have open roots and grow throughout life, enabling the animal to cope with very abrasive food. A few *Mimomys* persisted in the C-Cromerian and later, but the *Arvicola* type was already predominant.

In France, the mammal-bearing beds of Sainzelles at Le Puy [37] and of Saint-Prest northeast of Chartres [40] (figure 8) may perhaps be B-Waalian in date [39], while the Abbeville fauna of the Somme, northern France, appears to date from the C-Cromer [318].

An important group of sites is recorded from the Villány Mountains in Baranya, southern Hungary. Besides Astian and Villafranchian fissure fillings there are numerous Middle Pleistocene sites dating from 1-Günz I, B-Waalian and 1-Günz II [162]. The B-Waalian interglacial seems also to be represented by the rich fauna from Episcopia (formerly Püspökfürdö) at Betfia in the neighbourhood of Oradea Mare on the Roumanian side of the boundary between Hungary and Roumania [146]. Brassó further to the east, in Transylvania, is another well-known site of slightly younger date than Episcopia, perhaps approximately of 1-Günz II date, or transitional between B-Waalian and C-Cromerian.

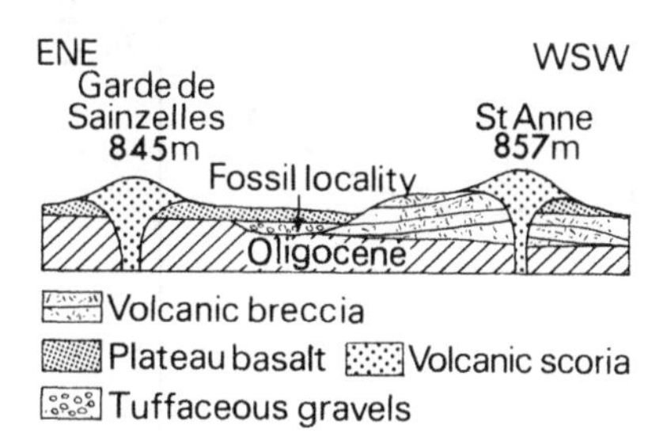

Figure 8. Section showing the stratigraphic position of the fossiliferous gravels at Sainzelles in the Puy-en-Velay Basin, Haute-Loire. After Boule & Movius.

In a cave by Stránská Skála near Brno in Czechoslovakia there is a long faunal sequence partly of this date [205; 263]. Several other cave fillings in Czechoslovakia may also be dated as 1-Günz, C-Cromer and 2-Mindel [82]. A rich fauna from fissure fillings at Gombasek (Gombaszög) contains both large and small mammals [160]. Unfortunately there are several different fissures and some age variation seems likely, but the modal age would appear to be 1-Günz II. This is also the probable age of the famous karst fissure of Hundsheim [314], where a vertical shaft formed a trap for innumerable animals. The Windloch at Sackdilling in upper Pfalz [97] may also be of about the same age, and the same probably holds for the steppe fauna from the gravels of Süssenborn (upper terraces of the Ilm, Thuringia, Germany) [123], where the 'cold' elements of the Ice Age make their first appearance: reindeer and musk ox.

A recently discovered fissure filling at Schernfeld near Eichstätt in Bavaria contains a rich fauna [56] which can hardly postdate the B-Waalian, since *Mimomys* is plentiful. In Poland there are also fissure fillings of early Middle Pleistocene date, i.e. at Kamyk [154]; the fauna suggests a cool climate and steppe vegetation and so perhaps points to a stage of the 1-Günz.

The continental faunas of C-Cromer age give good evidence of climatic oscillation. The sequence begins with steppe faunas of the

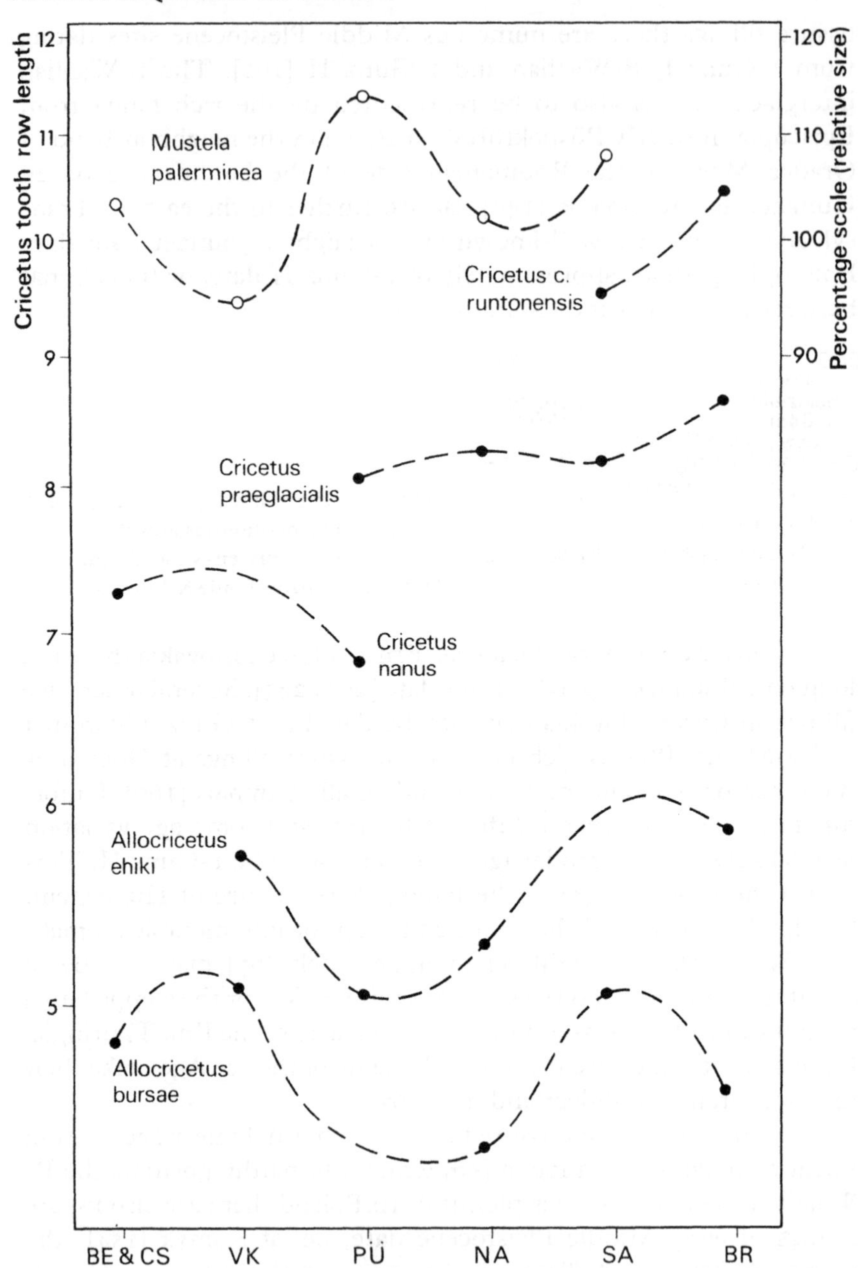

Figure 9. Fluctuation of mean size in five species of Middle Pleistocene Cricetidae and in *Mustela palerminea*. Localities abbreviated as follows: BE, Beremend; BR, Brassó; CS, Csarnóta; NA, Nagyharsány; PÜ, Episcopia (Püspökfürdö); VK, Villány Kalkberg. After Kurtén.

type of Süssenborn and Mosbach 1, continues with temperate forest faunas like that at Mauer, and then reverts to the steppe fauna type of Mosbach 2. Finally, the first great glaciation (the 2-Mindel) makes its advance, and the uppermost Mosbach sands (Mosbach 3) contain a tundra fauna.

At the preceding level, that of the B-Waalian, the oscillation is more difficult to verify. Studies of size fluctuation in some animals may give a clue [173]. Many living species of animals conform to Bergmann's rule, according to which the average body size tends to increase as the climate gets colder. The effect is to reduce the ratio between surface and volume, so that heat loss is minimized. This tendency is found in the numerous species of B-Waalian hamsters (*Cricetus* and related genera). The interglacial hamsters of Episcopia tend to be markedly smaller than members of the same species in the cool phases of 1-Günz I and 1-Günz II. Additional evidence is found in the primitive stoat, *Mustela palerminea*. Recent European *Mustela* vary inversely to Bergmann's rule (size increase to the south) and similarily the fossil stoat varies inversely to the hamsters. The combined evolutionary performances of these two groups are set forth in figure 9. The changes probably indicate a considerable shift of climatic zones; for instance, the size fluctuation in the hamsters is of the same magnitude as the difference between central European *Cricetus cricetus* in the 4-Würm and the G-Postglacial. Again, if the regression of size on geographic latitude in living populations is computed, the changes in the early Middle Pleistocene species are found to correspond to latitudinal shifts of between ten and twenty degrees, or from Hungary well up into Scandinavia [173].

The Middle Pleistocene up to and including the 2-Mindel may be regarded as a separate unit or sub-stage, the early Middle Pleistocene. In the oscillation between cold and warm phases of this sub-stage, the interglacials stand out as the main events while the glaciations of the Günz complex are limited, local phenomena. With the 2-Mindel this sequence – the Age of Interglacials – comes to an end and we enter a world dominated by the great glaciations.

Chapter 5

The Age of Glaciations

THE later Middle Pleistocene presents a strangely impoverished fauna. It is almost as if the great 2-Mindel glaciation had swept everything away. Actually, fossiliferous sites of this age may contain quite a respectable roster of animal species; the problem is that relatively few sites are known, in contrast with the profusion in the early Middle Pleistocene.

The later Middle Pleistocene comprises the D-Holstein interglacial and the 3-Riss glaciation, the latter perhaps including the incompletely known E-Ilford interglacial, probably intercalated between 3-Riss I and II. Most fossils from this time come from river deposits, of which those of the Thames in the London area are of special importance. They lie within the zone dominated by the fluctuations of sea level, so that the terraces here represent warm climatic phases (figure 10). The faunal succession has recently been summarized by Sutcliffe [271].

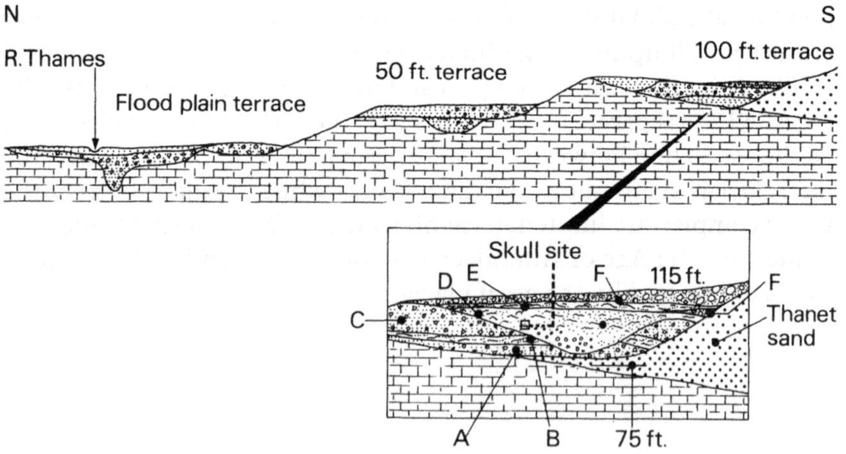

Figure 10. Sequence of terrace relationship in the lower Thames Valley. Inset, filling of Barnfield Channel at Swanscombe with the human skull site. After Howell.

The oldest fossiliferous terrace is the 100-foot terrace, dating from a time when the sea level was about a hundred feet higher than at present. By comparative studies of the ancient shore lines around the world it has been shown that this occurred during the D-Holstein interglacial (sometimes called the Great interglacial, as it was assumed by Penck and Brückner to have been much longer than the C-Cromerian and F-Eemian). The fossils in the gravels of this terrace are remains of animals that lived in D-Holsteinian times, such as the straight-tusked elephant, horse, and Merck's rhinoceros; there is no hippopotamus. In England, the steppe mammoth also occurs occasionally and this like the absence of Hippopotamus may suggest that the climate never became as warm as, for instance, in the C-Cromer interglacial. Important sites include Swanscombe in Kent [212], Grays Thurrock and Clacton in Essex [210; 304] and, outside the Thames area, Hoxne in Suffolk [309]. The last-mentioned site consists of a series of lake deposits and has yielded a complete pollen analytic history of the interglacial. The human skull from Swanscombe is one of the most interesting fossils of D-Holstein age.

The Ilford Terrace at 40 ft. has a fauna that resembles that from Swanscombe in some respects, but it lacks the fallow deer (*Dama clactoniana*) so common in the D-Holsteinian and includes an early form of the true woolly mammoth (*Mammuthus primigenius*) rather than the steppe mammoth (*M. trogontherii*). Merck's rhinoceros is rare, and instead we find the steppe rhinoceros (*Dicerorhinus hemitoechus*) in large numbers [271]. Plant remains from Ilford, however, are of Eemian type, and West [308] refers this terrace to an early stage of the F-Eemian. Whatever the ultimate solution, it can be safely asserted that the E-Ilford fauna is distinct from, and earlier than, the typical (or optimal) F-Eemian fauna.

The 27-ft. terrace takes us to the F-Eemian interglacial and into the Late Pleistocene [271]. In England, this phase is characterized by the complete absence of man (there are no artifacts, at least not from the optimal part of the interglacial) and of horse, while the hippopotamus, absent since the C-Cromerian, is back in great numbers. Plant remains confirm that this interglacial was warmer than the D-Holsteinian [305]. Finds of this age are not uncommon in the centre of London; excavations for a house foundation at Trafalgar Square uncovered a rich fossil fauna including remains of lions that could almost have competed with Lord Nelson's. (Some other European capitals also rest on fossiliferous deposits of F-Eemian date: in Warsaw, for instance, the remains of a great straight-tusked elephant were recently discovered.) Outside London a very rich F-Eemian fauna was found in river gravels at Barrington, Cambridge [113]. Besides Trafalgar Square, localities in the Thames estuary include Brentford, Upnor and Swalecliff (Kent); Brown's Orchard, Acton; and East Mersea (Essex). An early phase of the F-Eemian interglacial (pollen zones b and c of Jessen) is represented in the beach at Selsey (Sussex); here, both horse and human artifacts were found, but the hippopotamus, straight-tusked elephant and steppe rhinoceros had apparently not yet arrived.

The most recent terrace reaches about 15 ft. and dates from the interstadial between 4-Würm I and II [318]. Its fauna, with horse, mammoth, giant deer, aurochs and woolly rhinoceros, indicates a cold climate.

The long sequence of the Thames estuary may be correlated with shorter sequences in other areas. The earliest phase is well represented by the river deposits at Steinheim on the Murr (a tributary of the Rhine) near Marbach in Würtemberg [4]. The basal layers contain a cold fauna with steppe mammoth of characteristic 2-Mindel type, so that they may be correlated with the cold Mosbach 3 horizon. They are overlain by richly fossiliferous interglacial gravels with a forest fauna of the same type as that of Swanscombe (see figure 11): here, too, human remains have been found. Finally there is another cold fauna, dating from the 3-Riss, and showing the transition from steppe mammoth to woolly mammoth.

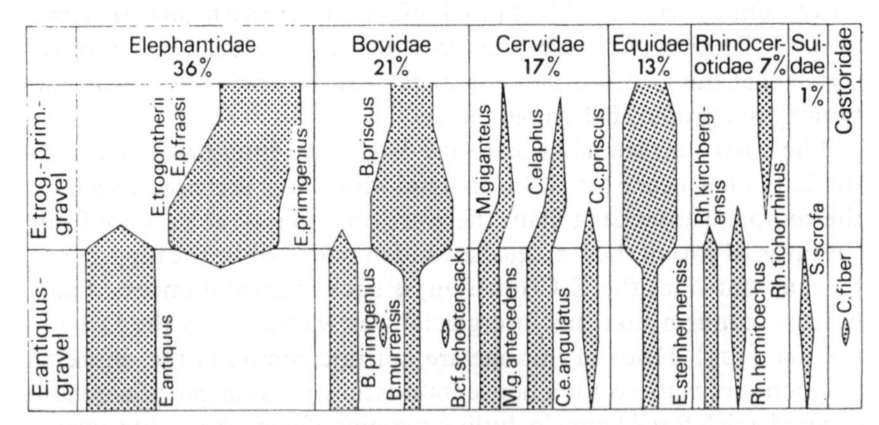

Figure 11. Quantitative distribution of proboscideans, ungulates, and beavers in the interglacial *antiquus* gravels and the glacial mammoth gravels at Steinheim. After Adam.

There are many other instances of fossiliferous D-Holsteinian river terraces, for instance at Châtillon-Saint-Jean on the Isère, northeast of Romans in Drôme [39; 47].

The D-Holsteinian interglacial is also represented in some caves, for instance the cave at Lunel-Viel in southern France [35; 91] and the Heppenloch near Gutenberg in Wurttemberg [5]. Both sites, like Grays Thurrock in Essex, have yielded the otherwise rare Gibraltar macaque.

The stage of E-Ilford is more difficult to identify on the continent, but it may be a correlative of the interstadial between 3-Riss I and II. The glacial terraces in the Saale area [291] have yielded rich faunas dating from both phases of the 3-Riss. It is possible that the earliest deposits in the caves of Montmaurin near Saint-Gaudens in Haute-Garonne [38; 231], made famous by the discovery of a jaw of a primitive human, may be of this age. These deposits fill a shaft, La Niche, and contain a pre-Mousterian industry and a temperate fauna which on stratigraphic grounds must antedate the F-Eemian. The lower terrace at Chelles and Mortières contains a temperate-cool fauna with straight-tusked elephant, steppe mammoth, and Merck's rhinoceros, apparently also of 3-Riss interstadial age.

Late Pleistocene fossiliferous sites, dating from F-Eemian and 4-Würmian times, are numerous. Both river deposits and loess formations in various parts of Europe contain Late Pleistocene fossils. Especially rich finds have been made in the calcareous tufas or travertines which were formed by warm springs during the last interglacial. Examples are found at Cannstatt, a suburb of Stuttgart [255], and Ehringsdorf and Taubach near Weimar [123].

Even more important for our knowledge of the faunal history in the Late Pleistocene, however, are the cave deposits. Bone caves may be formed in various ways. Primitive man was an important bone collector as he used caves for shelter and left the remains of his meals in them, giving us a sample of the animals he used for food. Other important bone collectors are carnivores and scavengers; the role of the raptorial birds, especially owls, in the accumulation of small bones is particularly noteworthy. Finally there are the animals that used caves as dens and, dying there, left their own bones to be buried in the cave sediments; the most important of these are the cave bear and the cave hyena.

Certain caves of this type may have a very long sequence. One of the most remarkable is Tornewton Cave in South Devon [272], where

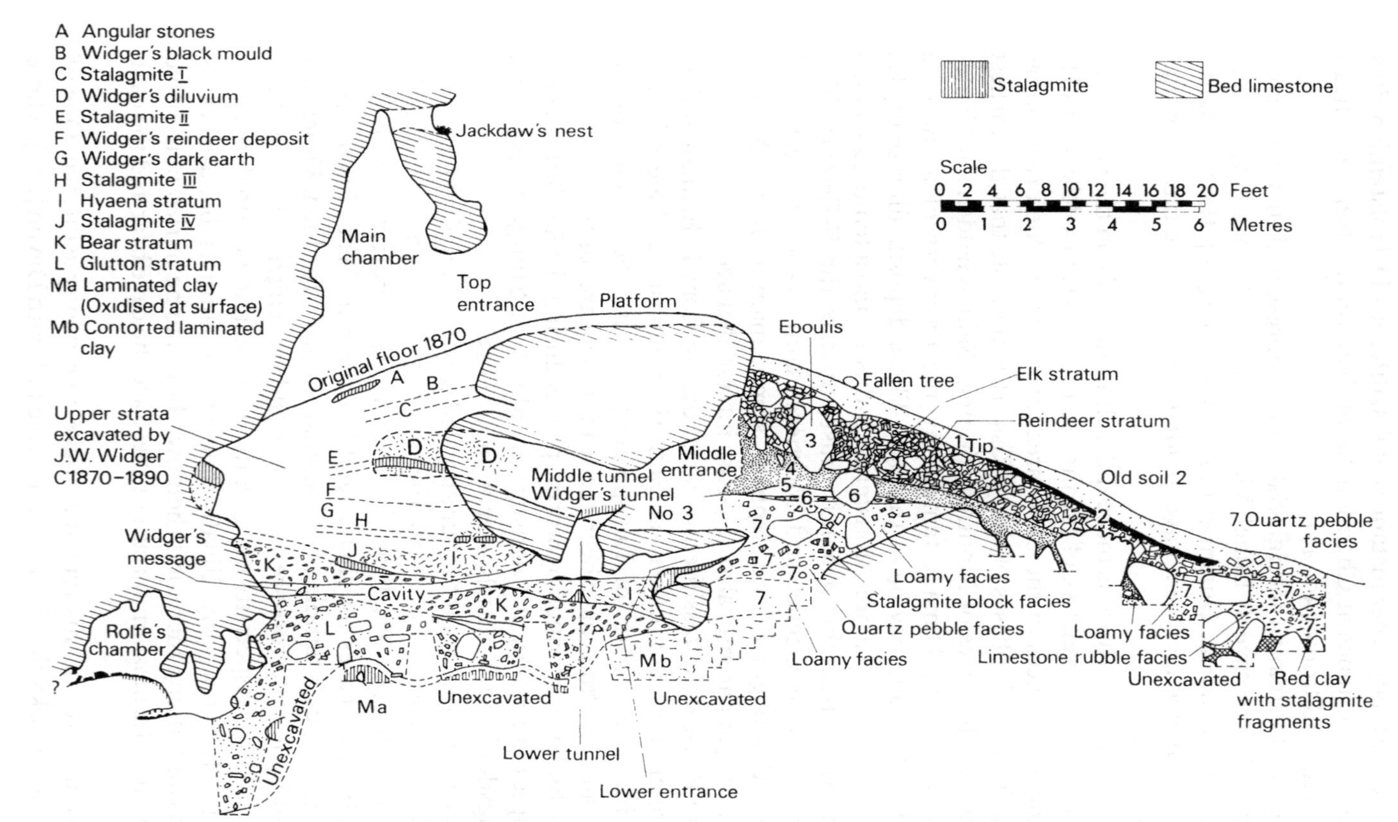

Figure 12. Deposits in Tornewton Cave, S. Devon. Layers in the cave: I, Hyena Stratum, F-Eem interglacial; K-L, Bear and Glutton Strata, 3-Riss phases. In the talus outside the cave are mostly found deposits from the 4-Würm. After Sutcliffe

excavations a century ago disclosed the remains of enormous numbers of hyenas (the excavator, J. L. Widger, estimated some 20,000 specimens). When excavations were resumed after World War II it was found that the Hyena Stratum was but a link in a long sequence going well back into the 3-Riss. That is the age of the two lowermost deposits, the Glutton Stratum and Bear Stratum (glutton is only found in the lowermost stratum, but the bear is common in both). The Glutton Stratum accumulated at a time of severe frost climate in the periglacial of the 3-Riss, whereas the Bear Stratum represents a somewhat milder end phase of the glaciation. Next follows the Hyena Stratum with a typical F-Eemian fauna: hippopotamus, steppe rhinoceros, fallow deer, red deer. Finally there are strata from a series of phases of the 4-Würm, with mammoth, reindeer, woolly rhinoceros, and traces of human occupation; the interstadial Elk Stratum carries, in addition, the woodland cervids elk and red deer indicating a cold-temperate break. (See figure 12.)

Other caves have acted as natural traps, like the Hundsheim fissure already described. There is a good example of this type in south Devon, the Joint Mitnor Cave at Buckfastleigh [271]. Here, among an intricate network of cavities and passages, there is a large cavity containing a great pyramid of debris interlarded with bones. They come from animals that fell down a chimney and were killed by the fall or, unable to scramble out, starved to death (figure 13). It may seem surprising that animals will be clumsy enough to fall into a yawning chasm fully visible to them, but this has actually been observed in many cases. Dr C. K. Brain [41] tells of suddenly coming face to face with a brown hyena in South Africa; the frightened hyena tried to jump the 10-ft.-wide opening of a deep shaft, missed its footing and was dashed to death. The floor of the cave turned out to be littered with the bones of small buck, hares and rock-rabbits. In the same way the great talus cone of Joint Mitnor Cave was built up, filling the cavity gradually, like a gigantic hourglass. The fauna, again, is the well-known F-Eemian assemblage.

A third type of bone cave is exemplified in south Devon by Eastern Torrs Quarry Cave east of Plymouth [271]; in this case the F-Eemian deposits were brought into the cave by running water.

On the continent bear caves are particularly common. Up to 99 per cent of the fossils in a bear cave may be bones and teeth of the extinct cave bear, which used the caves as winter quarters. Famous caves, containing almost incredible amounts of bones of this animal, include the

Figure 13. Restoration of scene at Joint Mitnor Cave during the F-Eemian. In the forest clearing may be seen straight-tusked elephants, cave lion and giant deer. A cave hyena is seen on the overgrown edge of the shaft of the cave, looking down at the decaying body of a bison. Other débris on the accumulating cone include branches of trees, skulls of a hippopotamus and a bear, and isolated bones. Restoration by M. Wilson; after Sutcliffe.

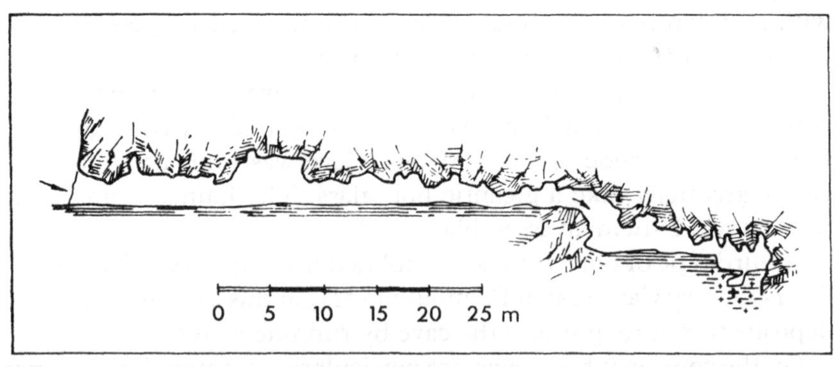

Figure 14. Section of the Gailenreuth Cave, Franconia. Bones of cave bear accumulated especially in the innermost recess marked with a cross. The shaft presumably acted as a trap for bears that entered the cave to hibernate and ventured too far in search of a lair. After Neischl & Zapfe.

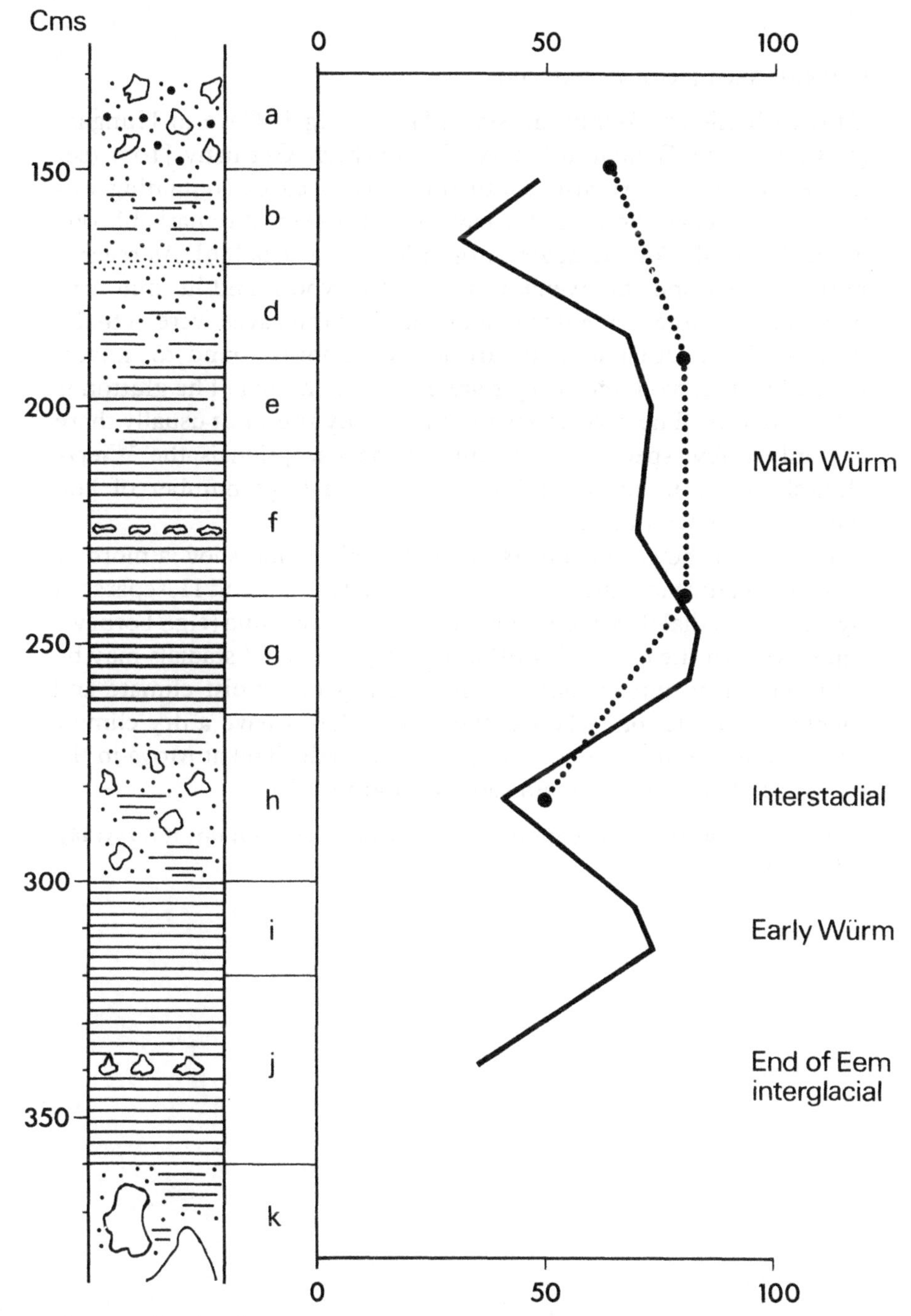

Figure 15. Profile of cave sediments in Cueva del Toll near Moyá, Barcelona, showing lithology (alternating between clayey and sandy-gravelly facies correlated with humidity), pine pollen curve (continuous line) and curve for woodland animals (broken line), as well as suggested correlation. After Donner & Kurtén.

Drachenhöhle at Mixnitz in Styria [1], the Igric Cave in Hungary [198], and the Gailenreuth Cave in southern Germany [100] (see figure 14). There are also continental hyena caves containing immense numbers of bones of the cave hyena; the most remarkable one is the Teufelslucken at Eggenburg in lower Austria [69]. As regards both the bear and the hyena, remains of newborn and juvenile animals form a large percentage, showing that the caves were actually used as dens. Other animals are less commonly found to inhabit caves, but fox, wolf and badger are common enough. The glutton is relatively rare. The cave lion is found at many sites, but usually there are only a few specimens; an interesting exception is the Wierzchowska Cave in southern Poland, where a large number of lion remains have been unearthed [311].

In general the cave deposits of the last glaciation show a more or less distinct division into two cold phases, 4-Würm I and II, separated by an interstadial deposit with a more temperate fauna (e.g. Tornewton Cave). In the caves of southern Europe the cold stadials may be represented by pluvial phases with a temperate-moist climate and increased forestation, whereas the warm phases have a dry climate and a steppe fauna; a good example is Cueva del Toll at Moyá in the province of Barcelona, Spain [60] (see figure 15).[1]

[1] For general surveys of the Pleistocene faunal succession, see [7; 50; 162; 187; 265; 285; 293; 297; 318].

Part Two
Pleistocene Mammal Species

Chapter 6

Introducing the Pleistocene Mammals

THIS part of the book is a survey of the Pleistocene mammalian species of Europe, especially central and western Europe, or approximately the area covered by Van den Brink [269]. The aim is to list every species considered to be valid, with its distribution in time and space, its evolutionary history, its roads of migration and its systematic relationships. It is obvious that this ideal can only be realized for a limited number of species. The larger mammals have been studied for a long time and the Pleistocene record of many of them is tolerably well established. As regards the small mammals, on the other hand, palaeontological study is still in the pioneer stage. We are still in the first flurry of species discovery and description and this probably means that a great number of synonyms are being created. These will be pruned off in time. Meanwhile it is hardly possible to achieve a balanced systematic account. As regards such forms, then, the present contribution is little more than an interim report of the main outlines.

Marine mammals (whales and seals) have not been considered in this book.

Some of the illustrations represent attempts to portray extinct animals in the flesh. The restorations are usually based on complete skeletons, so that the relative proportions of head, neck, body, and limbs should generally be correct. The skeletal anatomy also gives information on the characteristic poise of the animal; but the colour, the thickness of the fur and the presence or absence of fat humps and the like remains hypothetical for most extinct forms. The only exceptions are the occasional but very rare discoveries of cadavers with soft parts preserved and the animals pictured by Stone Age artists; some of the latter are reproduced here. Unfortunately the Palaeolithic hunters were interested only in a highly selected part of the fauna, mainly the large game and the more dangerous predators.

A few words about nomenclature may be useful. The scientific naming of animal species is subject to detailed legislation, aiming at nomenclatorial stability. A species name consists of two parts. The first part, in which the initial letter is capitalized, is also the name of the genus (group of related species) to which the species belongs. The second part, which should never be written with a capitalized initial letter, is termed the trivial name. The two together form the species name, followed by the name of the first describer of the species, thus *Homo sapiens* Linnaeus.

The valid name of a species is the earliest published name, beginning with the tenth edition of Linnaeus' *Systema naturae*, 1758. The result of this rule has occasionally been that names used for many years and well known among specialists have had to be dropped in favour of other names that were published at an earlier date but have been forgotten. In such cases I have given the discarded name, or synonym, in brackets after the valid name.

This procedure, however, has only been followed in the case of well-known synonyms frequently occurring in the literature. Not long ago, when the degree of morphological and size variation occurring in biological species was imperfectly understood, new species names were frequently bestowed on what we would now call local or temporal subspecies, or perhaps only individual variants. The procedure soon resulted in very long synonymies for certain species, thus creating a veritable jungle of meaningless names, through which only the seasoned specialist was able to hack his way. The resulting discredit to palaeontological systematics is not easily forgotten and hasty species-making is now in strong disfavour among most modern students. Instead the trend has now been to recognize broad, inclusive species with a span of morphological variation similar to that found in related species that live today. It should be emphasized that the description of a new species is a responsibility not to be taken lightly.

Subspecies are formally recognized subdivisions of species, representing temporal and/or local populations that are more or less clearly distinguishable morphologically from other populations of the same species. The subspecies name is a trinomial with the subspecific trivial name added to the species name. Individual variants do not have formal names. In this book a few important subspecies are mentioned, but no attempt has been made to list all the probably valid subspecies.

All the domestic mammals are specifically identical with some wild species, even though the latter may now be extinct, as in the case of domestic cattle (the aurochs, *Bos primigenius*). In accordance with recent recommendations [31] scientific names based on the domestic forms have not been used. Hence the use of *Equus przewalskii* rather than *E. caballus*, *Canis lupus* rather than *C. familiaris*, and so on.

There are few vernacular names for extinct species. The ones given here are selected so as to give the reader some idea of what kind of animal is being discussed. It might be thought that a direct translation of the Latin name would do, but that is not always advisable. The first namer of an animal may have been completely mistaken as to the characters of the creature he described, as when a fossil whale was named *Basilosaurus* or Imperial lizard. Despite the error this is a valid name under the rule of priority and cannot be changed. Zoologists have by now learned to use the scientific name as an identification tag only, no matter how it was derived. Still, it must be admitted that we are sometimes bothered by the sad necessity of using the generic name *Mammut* for mastodonts which are not mammoths, or the peculiar fact that the homeland of *Elasmotherium sibiricum* is southern Russia rather than Siberia.

Chapter 7

Order Insectivora

THE Insectivora are the most primitive order of living placental mammals and stand close to the ancestry of all other placental orders. The earliest members of the order appeared in the Cretaceous. In the Pleistocene, insectivore fossils are common in some types of fissures, especially those used as roosting places by owls. The study of these fossils, however, is still in an early stage and much work remains to be done to arrive at a proper understanding of their taxonomy and evolutionary history.

Family Erinaceidae, Hedgehogs

Hedgehog fossils are not particularly common in Pleistocene deposits. All have been referred to the modern genus *Erinaceus*, which arose in the Early Miocene.

The Hedgehog, *Erinaceus europaeus* Linné. Remains of hedgehog are occasionally found in Late Pleistocene cave deposits. There are also reports from the D-Holsteinian (Tarkö) and the early Middle Pleistocene (Schernfeld), but in the latter case the material is fragmentary and specific identification uncertain. Two related species have been described from the Villafranchian and early Middle Pleistocene: *E. praeglacialis* Brunner (Sackdilling; Villány) and *E. lechei* Kormos (Beremend; Chlum). The former was somewhat larger and the latter somewhat smaller than the modern species.

The European hedgehog now inhabits all of Europe except northern Fennoscandia; it ranges eastward through Asia to China. It is highly eurytopic, to be found in the most varied types of environment, but its habits are solitary and the population is always sparse [115].

Family Soricidae, Shrews

The great majority of living and fossil Insectivora in Europe belong to this family. Like the moles they have an elongate skull but the dentition is less complicated and the zygomatic arch is reduced or absent. The soricid family may be divided into two subfamilies, the Soricinae and Crocidurinae; the former have reddish-brown cusps on their teeth, while the latter are white-toothed.

The Pygmy Shrew, *Sorex minutus* Linné. The earliest fossils that have been ascribed to this species date from the Astian (Csarnóta; Podlesice; Weze); if the identifications are correct it is one of the oldest species of mammals living today. It has an almost continuous record throughout the Villafranchian, Middle and Late Pleistocene. At the present day it is distributed over almost all of Europe except the Mediterranean islands and Iberian peninsula. It prefers dry, comparatively open or shrubby ground. It has a wide range in northern Asia [269].

The Common Shrew, *Sorex araneus* Linné. The history of this species in Europe extends back at least to the D-Holsteinian (Breitenberg). Late Pleistocene records are plentiful in central, western and southern Europe. The species has almost the same distribution in Europe as the pygmy shrew but is not found in Ireland. It ranges eastward into Siberia.

This is an unusually hardy and adaptable species and may be found in great numbers in the most varying biotopes including forest, grassland, swampy areas and mountains above timber line.

A number of related species have been described from the earlier part of the Pleistocene, among which may be mentioned *S. araneoides* Heller from the early Middle (Schernfeld; Erpfingen; Sackdilling) and late Middle Pleistocene (Heppenloch) [56].

The Masked Shrew, *Sorex caecutiens* Laxmann. This species is probably a Postglacial invader from the east, although its present-day patchy distribution in Europe is suggestive of relict status [296].

The Alpine Shrew, *Sorex alpinus* Schinz. This species is little known in the fossil state, but there are a few Late Pleistocene and Postglacial records within its present-day range. It has been recorded from the Gaisloch in a possibly 2-Mindel association. Related and perhaps ancestral forms have been described from the early Middle Pleistocene and Astian: *S. praealpinus* Heller (Sackdilling; Schernfeld; Erpfingen) and *S. alpinoides* Kowalski (Podlesice; Weze) respectively.

The alpine shrew occurs in mountainous areas (the Alps, Pyrenees, Carpathians, etc.), mainly in coniferous forests. It is restricted to Europe [149].

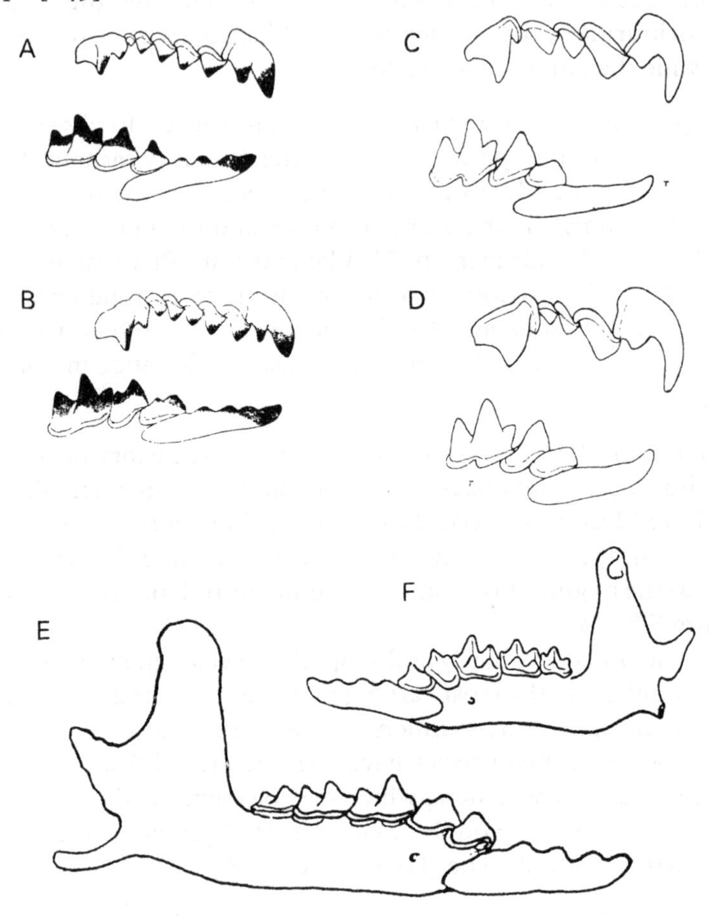

Figure 16. Anterior teeth of A, Common Shrew, *Sorex araneus*; B, Alpine Shrew, *S. alpinus*; C, Common White-toothed Shrew, *Crocidura russula*; D, Bicolour White-toothed Shrew, *C. leucodon*; all Recent. Mandibles of E, Savin's Shrew, *Sorex savini*; F, Runton Shrew, *S. runtonensis*; both from Hundsheim, Middle Pleistocene. All five times enlarged. A–C after Miller; E–F after Kormos.

Kennard's Shrew, *Sorex kennardi* Hinton, and the Runton Shrew, *S. runtonensis* Hinton. Two extinct, closely related species. *S. Kennardi*, a form of slightly smaller dimensions than the common shrew, occurs in the Middle and Late Pleistocene; the most recent finds evidently date from the 4-Würm, while other records are from F-Eemian and D-Holsteinian deposits. It has also been tentatively described from the early Middle Pleistocene (Erpfingen) but generally speaking *S. runtonensis* is the typical form of the Villafranchian and early Middle Pleistocene up to and including 2-Mindel (Koneprusy). It has been found at numerous localities from eastern Europe to England. The earliest finds are Astian (Podlesice; Weze). It differs from Kennard's shrew mainly by its smaller dimensions [104].

The Pearl-toothed Shrew, *Sorex margaritodon* Kormos. This is a comparatively large form, only found in the Villafranchian and early Middle Pleistocene in Roumania and Hungary. It forms a natural group with *S. savini* Hinton and *S. tasnadii* Kretzoi, two even larger forms (subgenus *Drepanosorex*). *S. savini* is found in the early Middle Pleistocene (C-Cromer at West Runton, 2-Mindel at Koneprusy) and perhaps the D-Holsteinian (Breitenberg), while *S. tasnadii* appears as a southern vicar,[1] recorded from Hundsheim, Erpfingen, Gombasek, etc., and the D-Holsteinian of Tarkö [143].

The Water Shrew, *Neomys fodiens* (Pennant). The species is rare in the fossil record but does occur at a few localities as far back as the F-Eemian [119]. It is also recorded from a possible D-Holsteinian association [45]. Its present-day distribution covers most of Europe except the Balkans, Spain and Portugal, the Mediterranean islands, Ireland and Iceland. The species is amphibious and thus a useful paleoclimatic indicator, but its mode of life makes it very rare in cave deposits, while its fragile remains are liable to destruction in river gravels. Several predecessors of this species have been named, for instance *N. browni* Hinton from the D-Holsteinian (Grays Thurrock) and *N. newtoni* Hinton from the early Middle Pleistocene. The genus is also represented in the Astian.

The Beremend Shrew, *Beremendia fissidens* Petényi. A large extinct form, originally described as a species of *Neomys*, which is very common in the Villafranchian and early Middle Pleistocene fissure fillings and so presumably differed from *Neomys* in its mode of life. The earliest finds are Astian (Weze; Gundersheim; Podlesice); it then occurs up to the C-Cromerian at Roumanian, Hungarian, Austrian, Italian, German, Czechoslovak, and Polish localities. Perhaps the

[1] Vicar or vicarious form: a species occupying the same ecological niche as another species not present in the area; for instance the coyote in North America acts as a vicar for the jackal.

youngest record of the species is that from Tarkö, which may date from the early part of the D-Holsteinian. The genus apparently also occurs in the Chinese Pleistocene.

Another relatively common form in the Villafranchian and early Middle Pleistocene is *Petenyia hungarica* Kormos, which ranges from the Astian to the C-Cromerian. *Soriculus kubinyi* Kormos is another much less common extinct form, which makes sporadic appearances from the Astian to perhaps the 1-Günz (Plesivec). Both the genera *Beremendia* and *Petenyia* are now extinct, while shrews of the genus *Soriculus* still live in Asia [142; 149].

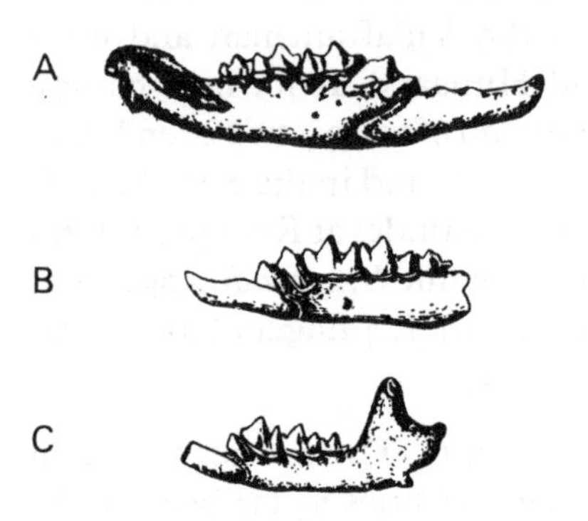

Figure 17. Mandibles of A, *Petenyia hungarica*; B, *Soriculus kubinyi*; C, *Suncus pannonicus*; from Podlesice, Astian. Four times enlarged. After Kowalski.

The Etruscan Shrew, *Suncus etruscus* Savi. This is the smallest living mammal known. The genus is mainly of tropical affinities (Africa, South Asia) and curiously enough contains not only the smallest of all shrews but also the largest (*S. indicus* Geoffroy). The European species belongs to the Mediterranean area. Its fossil history is unknown, but the related *S. pannonicus* (Kormos) is found in Pliocene and Villafranchian deposits in Hungary and Poland. The living form is mainly found in the cork oak forests and neighbouring glades and fields. With the genus *Suncus* we come to the crocidurine shrews, in which the teeth are white [142].

The Bicolour White-toothed Shrew, *Crocidura leucodon* Hermann. A few scattered records from France in the west to Roumania, Hungary and Czechoslovakia in the east indicate the presence of the species in the Postglacial, the 4-Würm, the F-Eemian and early D-Holsteinian (Tarkö). It is now found in a belt extending from France eastward through central and eastern Europe to Turkestan and Persia in the southeast, central Siberia in the northeast. It frequents shrubs and the forest edge, keeping to the dry ground [117].

The Common White-toothed Shrew, *Crocidura russula* Hermann. Separation of the two species *C. russula* and *C. leucodon* is often difficult, but geographic considerations perhaps make it most probable that the fossil forms found in Malta and on the Iberian peninsula represent this species. Other records, mostly uncertain as to species, date from the Late Pleistocene in central and eastern Europe, including F-Eemian finds at Lambrecht Cave and the Binagady asphalt deposits. The species now inhabits western, southern and central Europe and neighbouring parts of Africa as well as Asia Minor. Its biotope is much the same as for the bicolour shrew except that it favours somewhat more open ground [119].

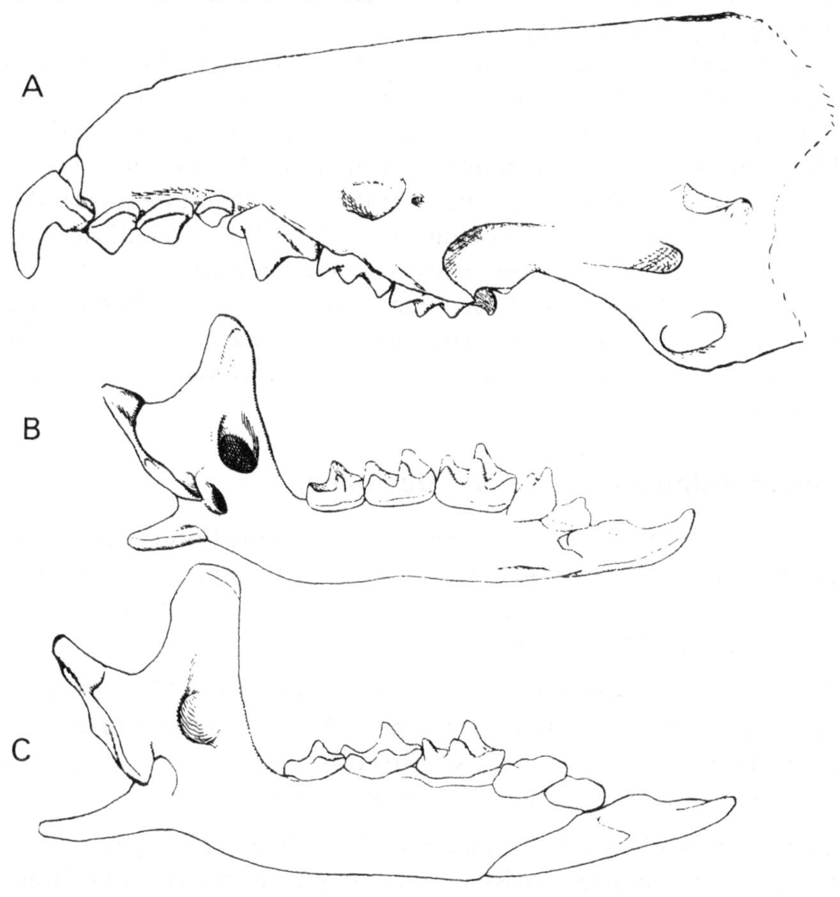

Figure 18. Tyrrhenian Shrews, genus *Nesiotites*. A, C, skull and mandible of *N. hidalgo*, Pleistocene, Mallorca; B, mandible of *N. similis*, Pleistocene, Sardinia; four times enlarged. After Bate.

The Lesser White-toothed Shrew, *Crocidura suaveolens* Pallas (*C. mimula* Miller). This form is very rare in the fossil state, but there is a tenuous record extending back as follows: Drachenhöhle at Mixnitz (4-Würm); Lambrecht Cave (F-Eemian); Breitenberg (D-Holsteinian). It is now found in northwest Spain, southern France, Italy, southern central and eastern Europe, and ranges widely in Asia to the Pacific Ocean [45].

The genus *Crocidura* is also represented in Europe by some extinct species, of which *Crocidura kornfeldi* Kormos (*C. praeglacialis* Kormos) is the most important; it ranges from the Astian to the early Middle Pleistocene. Sites include Weze, Csarnóta, Villány, Nagyharsányhegy, Plesivec [142].

The Mediterranean islands Malta, Sicily, Crete and Cyprus also had *Crocidura* in the Pleistocene, but the genus was then absent in the Balearics, Corsica and Sardinia. The Pleistocene shrews of these islands belong to an extinct genus, *Nesiotites*, which is most closely related to some living genera in central and southeast Asia [24]. Although these latter are amphibious, the relatively deep skull and unspecialized humerus suggest terrestrial habits in *Nesiotites*. The species found on Mallorca and Menorca is *Nesiotites hidalgo* Bate, while Corsica was inhabited by *N. corsicanus* Bate and Sardinia by *N. similis* Hensel. The exact date of these forms is still uncertain. *Crocidura* is now found on these islands.

Family Talpidae, Moles and Desmans

The mole family has been recorded as early as the Eocene. Its European representatives belong to two distinct subfamilies, the Talpinae or true Old World moles, and the Desmaninae or desmans; the former have a very good fossil record.

The Pyrenean Desman, *Desmana pyrenaica* (Geoffroy). The history of this relict species, now found only in the Pyrenees and northern part of the Iberian peninsula, is unknown. It is mainly a burrowing form but also, like other desmans, it is a good swimmer.

The European Desman, *Desmana moschata* Pallas. This species, now found only in southeastern Russia and in Asia, has been recorded from various Hungarian and German cave deposits apparently dating from the late 4-Würm and early Postglacial. An earlier incursion in Europe is represented by material from the C-Cromer and beginning of the

2-Mindel (Forest Bed; Mosbach); these specimens have been referred to extinct subspecies of *D. moschata*. Of slightly greater age is material referred to the extinct species *D. thermalis* Kormos, a small form from Episcopia and Hundsheim; while a fossil desman from Tegelen has been described as *D. tegelensis* Schreuder. In the Astian and perhaps the earliest Villafranchian occurs a species named *D. nehringi* Kormos. The living *D. moschata* is a highly aquatic form, preferring stagnant or sluggish water with a rich reed vegetation [137].

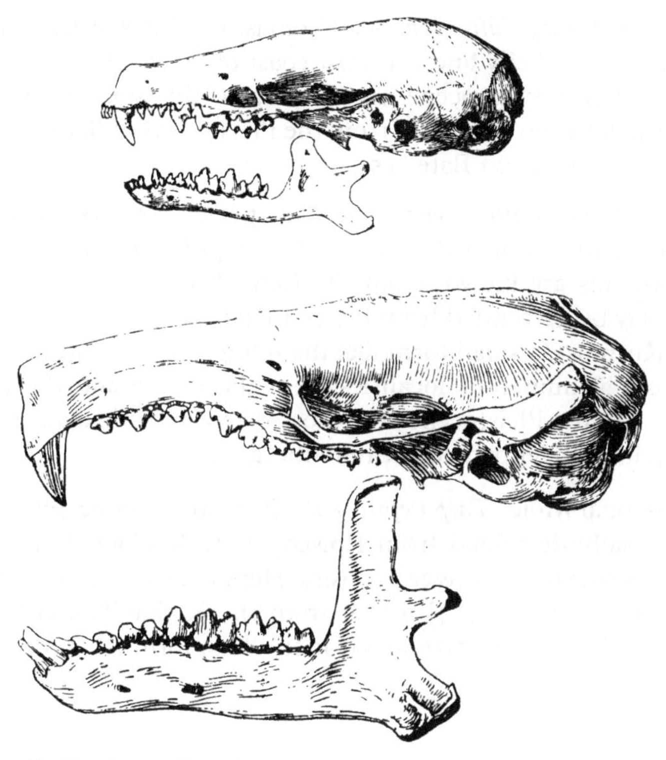

Figure 19. Skull and mandible of Mole, *Talpa europaea* (top) and European Desman, *Desmana moschata* (bottom). Recent, 1½ times enlarged. After Gromova.

The Mole, *Talpa europaea* Linné. This species has a long and well-documented history in Europe. Material from the 4-Würm and F-Eemian is plentiful. Specimens from the 3-Riss and D-Holsteinian are also referred to the living species (Achenheim; Breitenberg; Tarkö). The ancestral form is evidently *T. fossilis* Petényi (*T. praeglacialis* Kormos), which resembles the living form so much in morphology and size that specific separation appears doubtful. This form has a continuous record from the Astian up to and including the 2-Mindel and has been found at nearly all the localities in Europe with faunas of these dates.

The living species is also widely distributed in Europe, excluding only Ireland, northern parts of Scandinavia and the southern parts of the Mediterranean peninsulas; it ranges eastward to the Pacific, reaching Nepal and Sikkim in the south. It is an extremely specialized burrowing form, living mainly on earthworms, insects and grubs [147].

The Roman Mole, *Talpa romana* Thomas. A relict species, now distributed only in Sicily and the west coast of Italy. Many specimens from the Italian Pleistocene may belong to this species. A related fossil form has been described from the Pleistocene of Sardinia under the name *T. tyrrhenica* Bate [25].

The Blind Mole, *Talpa caeca* Savi. This species has a typically Mediterranean distribution with some relict populations further north. Fossil remains are known from the Late Pleistocene of Italy. The species may be descended from the extinct *T. minor* Freudenberg (*T. gracilis* Kormos), a small form like the living species. The fossil species occurs at numerous localities from the Astian and throughout the early Middle Pleistocene; the most recent records date from the D-Holsteinian (Heppenloch; Breitenberg) [147].

The Episcopal Mole, *Talpa episcopalis* Kormos. A large extinct species, originally described from Episcopia (B-Waalian) but also recorded in somewhat younger deposits (Heppenloch and Breitenberg, D-Holstein). Brunner [44] lists it from the Kleine Teufelshöhle at Pottenstein in an F-Eemian association. This form is larger than *T. europaea*.

Chapter 8

Order Chiroptera

BAT fossils are very common in some fissure faunas, but their study is as yet in a preliminary stage. From an evolutionary point of view the bats are a conservative group; many species have a very long stratigraphic range and fully evolved bats have been found as early as the Early Eocene.

Family Rhinolophidae, Horseshoe Bats

The Horseshoe Bats are characterized by the peculiar horseshoe-shaped development of the skin around the nostrils. There is only one genus, *Rhinolophus*, which dates back to the Late Eocene and is thus an extremely ancient mammalian genus. Horseshoe bats are now found widely distributed in the Old World from West Africa and Europe throughout Asia to the Philippines and northern Australia.

Blasius' Horseshoe Bat, *Rhinolophus blasii* Peters, and Mehely's Horseshoe Bat, *R. mehelyi* Matschie. These two species, which now have a patchy distribution in the Mediterranean area, are not known in the fossil state.

The Mediterranean Horseshoe Bat, *Rhinolophus euryale* Blasius. There are only a few fossil records of this species, from Italy, Roumania and Hungary; apparently the earliest finds date from the late Villafranchian (Beremend). The species is now found almost everywhere in southern Europe, with northern extensions in France and along the Danube. It also inhabits southern Russia, Palestine and North Africa. It is more gregarious than the two other main European species. [145].

The Greater Horseshoe Bat, *Rhinolophus ferrum-equinum* Schreber. This species has a good fossil record extending well back into the Middle Pleistocene. However, Astian and Villafranchian material previously referred to this species is now regarded as *R. delphinensis* Gaillard [156]. This species was originally described from the Late Miocene and is thus very ancient. The Astian and early Middle Pleistocene forms (Schernfeld; Episcopia; Kövesvarad; Kamyk, etc.) are slightly larger than the Miocene ones but not quite as large as the living form. Whether they are in fact ancestral to it remains to be seen.

The modern species inhabits the southwestern half of Europe; the boundary runs from southern Wales through southern England in an approximately southeastern direction to the Black Sea. It ranges east through Asia into Japan. It is mostly found in forested and hilly areas and is usually solitary in habits.

The Lesser Horseshoe Bat, *Rhinolophus hipposideros* Bechstein. This species has approximately the same distribution in Europe as its larger relative but ranges a little further to the north; it is also found in Ireland. Outside Europe it ranges to central Asia and North Africa. There are several Late Pleistocene and Postglacial records within the present-day range. Some finds from the later Middle Pleistocene (Breitenberg; Gaisloch) may also represent this species. Astian material from Gundersheim, Podlesice and Weze, previously referred to this species, has now been identified as *R. grivensis* Depéret from the Late Miocene [156].

The modern species resembles *R. ferrum-equinum* in habits and morphology but reaches only two-thirds of its size.

Family Vespertilionidae, Bats

This family comprises the majority of present-day and Pleistocene European bats. The family dates from the Early Oligocene. Most species belong to the world-wide genus *Myotis*, of which the earliest representatives have been found in Middle Oligocene deposits.

The Water Bat, *Myotis daubentoni* Leisler. There are various Late Pleistocene records from the 4-Würm and F-Eemian. This or a related form has also been recorded from the D-Holsteinian (Tarkö) and B-Waalian (Episcopia). Its modern range covers the greater part of Europe and northern Asia to the Pacific Ocean. It usually keeps to streams and lakes and hunts insects near the surface [117].

The Pond Bat, *Myotis dasycneme* Boie. The earliest representatives of this species are from the Astian (Podlesice). Other finds date from the early Middle Pleistocene (Nagyharsányhegy; Kövesvarad), the F-Eemian (Lambrecht Cave) and the 4-Würm (Zuzlawitz, etc.). The species now inhabits central and eastern Europe and parts of Siberia. Its biotope is about the same as for the water bat [149].

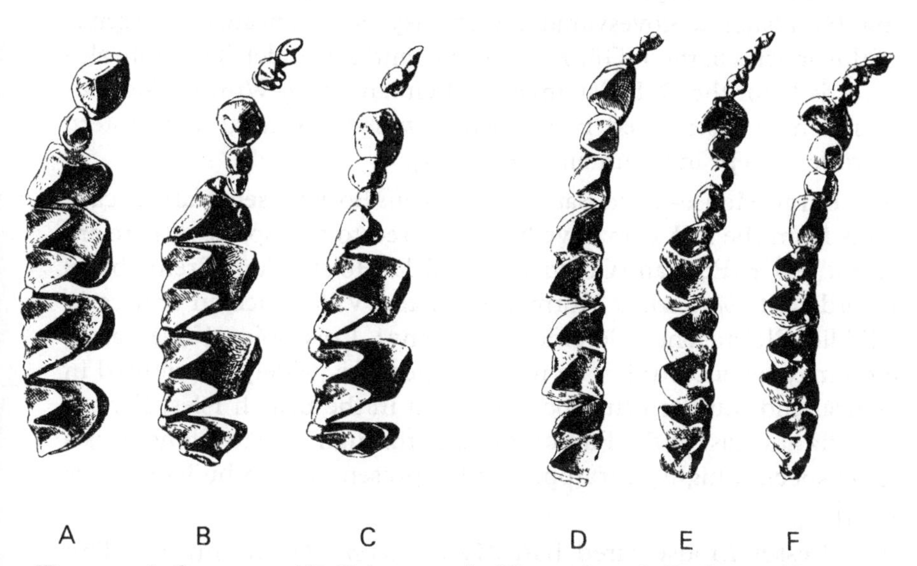

Figure 20. A–C, upper and D–F, lower teeth of Recent bats. A, D, Greater Horseshoe Bat, *Rhinolophus ferrum-equinum*; B, E, Whiskered Bat, *Myotis mystacinus*; C, F, Large Mouse-eared Bat, *M. myotis*. Ten times enlarged. After Miller.

The Whiskered Bat, *Myotis mystacinus* Leisler. This species makes its first appearance in the C-Cromer (Kövesvarad) and is also found in the D-Holsteinian, F-Eemian and 4-Würm. It now inhabits all of Europe except northern Fennoscandia and the southern part of the Iberian peninsula. In Asia it ranges to China and Japan. It is a woodland species [294].

Geoffroy's Bat, *Myotis emarginatus* Geoffroy. Tentative records date from the early Middle Pleistocene (Episcopia; Hundsheim; Koneprusy, etc.) and the D-Holsteinian (Tarkö). The species now has a somewhat scattered distribution in France, central and southeastern Europe, preferring wooded areas [117].

Natterer's Bat, *Myotis nattereri* Kuhl. The earliest finds come from deposits dated as late 1-Günz (Brassó; Sackdilling); later records are from 2-Mindel (Koneprusy), D-Holsteinian (Tarkö) and some Late Pleistocene cave deposits. The species is present in the greater part of

Europe except the southeast and most of Fennoscandia; it ranges on through Asia to Japan. It is less dependent on woods than the whiskered bat and Geoffroy's bat [117].

Bechstein's Bat, *Myotis bechsteini* Leisler. Specimens from Sackdilling, Hundsheim, Kövesvarad, Koneprusy, etc., indicate the presence of this species in the 1-Günz, C-Cromer and 2-Mindel. There are also records from the D-Holsteinian, F-Eemian and 4-Würm. The species is now distributed in temperate western, central and eastern Europe, except in mountain areas [294].

The Large Mouse-eared Bat, *Myotis myotis* Borkhausen. Various cave finds from the 4-Würm have been referred to this species; there are also a few F-Eemian (Cotencher) and D-Holsteinian (Breitenberg) records. *M. baranensis* Kormos from the Villafranchian and early Middle Pleistocene (Beremend; Villány; Episcopia; Kövesvarad, etc.) may be ancestral. The modern species is widely distributed in central and southern Europe but absent in the British Isles, Fennoscandia and east of the Baltic; it ranges through Asia to China. It is a large species, highly eurytopic, and is present also in built-up areas [294].

The Lesser Mouse-eared Bat, *Myotis oxygnathus* Monticelli. This form is closely related to *M. myotis* but somewhat smaller. Fossil remains are scarce but include material from Episcopia (B-Waalian), Hundsheim (1-Günz II) and Tarkö (D-Holsteinian). This species probably branched from a common ancestor with *M. myotis*, perhaps *M. baranensis*. Its modern distribution is Mediterranean [117].

Several extinct species of *Myotis* have been described from the Astian but only a few in the Pleistocene; three species were erected by Kormos for material from the late Villafranchian and early Middle Pleistocene.

The Long-eared Bat, *Plecotus auritus* Linné. A small form, easily spotted by its long ears. The fossil record is good and extends back to the B-Waalian (Episcopia); other finds date from 1-Günz II (Brassó), C-Cromer (Kövesvarad), 2-Mindel (Koneprusy), D-Holsteinian (Tarkö) and late Pleistocene (many sites). It now inhabits almost all of Europe except northern Fennoscandia; it is also found in temperate Asia to Japan and in Egypt, Palestine and northern India. A fossil species, *P. crassidens* Kormos, has been recorded in the Astian (Podlesice; Weze; Gundersheim) and early Middle Pleistocene (Episcopia) [156; 294].

The Grey Long-eared Bat, *P. austriacus* Fischer. This species is slightly larger than *P. auritus* and of a slightly different colour. It has a Mediterranean–Mongolian distribution extending to central Germany. That it really is a separate species has only recently been proved, and its relationship to the fossil remains has not yet been worked out [27].

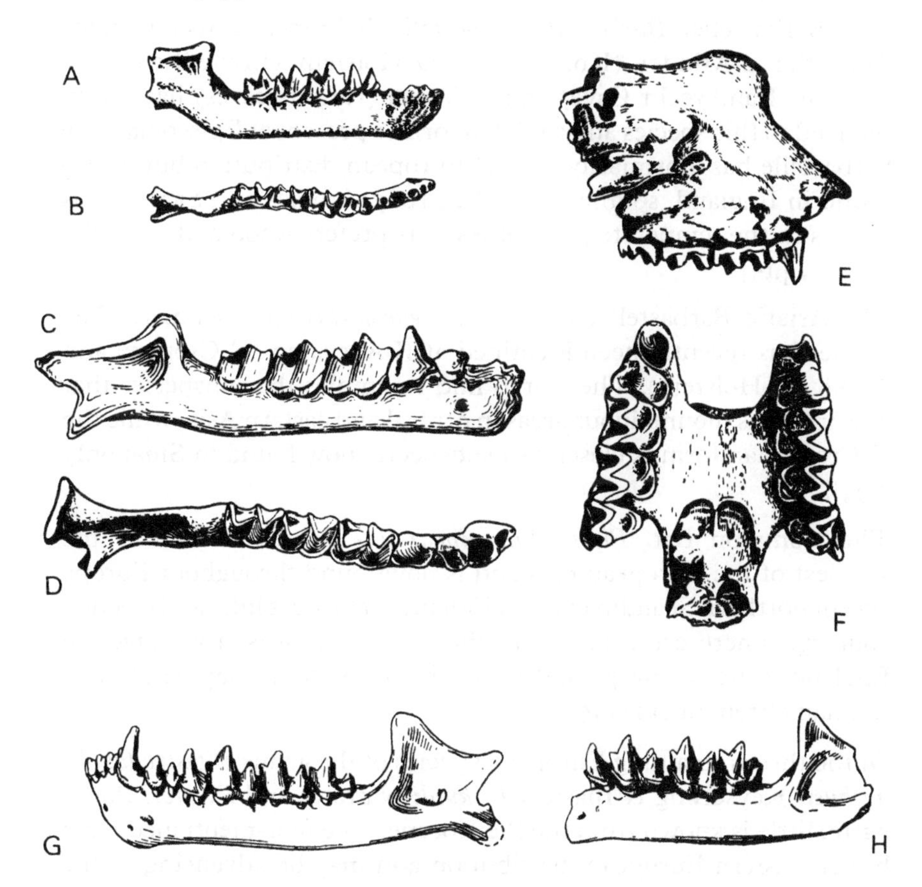

Figure 21. A, B, mandible of Long-winged Bat, *Miniopterus schreibersi*; C, D, mandible and E, F, skull fragment of extinct Horseshoe Bat, *Rhinolophus delphinensis*. G, H, mandibles of Long-eared Bat, *Plecotus auritus*. A–F, Astian, Podlesice; G, Recent; H, Late Pleistocene, Cotencher. A–F four times, G–H ten times enlarged. A–F after Kowalski; G–H after Stehlin.

The Long-winged Bat, *Miniopterus schreibersi* Natterer. A medium sized species with short, truncated ears and pointed wings; this is a fast, accomplished flyer. Its fossil record is a long and good one. The species is well represented in the Astian (Podlesice; Gundersheim; Weze). There are also records in the early Middle Pleistocene (Episcopia, etc.) but the species is then absent until the Late Pleistocene, when there are some cave finds. It now inhabits the Mediterranean area and also North Africa, south and east Asia and northern Australia [156].

The Barbastelle, *Barbastella barbastella* Schreber. Fossil remains from the 2-Mindel (Koneprusy), D-Holstein (Breitenberg), F-Eemian (Lambrecht Cave) and 4-Würm (various caves) have been referred to this species and a related form is present at Episcopia. The barbastelle has a primarily central European distribution but is also found in England, southern Sweden, Italy, Corsica and Sardinia, as well as in the temperate parts of Asia. It prefers woodland and hilly country [45].

The Asiatic Barbastelle, *Barbastella leucomelas* Cretzschmar. This species has recently been identified at Kövesvarad (C-Cromer) and Tarkö (D-Holstein). The Tarkö find resembles the subspecies that now inhabits mountainous areas in central and eastern Asia, while the C-Cromerian form is closer to a subspecies now found in Sinai only [294].

The Common Bat, *Pipistrellus pipistrellus* Schreber. This is the smallest of the European bats and is now found throughout Europe except northern Scandinavia and Finland. It favours hilly and wooded country. There are a few Late Pleistocene records from caves in England, central Europe and Italy; the oldest finds may be D-Holsteinian (Breitenberg) [45].

Of the three remaining European species of the genus *Pipistrellus*, *P. nathusii* Keyserling & Blasius, *P. kuhli* Natterer and *P. savii* Bonaparte, little is known from the Pleistocene. The first-mentioned form has an eastern European distribution and may be advancing to the west at the present time. The two latter species are Mediterranean in distribution; fossil finds from Salerno and Spezia have been referred to *P. savii*.

The Serotine, *Vespertilio serotinus* Schreber. Fossils are scarce but the species has been found in the 2-Mindel (Koneprusy) and D-Holsteinian (Breitenberg). It now inhabits southern England and continental Europe except north and east of the Baltic; it ranges eastward to China. It is found in spinneys, park woods and built-up areas [82].

The Northern Bat, *Vespertilio nilssoni* Keyserling & Blasius. This species has been tentatively identified in the C-Cromer (Kövesvarad) and D-Holsteinian (Tarkö; Breitenberg). It is also recorded in the F-Eemian (Lambrecht Cave). The species is now found in northern and eastern Europe, in the mountains of central Europe, and eastward to the Pacific Ocean. It is somewhat smaller than the serotine [294].

The Parti-coloured Bat, *Vespertilio murinus* Linné (*V. discolor* Kuhl). There are several Late Pleistocene records from France and Italy as well as central and eastern Europe, both within and outside the present range of the species, which is now found in central and eastern Europe and southern Fennoscandia and in Asia to Japan. Its earlier history is unknown [312].

Two extinct species of the genus *Vespertilio* have been described from the European Pleistocene: *Vespertilio majori* Kormos (Villány) and *V. praeglacialis* Kormos (Villány; Episcopia) [145].

The Common Noctule, *Nyctalus noctula* Schreber. This large species inhabits Europe except Ireland and most of Fennoscandia (it is present in south Sweden and Denmark) and in Asia to Japan. Fossil remains are not uncommon in Late Pleistocene cave deposits. The oldest may date from the D-Holsteinian (Breitenberg). The species inhabits woods and parkland. The related lesser noctule, *N. leisleri* Kuhl, is not known in the fossil state. Its present-day distribution is in central and eastern Europe, with isolated relict populations in Ireland, England, France, Portugal, etc. [45].

Family Molossidae, Free-tailed Bats, etc.

The European Free-tailed Bat, *Tadarida teniotis* Rafinesque. This is the only living species of the molossid family in Europe, but extinct members of the genus *Tadarida* are known as early as the Oligocene. The modern species, which is the largest of all the European bats, is found in the Mediterranean peninsulas and in Asia to Japan and Formosa; its fossil history is obscure.

Chapter 9

Order Primates

PRIMATES are comparatively rare in the Pleistocene of Europe. Most of the fossils belong to the hominid family, and if we include the artifacts of man, especially the stone tools, man may be regarded as the commonest primate of this period. But skeletal finds are exceptional; in the Middle Pleistocene and up to the F-Eemian they are extremely rare. Other fossil primates include only two or three types of monkeys whose range intermittently extended to Europe.

Family Cercopithecidae, Old World Monkeys

The cercopithecids are the only non-human primates known in Europe in the Pleistocene. In Tertiary times apes of the genus *Dryopithecus* lived here, but they were gone before the beginning of the Villafranchian. The history of the Cercopithecidae goes back to the Miocene.

The Florentine Macaque, *Macaca florentina* Cocchi. A macaque apparently quite closely related to the living Gibraltar 'ape', the only monkey living in Europe today in the wild state. The species is not uncommon in the Villafranchian; it is present from the very beginning of the age (Villafranca d'Asti) and occurs in various types of faunas (Saint-Vallier; Senèze; Val d'Arno; Tegelen; Beremend). The macaques are usually forest forms, but some of them, as the rhesus monkey and the Gibraltar 'ape', also like rocky ground.

The Florentine macaque is probably the direct ancestor of the living Gibraltar 'ape'. Finds from the B-Waalian and C-Cromer interglacials are often referred to the Villafranchian species. A possible forerunner of this species is *M. prisca* Gervais from the Astian deposits of Montpellier, Gundersheim and Csarnóta [302].

The Gibraltar 'ape', *Macaca sylvana* Linné. The classification of the early Middle Pleistocene finds from Episcopia (B-Waalian), Koneprusy and the Forest Bed (C-Cromerian) and Mosbach (early 2-Mindel) is uncertain; perhaps they are Florentine macaques or transitional forms. A macaque phalanx from Voigtstedt in Thuringia (C-Cromer) resembles rhesus monkey (*M. mulatta*) rather than Gibraltar ape, which may suggest that these animals should be classed within *M. florentina* [287]. The D-Holsteinian macaques from Heppenloch, Grays Thurrock and Montsaunès may probably be referred to the modern species. All the fossils date from interglacial or temperate (Mosbach) phases and the species is evidently a good indicator of comparatively warm climate.

A unique characteristic of the present-day species is the absence of a tail, as in true apes, although this animal is a monkey and not an ape. The fossils give no information on this trait in the Pleistocene form. The length of the tail varies in modern species of the genus *Macaca* [6].

The Auvergne Monkey, *Dolichopithecus arvernensis* Depéret. The extinct genus *Dolichopithecus* belongs to the colobine monkeys, often regarded as a separate family; it is related to such living forms as the guerezas of Africa and the langurs and proboscis monkeys of Asia, all of them leaf-eating and with highly specialized stomachs. It was originally described from the late Villafranchian of Senèze. Other finds from Budapest and Saint-Estève may be even later, while on the other hand a record from Vialette dates from the beginning of the Villafranchian. An Astian member of the same genus is known from Roussillon and may be ancestral to the Auvergne monkey [59; 60].

Family Hominidae, Men

The earliest known members of our own zoological family date from the Miocene of India and Africa and belong to the genus *Ramapithecus*. Their evolutionary level was that of the apes, but their teeth were human. Still more man-like were the Australopithecinae of the Villafranchian and Middle Pleistocene, but there is no evidence that they ever ranged into Europe. There are no certain artifacts from the Villafranchian in Europe, so that it seems reasonable to assume that tool-making hominids were not present here at that early date. The earliest artifacts in Europe are from the C-Cromer interglacial (Abbeville; ?Mauer) and may be regarded as the handiwork of Heidelberg Man.

The evolution of man and the life of primitive man is a great and fascinating theme, which can only be touched on in the briefest terms here.

Heidelberg Man, *Homo heidelbergensis* Schoetensack. The oldest human fossil in Europe comes from the C-Cromer interglacial sands of Mauer. Unfortunately, the specimen is only a lower jaw, albeit very well preserved. It seems to have characters allying it to the Neandertal type and fairly distinct from *Pithecanthropus*, which at that time flourished in Asia and Africa (not to mention the still more primitive australopithecines, of which relict tribes were still in existence in southern Africa). Heidelberg Man is probably ancestral to the European line of evolution which leads to Neandertal Man (and perhaps, as a side branch, to modern man). Whether he merits specific distinction from Neandertal Man on one side, and *Pithecanthropus erectus* Dubois on the other, may be disputed; more material will be needed to settle this question [112].

Recently a few teeth and a skull fragment have been found at Vértesszöllös, northwest Hungary, in deposits dating from the 2-Mindel I–II interstadial and in association with a pebble-tool industry. This material closely resembles *Pithecanthropus* [209].

Neandertal Man, *Homo neanderthalensis* King. The true or 'classical' Neandertaler with his dorsally flattened, very large and long head, big brow ridges, prognathous face, round eye sockets, very large braincase and a stockily built, powerful body, lived during the 4-Würm up to the great interstadial about 35,000 years ago, when this species became extinct. As regards the classification of human fossils from preceding interglacials (F-Eem and D-Holstein, perhaps also E-Ilford – the Montmaurin jaw), opinions diverge greatly. The D-Holsteinian fossils come from Steinheim and Swanscombe; they have many characters that seem to ally them to modern man and they are classified as *H. sapiens* by many authorities. Others prefer to regard them as primitive or unspecialized Neandertalers. If so, the F-Eemian hominids from various sites in Europe must obviously also be classified as *H. neanderthalensis*, since they appear to be transitional between the D-Holsteinian and 4-Würmian types. The stratigraphic span of the species would then extend from the D-Holsteinian to the interstadial Würm I–II. This is one way of stressing the unified character of the European evolutionary line during the entire time span from C-Cromer to 4-Würm.

Many authors regard Neandertal Man as nothing but an aberrant stock of *H. sapiens*. Available evidence may suggest that at least the end stage attained full specific distinction. The classification used here then requires that modern man be regarded as a side branch of the Neandertal stock [112; 135; 275].

Modern Man, *Homo sapiens* Linné. Our own species may be definitely identified in the 4-Würm interstadial, the time when men of modern type invaded Europe and ousted the Neandertalers. On the other hand, it has been thought that specimens like Fontéchevade Man (two skull caps without a trace of the Neandertal brow ridge) have to be counted in this species, even if Steinheim and Swanscombe are left as Neandertalers. The Fontéchevade fossils date from the F-Eemian, a time when there were certainly Neandertalers in Europe (Taubach; Ehringsdorf; Saccopastore; Gánovce). Are we to conclude that both species coexisted in Europe and actually behaved like good species in the F-Eemian – that is to say, that they did not hybridize? It is, of course, very probable that *H. sapiens* was in existence *somewhere*, in the guise of an early variant at any rate, during the F-Eemian; and no doubt Fontéchevade Man is anatomically acceptable as one of these. Could the find represent a temporary invasion of our species, later to become extinct or be swamped by Neandertalers?

Early extra-European finds include the at present insufficiently dated Kanjera skulls from East Africa and a skull from Niah Great Cave in Sarawak. The latter has a radiocarbon age of nearly 40,000 years, which makes it the oldest well-dated *H. sapiens* so far. The F-Eemian Fontéchevade Man must be at least twice as old [208].

Chapter 10

Order Carnivora

THE carnivores have an extremely good fossil record in the Pleistocene, especially as regards the larger species. The most serious lacunae occur among the mustelids. Some of the larger carnivores are among the most common fossils in Ice Age deposits.

Family Viverridae, Civets

The viverrids are the most primitive of living carnivores. In the past, this family gave rise to two other families: the cats and the hyenas. In our time the viverrids have a mainly tropical distribution, but in the Tertiary they were quite common in Europe. Today only two species are found here and both have a marginal distribution.

In the Astian, members of the genus *Viverra* existed in Europe; a large, hitherto undescribed species of this genus persisted in the earliest Villafranchian (Villafranca d'Asti).

The Genet, *Genetta genetta* Linné. This pretty little carnivore is found nowadays in part of France and on the Iberian peninsula, but it is probably a Postglacial invader. The main range of the species is in Africa; it has not been found in the Pleistocene of Europe but does occur in the Pleistocene of Africa. The habits of the genet are somewhat reminiscent of the wild cat; like the cats, the genet has sharp, retractile claws, and is an excellent climber [227].

The Ichneumon, *Herpestes ichneumon* Linné. Like the genet, this species is found only in the Holocene in Europe (though the genus *Herpestes* occurs also in the Tertiary in Europe). Coming from Africa, where it is known in the Pleistocene (as it is also in Palestine), it has invaded the southern part of the Iberian peninsula [227].

Family Hyaenidae, Hyenas

The hyenas are a young family of carnivores: they date from the Late Miocene. They arose from the Viverridae, and the genus *Viverra* may be close to the actual ancestry of the hyenas. Several types of hyenas are known from the Tertiary, but most of them belong to extinct genera. The Pleistocene hyenas of Europe belong to three quite distinct genera of which two, *Hyaena* and *Crocuta*, are still in existence, while the third, *Euryboas*, is extinct.

The hyenas are specialized carrion feeders, able to smash large bones in order to get at the marrow and the nutritious spongy bone tissues. As an adaptation for this way of life their cheek teeth have evolved into powerful, conical structures. The remains of a hyena meal are easily recognized, but the unwary student may mistake them for bone tools intentionally fashioned by man. A large proportion of the bones in a hyena den may be made up of remains of hyenas, ranging from juveniles to very old animals. The hyenas are probably not intentional cannibals, for eyewitnesses state that a hyena does not eat a dead hyena; however, a completely putrefied carcass will not be recognizable as hyena any more and will be eaten.

The hyena plays an important role as a bone collector, to the benefit of the palaeontologist. Although it will mutilate the bones, it always leaves the teeth intact and systematically they are the most important part of the skeleton. As a result, the activities of hyenas in the European caves have been of great importance for our knowledge of the Ice Age and its fauna. The hyena itself is one of the most commonly found fossil mammals and the evolutionary history of the family Hyaenidae is thus comparatively well known.

The extinct *Euryboas* deviated greatly from the orthodox hyenas and was one of the most remarkable animals in the Pleistocene fauna. The hyaenid family has also produced another aberrant animal in the aard wolf (*Proteles*), known only from Africa; it has vestigial teeth and is insectivorous.

The Perrier Hyena, *Hyaena perrieri* Croizet & Jobert (*H. arvernensis* Croizet & Jobert). This is the common hyena of the Villafranchian, lasting to the C-Cromer interglacial. Its record begins with the basal Villafranchian (Villafranca d'Asti; Etouaires) and later on it occurs at practically every Villafranchian site with a reasonably good macromammalian fauna, including the Crags in East Anglia. The geologically most recent finds come from Mauer (C-Cromer) and

Mosbach (probably from the basal sands). There is no certain record of the species outside Europe.

The Perrier hyena is so close to the living Brown Hyena (*H. brunnea* Thunberg) that it might even be questioned whether they are really distinct species. However, the relatively great geological age of the Perrier hyena, its somewhat larger size and the slightly more advanced construction of its carnassial teeth suggest that it should be regarded as a distinct species and not merely a European subspecies of the brown hyena. In *H. brunnea* the lower carnassial always carries the vestigial inner cusp (metaconid) inherited from the viverrid ancestors of the hyenas; this cusp is usually absent in *H. perrieri*, or if present it is very small.

Figure 22. Restoration of Short-faced Hyena, *Hyaena brevirostris*, based on a skeleton from Choukoutien and on the related living Brown Hyena, *H. brunnea*.

The brown hyena has been found in the fossil state only in Africa (the European reports I have checked have been erroneous) and its present range is confined to the southern part of the African continent. The oldest finds come from the early Middle Pleistocene of the Transvaal. The morphological resemblance between *H. perrieri* and *H. brunnea* indicates that the two species diverged from a common ancestor in the Late Pliocene; an incompletely known form in the Astian of France (*H. donnezani* Viret) may be that ancestor.

The present-day brown hyena is regarded as a somewhat more active carnivore than other living hyaenids. Its cheek teeth are in fact somewhat less specialized for carrion-feeding than those of the genus *Crocuta*, while the canine teeth are relatively large. The latter, of course, are important in the catching and worrying of the prey, but are of little use to a scavenger and in *Crocuta* they are rather feeble. It seems reasonable to conclude that the Perrier hyena, which had powerful canines, obtained part of its food by active hunting, just like its living counterpart.

The extinction of the species at the end of the C-Cromer may be due to competition from the species *Crocuta crocuta*, at that time a recent immigrant [166; 223; 302].

The Short-faced Hyena, *Hyaena brevirostris* Aymard (*H. robusta* Weithofer, *H. sinensis* Owen, etc.). This gigantic hyena was as big as a lion; the shoulder height of the largest specimens was three feet, the length of the head and body up to five feet. Apart from size, however, the species is so similar to the Perrier hyena that some authors even have regarded them as variants of a single species. While the two species are certainly closely related, the difference in size is so great that specific identity is out of the question.

It seems probable that the short-faced hyena evolved as a side branch out of a local population of Perrier hyena somewhere outside Europe and made its entry in Europe towards the end of the Villafranchian. As late as at Senèze it is still absent, but at Olivola and Val d'Arno the giant form is present together with the Perrier hyena. It persisted into the Middle Pleistocene (Sainzelles; Bacton) up to 1-Günz II (Gombasek; Süssenborn; Stránská Skála) but may have become extinct before the C-Cromer interglacial or early in the interglacial (Koneprusy).

The geographic distribution of the short-faced hyena was unusually wide. It is represented in Villafranchian strata in India and China; in China it survived up to 2-Mindel, for its bones were found in great numbers in the lower strata of the Peking Man site at Choukoutien. A Javanese local race of the species occurs in the same layers as *Pithecanthropus* (the Djetis strata); and finally, a fragment from a cave in Transvaal, of the same age, has been thought to represent this species also.

In the upper layers of the Peking Man site at Choukoutien, the short-faced hyena disappears and its place is taken by *Crocuta crocuta*. In Europe both species existed side by side for some time, then the short-faced hyena became extinct. There seems to be reason to suggest that competition between the two species was involved and the success of the more highly specialized scavenger (*Crocuta*) would not be too surprising. For the short-faced hyena probably was a less exclusive carrion-feeder and a more active hunter than *Crocuta*. Its canine teeth are enormous and in life the animal must have been an extraordinarily powerful, truculent beast. Anybody who has seen an angry, snarling brown hyena with its big shaggy mane on end will appreciate what a terrifying apparition its gigantic extinct ally must have been to the primitive men of its day [37; 166; 167].

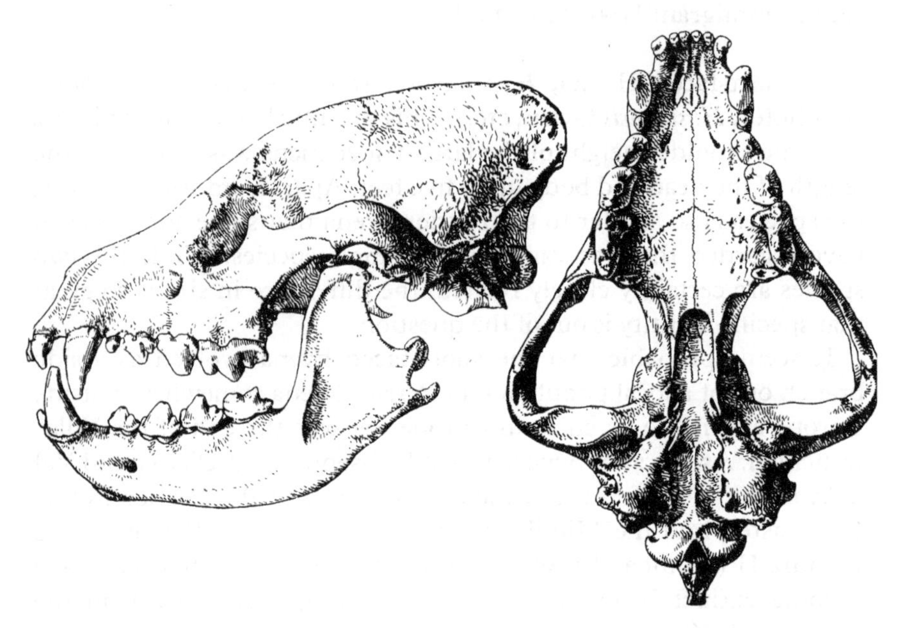

Figure 23. Skull and mandible of Striped Hyena, *Hyaena hyaena*, Recent. One-third natural size. After Gromova.

The Striped Hyena, *Hyaena hyaena* Linné. During the Pleistocene this species occurred for some time in Europe. In the D-Holsteinian it was widespread in France and Germany (Lunel-Viel; Montsaunes; Kreuznach). It is also found at Montmaurin (E-Ilford?) and at Hollabrunn, Austria, in the Furninha Cave in Portugal and the Genista Caves, Gibraltar (Late Pleistocene?). The European form of the striped hyena is quite similar to the living form morphologically but was somewhat larger and it has also been regarded as a distinct species, *H. prisca* De Serres (*H. monspessulana* Christol). The same form lived in Palestine during the F-Eemian interglacial and 4-Würm. This large form is of a size with the living brown hyena but differs from that species in various diagnostic characters (size of upper molar, etc.).

In Africa the striped hyena is not uncommon as a fossil. There are

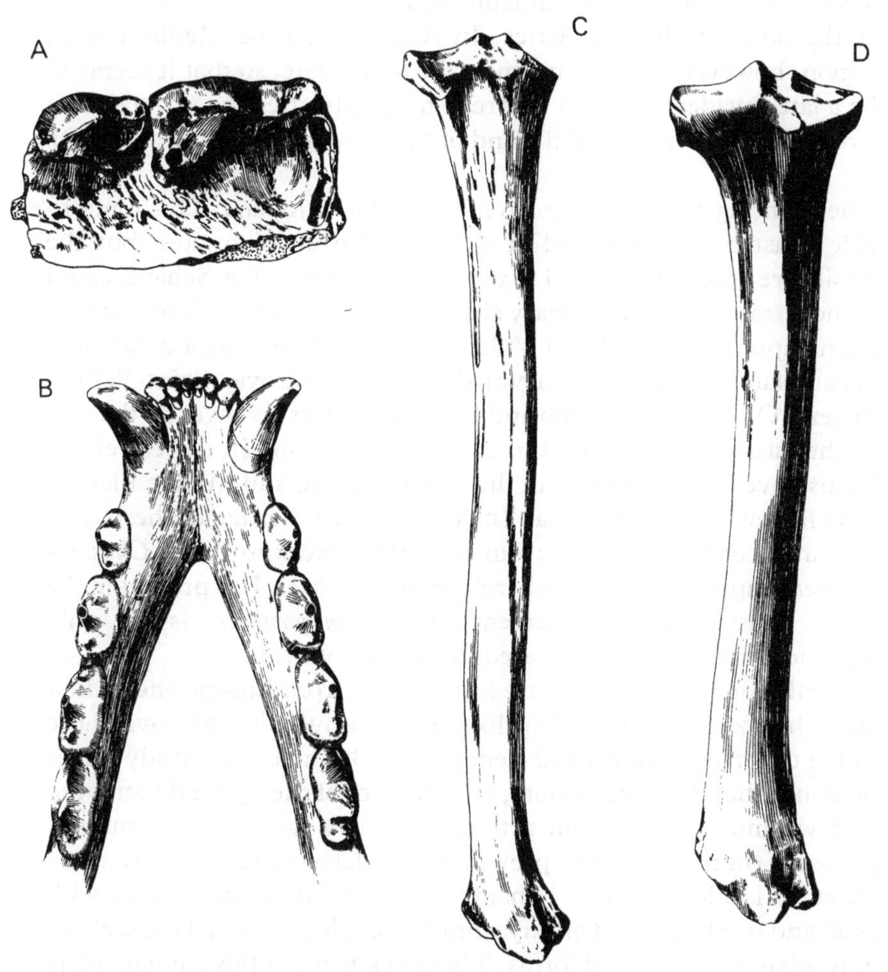

Figure 24. A, upper P^3–P^4, B, mandible and C, tibia of Hunting Hyena, *Euryboas lunensis*; D, tibia of Spotted Hyena, *Crocuta crocuta*. A, C, Pardines; B, Val d'Arno; Villafranchian. D, Recent. A one-half, B two-thirds, C–D four-ninths natural size. After Schaub.

records going back to the Middle Pleistocene (*H. makapani* Toerien from the Transvaal) and even to the Villafranchian (for instance Aïn Brimba, Tunisia; Serengeti, East Africa). The origin of the species probably lay in Africa; there is a possible ancestor (*H. namaquensis* Stromer) in the Pliocene.

H. hyaena is the smallest and most primitive of the living true hyenas and retains many viverrid characters that were lost in larger members of the genus *Hyaena*. It is now more widely distributed than other hyaenid species and inhabits south and southwest Asia as well as the northern half of Africa. In Asia outside the Mediterranean region, however, it is unknown in the fossil state, so that it seems to be a late invader; perhaps it spread here only after the extinction of *Crocuta crocuta* in Asia at the end of the Ice Age [91; 166; 179; 288].

The Hunting Hyena, *Euryboas lunensis* Del Campana. This remarkable beast has been found at several Villafranchian sites, but the fossils are fragmentary and it was not until 1941 that Schaub could demonstrate the extraordinary characters of the species. The hunting hyena appears in the basal Villafranchian at Villafranca d'Asti and occurs later on at Roccaneyra-Pardines, Villaroya, Saint-Vallier, Senèze, Olivola, Val d'Arno and Erpfingen; thus it ranges all the way to the late Villafranchian, but is rare throughout. In Africa related forms have been identified in the Villafranchian and Middle Pleistocene faunas of the Transvaal. In Asia, on the other hand, there is no certain record of *Euryboas*; but a North American form, *Chasmaporthetes*, appears to be closely related to *Euryboas*. It is present in the Early Pleistocene and, incidentally, *Chasmaporthetes* is the only hyaenid known to have migrated to the New World.

Limb bones of *Euryboas* are known only from Europe and at first sight they certainly do not look like those of a hyaenid; they rival those of the cheetah in length and slenderness. But a detailed study leaves no doubt that the bones belong to a member of the hyaenid family, so that we must visualize an active, extremely fast cursorial hunter, probably running down its prey by a swift dash in the same way as the cheetah. Logically, then, Schaub combined these limb bones with skull and tooth material of a hyaenid that evidently lacked the scavenging adaptation of allied forms. The cheek teeth in this animal, while basically hyaenid, are slender and sharp-edged like those of a cat; the dentition is reminiscent of that found in some Pliocene hyaenids (*Ictitherium*, *Lycyaena*, *Hyaenictis*). Unfortunately the limb bones of the Pliocene hyaenids are imperfectly known, so that we cannot say whether any of them had developed the fast-running habits of *Euryboas*. The limbs of the North American *Chasmaporthetes* are also unknown at present.

With *Euryboas* we have travelled very far from the traditional hyena type. Like the cheetah it probably inhabited open ground (bush steppes and savannas would seem ideal) and was a daytime hunter. In size it resembled the living cheetah; but it is part of the story that it lived together with the great Villafranchian cheetah, *Acinonyx pardinensis*, throughout the Villafranchian age, so that evidently it was able to hold its own in spite of powerful competition. This also gives some idea of the abundance of fast cursorial game on the Villafranchian plains. Of the species known to us, the Bourbon gazelle and the chamois-antelope may well have been the favourite prey of the hunting hyena [240; 268; 302].

The Spotted Hyena or Cave Hyena, *Crocuta crocuta* Erxleben. This is one of the best-known fossil mammals of the Ice Age. The species is represented in most European bone caves and in some, used as dens for thousands of years, bones of the cave hyena occur in profusion. A famous cave of this type is Kirkdale Cave in Yorkshire, described by William Buckland in 1823. Dean Buckland showed that the enormous accumulation of hyena bones resulted from continuous use of the cave as a living site by the hyenas over a very long period (the F-Eemian interglacial, as we now know). Here are found remains of hyenas of all ages from newborn to senile and in addition their characteristic faeces and the bones of their prey. Another British hyena cave with a rich hyena stratum of F-Eemian date is Tornewton Cave in South Devon (reckoned to have yielded more than 20,000 hyena teeth), while Kent's Cavern in Torquay contains an immensely rich hyena stratum of more recent date, 4-Würm. On the continent several hyena caves are known and at least one – the Teufelslucken at Eggenburg in Austria – is of the same class.

All the known mass occurrences of this type date from the F-Eem or 4-Würm, but the cave hyena entered Europe much earlier. The first immigrants were animals of about the same size as the present-day African spotted hyena; they appear in the 1-Günz II at Süssenborn and Gombasek. Slightly later, in the C-Cromer (Forest Bed; Mosbach) the European race reached its distinctive large size. It is also known from the D-Holsteinian (Grays Thurrock and

Figure 25. Skull and mandible of Cave Hyena, *Crocuta crocuta spelaea*, Pleistocene, USSR. One-third natural size. After Gromova.

Lunel-Viel; the latter site, recently reopened by Dr and Mrs Bonifay, is now yielding what may be another mass occurrence) and from 3-Riss (Archenheim; Tornewton Cave).

The ancestor of the spotted hyena is evidently the Indian *Crocuta sivalensis* Falconer & Cautley, which lived in the Villafranchian; it resembles *C. crocuta* but has a somewhat more primitive dentition (the upper molar, for instance, is still functional, whereas it is vestigial or absent in the modern species). The population probably spread out from its original area in the early Middle Pleistocene: at this time, spotted hyenas suddenly appear in a very wide area from Europe to southern Africa and China. Within most of this immense range, the spotted hyena remained the dominant hyaenid until the end of the Pleistocene. There was however some local fluctuation or alternation, as for instance in Palestine. Here *C. crocuta* was absent in the F-Eemian and in its place we find the large *H. hyaena prisca*. In the 4-Würm *C. crocuta* entered the scene and the striped hyena vanished. In Post-glacial times, with the development of agriculture, the spotted hyena became much dwarfed and finally extinct and the striped hyena (in its modern form) returned.

The European form was considerably larger than the living African spotted hyena. There is however a gradual transition in size over southern Russia and Palestine, indicating that there was once a continuous population extending all the way from Africa to Europe. In the Late Pleistocene the European form had peculiarly shortened and thickened bones of the fore- and hind-feet, while the humerus and femur were very long. This subspecies, *C. crocuta spelaea* Goldfuss, is the true cave hyena.

There are very few pictorial representations of the cave hyena in Palaeolithic art. The one from La Madeleine figured here is, however, a superb likeness (figure 26).

C. crocuta is the most specialized scavenger of the Hyaenidae. The premolars are powerful bone-cracking teeth, while the carnassials have been much elongated to form long, sharp blades to cope with tough pieces of hide. The canines, on the other hand, are relatively feeble.

At the end of the Ice Age this species became extinct both in Asia and Europe. From a zoogeographical point of view it is noteworthy that this species, which today is restricted to Africa, appears to be of Asiatic origin; while the striped hyena, which now has the main part of its range in Asia, almost certainly originated in Africa! This shows how misleading zoogeographic conclusions may be if based on Recent animals only rather than on fossil evidence [69; 166; 169; 170; 203].

Figure 26. Ivory sculpture of Cave Hyena, *Crocuta crocuta spelaea*, from La Madeleine, Dordogne. After Maringer & Bandi.

Family Felidae, Cats

The earliest felids appeared in the Late Eocene, about 50 million years ago. The family has a good fossil record. Apart from cats with normally developed teeth, the family also contains the extinct sabre-toothed 'tigers' or sabre-tooths, which are often set off as a distinct subfamily of the Felidae. They do not, however, form a homogeneous group, but simply a recurrent adaptive type which evolved in several different lineages out of 'normal' cats. This may sound surprising, but it so happens that pseudo-sabre-tooths evolved even within groups of carnivores other than the cats, thus showing that this type of adaptation is highly successful. Both the meat-eating marsupials of South America during the Tertiary and the creodonts (primitive pre-Carnivora) evolved sabre-toothed types. A closer study of the sabre-toothed felids reveals that they form quite divergent lines of evolution, with the enlarging of the upper canine teeth as the only significant common character. The two groups of Pleistocene sabre-tooths are good examples of this divergence.

That the true cats may well give rise to sabre-toothed forms is clearly shown by a living species, the clouded leopard (*Felis nebulosa* Griffith) in the East Indies. In this animal the upper canine teeth are very large, forming great dirks fully comparable to those of many extinct sabre-tooths. But unlike the real sabre-tooths, the clouded leopard has suffered no decrease in the size of its lower canines so that it is still able to bite like a normal cat, while the extreme sabre-tooths were specialized for stabbing.

All the Pleistocene felids may be referred to the subfamily Felinae, characterized by a reduction of the number of teeth and a complete ossification of the inner ear. They may be divided into three different groups with the systematic rank of tribes: Tribe Smilodontini, the dirk-toothed cats; Tribe Homotheriini, the scimitar cats; and Tribe Felini, the true cats – the only one to survive to the present day.

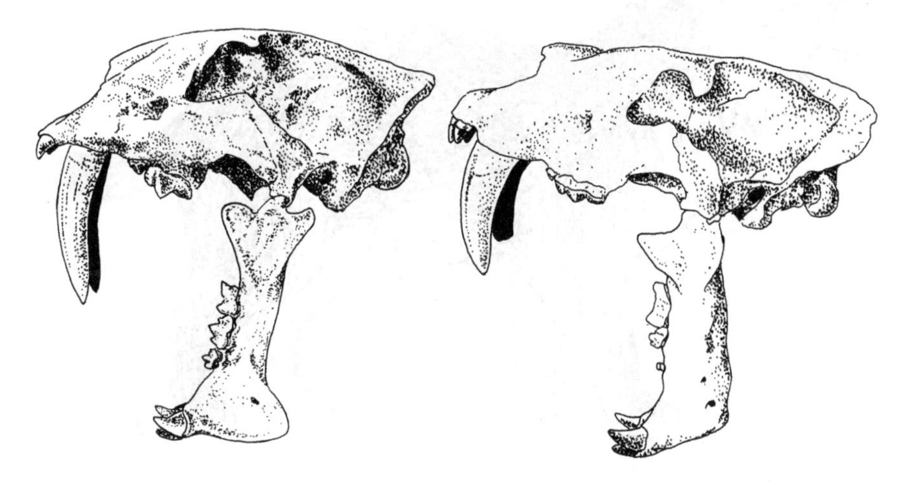

Figure 27. Skull and mandible of Dirk-tooth, *Megantereon megantereon*, Senèze (left) and Greater Scimitar Cat, *Homotherium sainzelli*, Perrier (right).

The Dirk-Tooth, *Megantereon megantereon* Croizet & Jobert. All the European members of the Smilodontini may be referred to a single species, which was originally described from Mount Perrier and is a Villafranchian guide fossil. It appears in the basal fauna (Villafranca d'Asti; Etouaires) and ranges all the way up to the Val d'Arno. It is also present in Spain (Villaroya). It seems to be mainly Mediterranean in distribution, although there are a couple of records north of the Alps (Schernfeld; Hajnacka). A closely related form, perhaps just a subspecies of the same species, is present in the Villafranchian of China. A complete skeleton of a dirk-tooth from Senèze may be seen in the Basel Museum.

The smilodont cats are among the most extreme of all known felids as regards the development of the canine teeth. The upper canines were greatly elongated, slender and slightly curved; evidently they were employed for the sole purpose of stabbing the prey with a rapidly killing blow. In the Late Pleistocene terminal form in America, *Smilodon*, the canines are enormous sabres, the largest tusks known in the cat family. In *Megantereon* they were smaller but otherwise of the same type. The skull is rather high and short with a triangular profile. The jaw joint was so constructed as to permit the lower jaw to swing down through an arc of 90 degrees, clearing the points of the big tusks.

Figure 28. Restoration of Dirk-tooth, *Megantereon megantereon*, based on a skeleton from Senèze.

The smilodontine tribe originates with the small *Megantereon* of the early Villafranchian in Eurasia (there is a possibly ancestral form in the Pliocene of India). The earliest members of the European species were no larger than a puma, but they tended to grow larger during the Villafranchian. At the end of the Villafranchian the genus died out in Europe, but it survived into the Middle Pleistocene both in Asia and Africa. The last Eurasian representative of the genus, *Megantereon inexpectatus* Teilhard from Choukoutien (2-Mindel), was as large as a panther. At this time the line apparently invaded North America across the Bering bridge, later to advance into South America as well. The Old World line meanwhile became extinct but the American forms, which are classified in the genus *Smilodon*, grew still larger and culminated in the monstrous South American *Smilodon neogaeus* Lund, larger than a lion.

Megantereon was a typical smilodont with a very powerful neck (strong neck muscles, necessary for stabbing), short but massive front legs and relatively feeble hind quarters. The animal obviously was not a fast runner but relied on its tusks and the great strength of its front paws, so that the prey may be visualized as a relatively large, slow-moving animal. In the Villafranchian fauna the rhinoceros may be a possibility, or perhaps young mastodonts and elephants.

In *Megantereon* the lower jaw carried a large chin process or dependent flange, which formed a kind of sheath for the tusk when the mouth was closed. In the earliest American *Smilodon* there is also a flange, but it is somewhat reduced so that the points of the teeth extend beyond it when the jaws are closed. It is probable that the large tusk bit outside the lower lip in this form. The flange became almost completely reduced in the later smilodonts [177; 234].

Figure 29. Restoration of Greater Scimitar Cat, *Homotherium sainzelli*, based on a skeleton from Senèze; coloration hypothetical.

The Greater Scimitar Cat, *Homotherium sainzelli* Aymard (*Epimachairodus crenatidens* Fabrini). This is the large sabre-tooth of the Villafranchian and early Middle Pleistocene in Europe. It is present from the Etouaires level onwards and persists to B-Waalian times (Sainzelles; Episcopia). In the French and Italian Villafranchian faunas it is the constant companion of *Megantereon*, but unlike the dirk-tooth it ranged far north and even occurs in a Villafranchian cave at Doveholes in England. The best specimens so far known are a complete skeleton from Senèze (Lyon) and a perfectly preserved skull from Mount Perrier (Clairmont).

The homotheriine sabre-tooths are rather different from the smilodontines. The sabres are shorter, more curved and much flatter; they are thin, twin-edged scimitars with razor-sharp, crenulated edges. The lower canines are not much reduced in size and the homotheres often bit in the normal way, as proved by the presence of wear facets on the canines. The cheek teeth are highly modified to form thin, sharp-edged blades.

The strong curvature of the tusks indicates that the homotheres did not make a deep stab into the body of their prey but used the canines for slashing and slicing. In contrast with the smilodonts, the head is elongated with long jaws and the skeleton is more rangy; the front limbs were high and the lower arm particularly long. But like the smilodonts, the homotheres had a short tail.

The distribution of the greater scimitar cat outside Europe has not been definitely settled, mainly because of the difficulty in distinguishing this species from the lesser scimitar cat. There is a Villafranchian form in China which may be the same species, and the same may be true for some incomplete material from the Siwaliks in India.

The origin of the homotheres is uncertain. Possibly it has some connection with an Astian genus, *Therailurus*, which is rather less specialized. On the other hand some authors think the scimitar cat may have been descended from the characteristic Pliocene sabre-tooth genus, *Machairodus* [19].

The Lesser Scimitar Cat, *Homotherium* (*Dinobastis*) *latidens* Owen. The discovery of sabre-tooth canines in British caves was a sensation a hundred years ago; they were described by Richard Owen under the name *Machairodus latidens*. Recently, complete skeletons of adult and juvenile specimens of a closely related scimitar cat, *Homotherium serum* Hay, were discovered in a cave in Texas (Friesenhahn Cave) and they have turned out to be closely similar to the European form. The lesser scimitar cats differ from *H. sainzelli* in the smaller size of the scimitars, but otherwise hardly at all. Like *H. sainzelli* this form was of the size of a lion and was rather long-legged, with a particularly long lower arm. A smaller form, *H. ultimum* Teilhard, has been found in the Middle Pleistocene of China at Choukoutien; and another form is known from the Djetis beds of Java.

The European species made its first appearance in the B-Waalian interglacial (Bacton, Forest Bed); other early finds date from 1-Günz II (Hundsheim) and C-Cromer (Cromer Forest Bed; Abbeville). The earliest forms are probably contemporaneous with the last representatives of the greater scimitar tooth, which at that time was still to be found in southeastern Europe. A definite classification of material from Gombasek, Koneprusy, Süssenborn, Stránská Skála, etc. will be most informative in this respect.

In the later Middle Pleistocene the lesser scimitar cat could still be found in Europe, though it must have been rare; it occurs in the D-Holstein (Steinheim; Montmaurin; Grotte de la Baume). In England, however, the species is found in still younger deposits. It has been found in strata that evidently can be no older than the 4-Würm at no less than three sites – Kent's Cavern, Torquay, Pin Hole and Robin Hood Cave at Creswell, Derbyshire. Thus the lesser scimitar cat was still in existence during part of the last glaciation in Europe. Its rare occurrence probably indicates a sparse population and solitary habits. Perhaps the 4-Würmian British finds represent a relict population and the species may have been extinct on the continent at that time.

H. latidens has not yet been found in southern Europe, but its presence in North Africa (e.g. Ternifine, probably 2-Mindel) suggests that its range included the Mediterranean coasts at least in the Middle Pleistocene.

The fossils found in Europe are mainly isolated teeth and jaws, but the partial skeleton from Hundsheim resembles the Texan form. The mode of life of the latter has been illuminated by the discovery of great numbers of teeth and bones of young elephants in Friesenhahn Cave. It suggests that juvenile elephants formed the staple diet of *H. Dinobastis*, while adult elephants were immune to their attacks. How the great cat managed to avoid the adult elephants while preying on the young is a matter for fascinating but perhaps futile speculation. However this may be, the extinction of the scimitar cats might be connected with the vanishing of their special prey in both Europe and America at the end of the Ice Age [6; 196; 207].

With Martelli's Wild Cat, *Felis lunensis* Martelli we come to the tribe Felini, which includes all the living felids. The larger species of this group have a fairly good fossil record, but the small cats are rare everywhere except in the very youngest cave deposits.

This rule certainly holds for the Villafranchian *F. lunensis*. The description is based on a jaw from Olivola of a cat the size of a modern wild cat, but with somewhat different dental characters. Other remains of cats, probably of the same species, are found sporadically at different levels, beginning with the Pardines. This may also hold for some Astian or Villafranchian remains from Csarnóta, referred tentatively to the steppe cat, *F. manul* Pallas.

The species may have persisted in Europe into the Middle Pleistocene; a fossil (unfortunately but a single tooth) from the Forest Bed at West Runton (C-Cromer interglacial) resembles Martelli's cat rather than modern wild cat. Perhaps this is a transitional form. There are other finds demonstrating the presence of small cats in Europe during the early Middle Pleistocene, but either they are indeterminable as to species (an incomplete radius from Mauer) or else they have not been studied from this point of view (e.g. Schernfeld) [181].

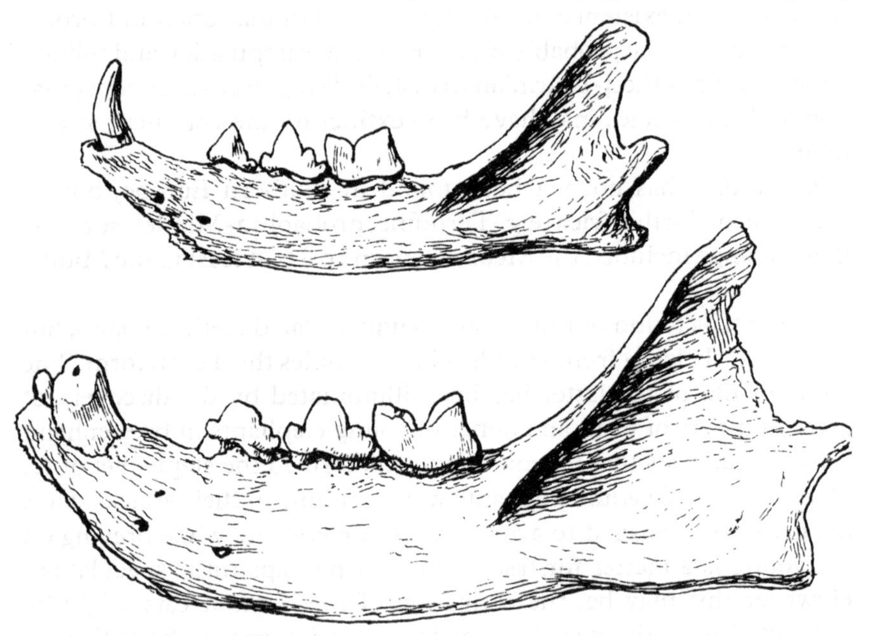

Figure 30. Mandibles of domestic tabby (top) and European Wild Cat, *Felis silvestris* (bottom). Both from Ravencliffe Cave, Late Pleistocene or Holocene. $1\frac{1}{2}$ times natural size.

The Wild Cat, *Felis silvestris* Schreber. The wild cat is fairly common in cave deposits dating from the last glaciation and Postglacial. Somewhat more sporadic finds in older strata back to the D-Holsteinian are also referable to this species, but C-Cromerian fossils may be *F. lunensis* or transitional forms. Apparently the Villafranchian species was directly ancestral to the living one; only a complete fossil record can show just where the species boundary should be drawn.

The material from the D-Holstein represents a large form of the modern wild cat (Heppenloch; Lunel-Viel; Tarkö). The F-Eemian form is only slightly smaller (various travertines and caves).

The wild cat today has a very wide geographic range and inhabits most of Africa down to the Cape, large parts of Europe (although exterminated in many areas) and a great tract of Asia to Turkestan and India. In the fossil state the species has an excellent record in Palestine, especially in the 4-Würm.

The Pleistocene wild cat is remarkable for its size. Nowadays there is a very large form of wild cat in southern Spain, *F. silvestris tartessia* Miller. In other parts of Europe the wild cats are somewhat smaller, although larger than the domestic tabby. The Pleistocene forms, both in Europe and Palestine, are as large as the Spanish subspecies. Towards the end of the 4-Würm and in the Postglacial there occurred a gradual size decrease, leading to the type seen in central Europe and Scotland today.

The hunting method of the wild cat is similar to that of the house cat: it lurks in ambush or stalks its prey, makes a dash, beats down the prey with its paws and bites to kill. The staple diet consists of small rodents. The living European form is a woodland animal preferring mixed forest, but other subspecies are steppe forms. One of the latter, the Egyptian *F. silvestris lybica* Forster, is probably ancestral to the domestic cat; domestication dates from about 2500 BC.

The European wild cat is easy to identify externally from its thick, grey and black banded fur and its moderately long, thick tail, which does not taper as in the tabby. The identification of skeletal remains is much more difficult, especially in the case of fragmentary material – which, of course, is the usual state of fossil bones. Often, however, size alone may be sufficient to distinguish wild cat from the domestic form [181].

The Steppe Cat, *Felis manul* Pallas. This species has been identified in the 4-Würm in Europe (Kesslerloch; Schweizersbild) but the identifications have been seriously doubted. In the early Middle Pleistocene (Kamyk) there are however remains of a forerunner of this species, clearly identifiable on the very characteristic carnassial; the material is at present undescribed. The modern steppe cat inhabits central Asia in a belt from Transcaucasia to western China.

The Jungle Cat, *Felis chaus* Gueldenstaedt. This widely distributed Asiatic and African felid ranges into eastern Europe in a limited zone on the west coast of the Caspian Sea south of the Volga estuary. It is larger than the wild cat but smaller than the true lynxes; some external characters such as the tufted ears and the relatively short tail are reminiscent of the latter, but the dentition is not lynx-like and the second upper premolar, characteristically absent in lynxes, is present in *F. chaus*.

There are many records in the Pleistocene of Europe of felids larger than the wild cat but smaller than any of the lynxes, but the material has mostly been too fragmentary for definite identification. There appears to be only one reliable record [95]; this is from the F-Eemian interglacial travertines at Untertürkheim near Stuttgart. That this tropical and subtropical animal ranged so far north may be due to the warm climate during the optimum of the last interglacial. Unfortunately the Untertürkheim specimen consists only of parts of the fore and hind limb, while skull and tooth remains are missing.

In Palestine the species is present in the 4-Würm [179].

The Issoire Lynx, *Felis issiodorensis* Croizet & Jobert. The characteristic lynx of the Villafranchian is represented by numerous finds ranging throughout that stage: Etouaires, and later at nearly all the sites mentioned from Italy, France and Spain. The species has also been found in eastern Europe and a close relative, *F. shansius* Teilhard, occurs in Villafranchian strata in China; it may be simply a local subspecies of the same species.

The entire skeleton of the European form is known, mainly on the basis of finds from Mount Perrier. The Issoire lynx is a well-defined species. The dentition is of the lynx type, but otherwise the species differs from typical lynxes. The head is quite large with a long face, the body is long and the legs short, so that the Issoire lynx may have resembled a small puma rather than a lynx; the tail, however, appears to have been short as in the lynxes. Evidently the Issoire lynx is close to the starting point of the evolution of lynxes. This shows that the small-headed, long-legged and short-bodied lynxes of today have evolved out of ancestral forms with the body proportions of normal cats.

The evolution of the northern species of lynx (*F. lynx*, *F. pardina* and *F. canadensis*) may probably be interpreted as an adaptation to the

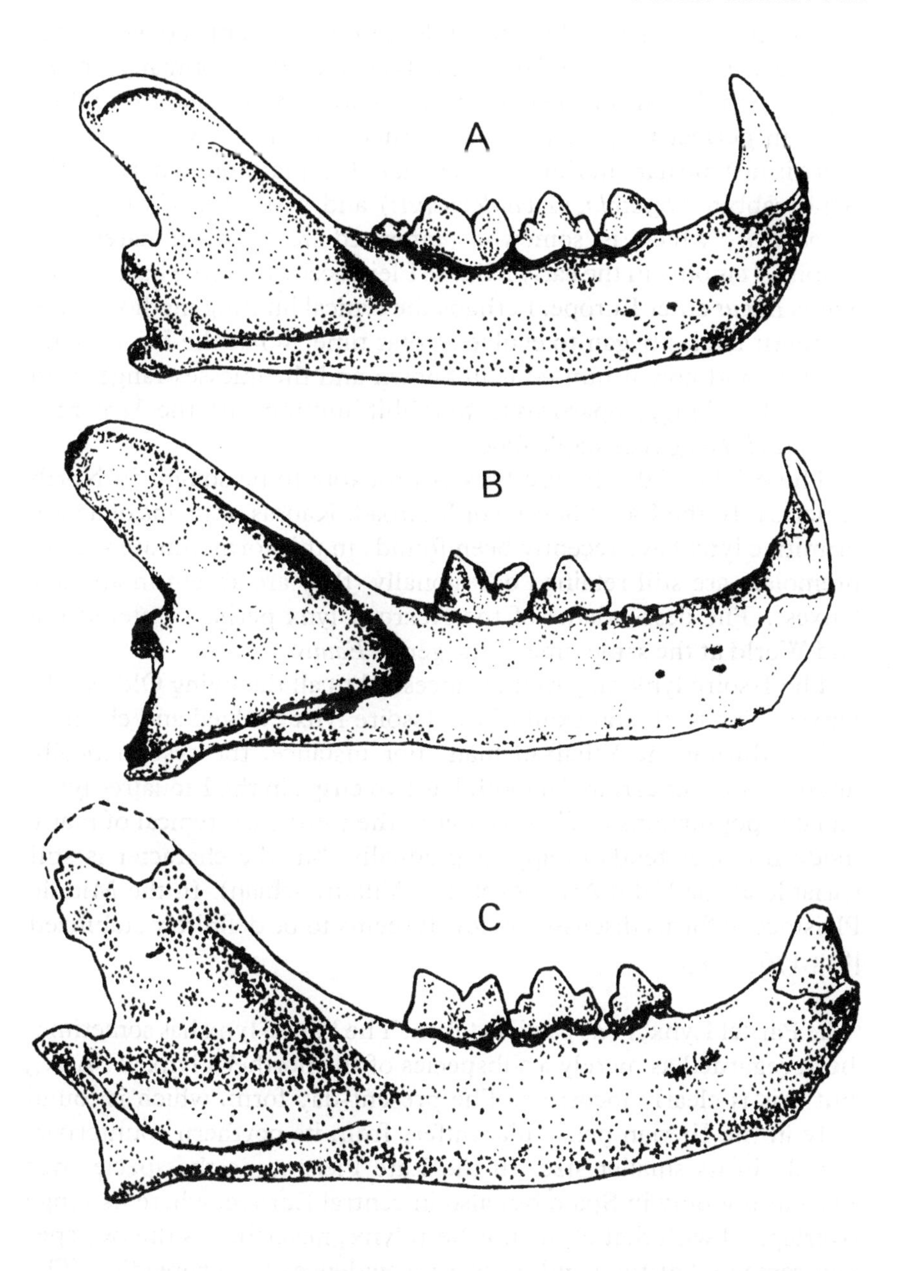

Figure 31. Mandibles of A, Northern Lynx, *Felis lynx*, Finland, Recent, specimen with supernumerary molar; B, Issoire Lynx, *F. issiodorensis*, middle Villafranchian, Saint-Vallier, with trace of third cusp on molar; C, Issoire Lynx, *F. issiodorensis*, early Villafranchian, Etouaires, with only two cusps on molar. Natural size.

hare as the main prey. The dependence on hares appears very convincingly in the trapping data of the Hudson's Bay Company, showing the populations of the two species – prey and predator – to fluctuate in perfect rapport with each other. But in the Villafranchian there is no true hare in Europe; the related animals found here are the Arno rabbit (*Oryctolagus valdarnensis*) and a species of the genus *Hypolagus*, which is somewhat intermediate between hares and rabbits. It is only in the early Middle Pleistocene that true hares of the genus *Lepus* enter Europe. Perhaps the typical hunting method of the northern lynxes began to evolve at that time. What the Issoire lynx lived on and how it hunted is unknown and the guesses range from arboreal stalking, leopard style, to rabbit hunting with the Arno rabbit and *Hypolagus* as staple food.

The origin of the Issoire lynx is probably to be sought in North America. In the Late Pliocene of Rexroad, Kansas, remains of a very primitive lynx have recently been found; in this form minute second premolars are still retained occasionally (they are absent in all later lynxes). The descendants of the Rexroad lynx perhaps entered the Old World at the same time as the genus *Equus*.

The Issoire lynx may well be ancestral to all the living Old World lynxes. The teeth and skull of the Issoire lynx evolved and changed visibly during the Villafranchian. For instance, the systematically important lower carnassial tooth has two cusps in the Etouaires lynx. In later populations of *F. issiodorensis* the third cusp typical of many modern lynxes tends to appear gradually, but the character is still variable at the Val d'Arno level (late Villafranchian). In the Middle Pleistocene forms discussed below it seems to be definitely stabilized [81; 176; 302].

The Pardel Lynx, *Felis pardina* Oken. The Pardel lynx has sometimes been regarded as merely a subspecies of the northern lynx (*F. lynx*), but this is clearly incorrect. The present-day form, which is found only in the Iberian peninsula, differs from its northern counterpart mainly in its smaller size. In the Late Pleistocene the species was present not only in Spain but also in central Europe, where its range overlapped with that of the northern lynx; nevertheless the two species remained distinct and there is no evidence of interbreeding. The modern Pardel lynx is a less exclusive hare hunter than the Canada lynx and northern lynx and rabbits and small rodents form a large part of its food.

Lynxes that have usually been referred to *F. issiodorensis* occur at several Middle Pleistocene sites in Europe (Episcopia; Mauer; Solilhac-Blanzac); they might perhaps as well be regarded as early representatives of the Pardel lynx. Unfortunately we have only a few teeth and jaws of these animals, so that no definite decision appears possible at present. In size the Middle Pleistocene form exceeds the modern Pardel lynx – it is as big as the Issoire lynx or the modern northern lynx – but this is also true for the fossil Pardel lynx from the 4-Würm in Spain, of which excellent material is known from the Gibraltar caves and Cueva del Toll near Moyá (Barcelona). These animals are nearly as large as the northern lynx, so that it would seem that the body size of this species has diminished since the Ice Age [176; 302].

The Northern Lynx, *Felis lynx* Linné. The earliest certain finds of this species date from the F-Eemian; it occurs, for instance, in the travertines near Weimar. It is often found in deposits from the 4-Würm, when it ranged into Italy and the Balkans; alleged Spanish finds are more likely to be the large Pleistocene *F. pardina.*

The fossil specimens are quite similar to the modern form, which indicates that there has been no size reduction in this species after the Pleistocene. *F. lynx* is the largest of the living lynxes and has the lynx characters in greater measure than any other species. It is capable of standing leaps up to 15 feet; this is the basis of its hunting method. The lynx is dependent on woodland and hence absent in the tundra phases of the 4-Würm.

Although it has been exterminated in most of its European habitat, the lynx still has a great geographic range; it inhabits all of the northern Asiatic taiga belt and ranges south to Asia Minor, Persia and Tibet. The American *F. canadensis* Kerr is a related form but of smaller size, approximately equal to the Pardel lynx.

The third cusp of the lower carnassial is very well developed in *F. lynx*, and some specimens – about ten per cent of the individuals – carry a small second molar behind the carnassial. This is an odd reversion of the normal trend in felid evolution, which tends to elimination of the tubercular cheek teeth behind the carnassials [176; 281].

The Tuscany Lion, *Felis toscana* Schaub. This is the large cat typical of the Villafranchian. About the size of the small lion, it differs from *F. leo* in various details in the dentition. Whether its body build resembled that of the lion we cannot tell at present, for very little is known of the skeleton and even the sole known skull (from Val d'Arno) is badly crushed and deformed.

The Tuscany lion makes sporadic appearances at various sites from Etouaires times on; it has not been found in Spain. On the other hand, it did range north of the Alps and is very well represented at A-Tegelen, so that it seems to have been quite common at the end of the Villafranchian. Perhaps *F. gombaszögensis* Kretzoi, a rather little-known species from Gombasek, Kövesvarad and Koneprusy, is identical with the Villafranchian form; that would extend its range to the 2-Mindel.

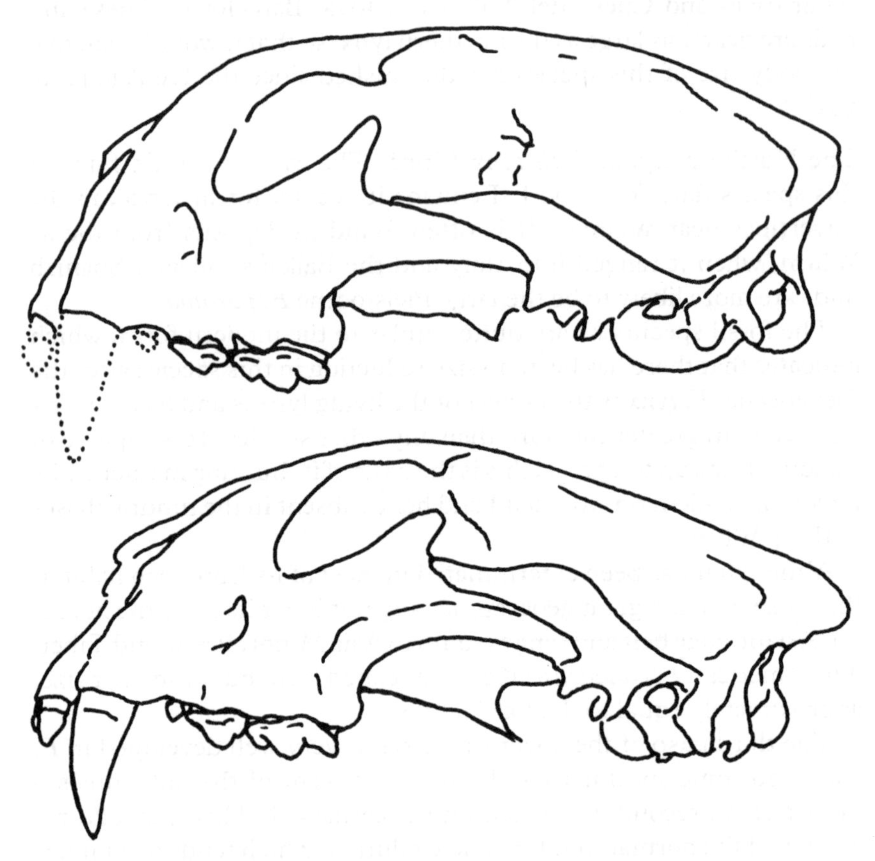

Figure 32. Skulls of Villafranchian Leopard, *Felis schaubi* (above) and modern Leopard, *F. pardus* (below). Not to scale.

The Tuscany lion does not seem to qualify as an ancestor of the true lion; in some respects it shows closer affinity to the leopards. The same is true for a Chinese great cat (*F. palaeosinensis* Zdansky) of the same date; possibly all these forms are really a single species, in which case Zdansky's name has precedence. In India there is also a large true cat in the Villafranchian, *F. cristata* Falconer & Cautley, but it is aberrant and does not resemble any modern species. Perhaps we must look to Africa for the ancestors of both the lion and the tiger [243].

Owen's Panther, *Felis pardoides* Owen. This leopard-like cat was originally described on the basis of a few teeth from the East Anglian Crags, probably from the late Villafranchian. It is possible that a small leopard from Saint-Vallier, *F. schaubi* Viret, may belong to the same species; the size is the same. Again, the same may be true for some fragmentary specimens from Villány and Csarnóta, but knowledge of this form is still quite incomplete [213; 302].

The Lion or Cave Lion, *Felis leo* Linné. This great cat is the best known of all the Pleistocene felids of Europe. Yet generations of paleontologists have tried to decide whether the cave lion was a real lion, or perhaps a tiger, or a species distinct from both. There is still no consensus of opinion on this point. Most German authors regard the cave lion as a species of its own, *F. spelaea* Goldfuss, while most English and French students favour the lion alternative. The cave tiger solution nowadays has few advocates, although one author has advanced the original idea that both species existed in Europe in the Pleistocene – the tiger during glaciations, the lion during interglacials.

The tiger alternative need not detain us. There are gigantic tigers in the Pleistocene of China and they show what a tiger looks like when blown up to this format [48]. It is quite different from the European great cat: it has much larger canine teeth in relation to the premolars and the nasal opening has a different shape. In these and many other characters the cave lion resembles the true lion. Differences from the living lion are minor and many of them are simply by-products of the larger size in the European form. There is thus insufficient evidence for a specific separation between *F. leo* and *F. spelaea*.

The lion entered the European scene in the C-Cromer (Mauer; Forest Bed; Mosbach) with a gigantic form. African lions of the same date resemble the Mauer lion, though they do not reach such an enormous size. The Cromerian lion in Europe may be the largest felid that ever existed. In the D-Holsteinian, the lion is common (Swanscombe; Steinheim; Heppenloch; Lunel-Viel, etc.); it has also been found at E-Ilford in large numbers. 3-Riss finds are numerous (Châtillon Saint-Jean; Achenheim; Saale terraces; Tornewton Cave,

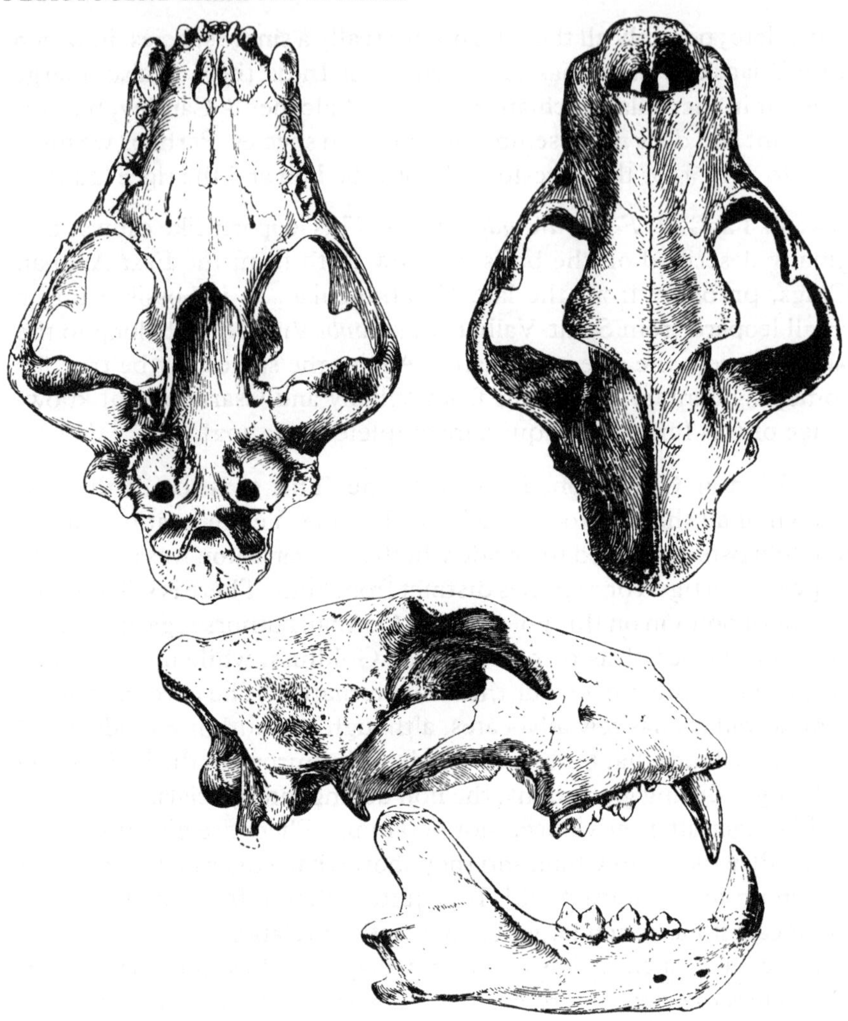

Figure 33. Skull and mandible of Cave Lion, *Felis leo spelaea*, Pleistocene, USSR. Two-ninths natural size. After Gromova.

etc.). In cave deposits from F-Eemian and 4-Würmian times the lion is frequently encountered, though rarely in great numbers. An interesting exception is the Wierzchowska Cave in southern Poland, where a large number of lion bones, representing about twenty animals, have been found. Two size groups, representing males and females, may be distinguished in the material from this cave. The sexual size variation has often caused confusion and led to the erroneous conclusion that there were two different races or species of cave lion. The

Late Pleistocene cave lion was not as large as its forerunners in the C-Cromer, but was distinctly larger than modern African lions.

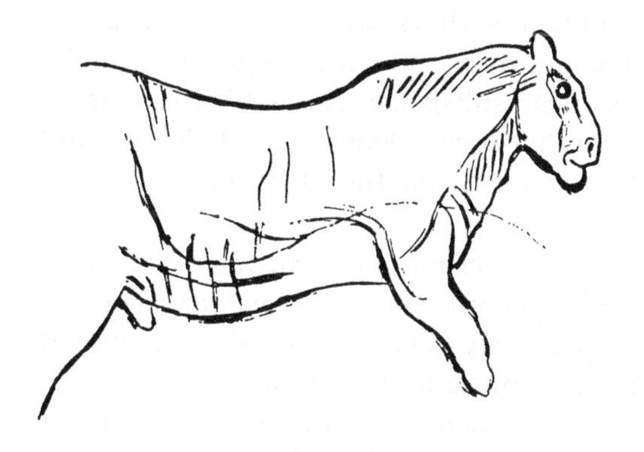

Figure 34. Engraving of Cave Lion, *Felis leo spelaea*. Les Combarelles, Dordogne. After Breuil.

The lion (probably the cave lion, *F. leo spelaea*) survived in historical times in the Balkans and Asia Minor. Probably there was a continuous population extending into India, but now only a small group of the Asiatic lion survives. The Indian population was connected by way of Arabia with the African stock, which has also been decimated in recent years. The famous Berber lion and other North African forms are gone, and so is the Cape lion.

The lion is the only living felid to hunt in groups. Whether this method was practised by the European cave lion is of course unknown to us. Perhaps group tactics were necessary to cope with such powerful beasts as the aurochs and the bison. Of course the cave lion may have preyed on other forms of large game, for instance the giant deer, red deer and elk.

The comparatively large number of finds may indicate that the lion was a rather common animal all over Europe, perhaps more so than the leopard and almost certainly more so than the scimitar-tooth (*Homotherium*).

A possibly ancestral form combining tiger-like and lion-like characters has recently been described from the Villafranchian of East Africa [80; 128; 187].

The Leopard, *Felis pardus* Linné. Like the lion, this species entered Europe in the C-Cromerian; the best early specimen is a lower jaw from Mauer. The leopard already had a wide distribution at this early date: it is found in Africa (Transvaal Caves) and in Asia (Choukoutien; Java). For all of the Pleistocene in Europe, Stehlin in 1933 was able to list 69 sites with fossil leopard and many more have been added since then. Though it seems never to have been so abundant in numbers as the lion (lion sites are more than three times as numerous as leopard sites), the leopard ranged with its larger cousin throughout the Middle and Late Pleistocene, dying out in Europe towards the end of the 4-Würm. The European subspecies is notable for its large size, but this also holds for the Pleistocene leopards of Palestine and China. The African Pleistocene form, on the other hand, was no larger than the living one.

The ancestry of the leopard is uncertain. Descent from the Villafranchian *F. pardoides* is a distinct possibility, *F. palaeosinensis* of China is another, and what Africa may have harboured in the line of Villafranchian proto-leopards is at present little known.

The geographic range of the leopard in Europe was found by Stehlin to be more restricted than that of the lion. Its northern boundary passes from southern England through Liège, Thuringia, Moravia and the Transylvanian Alps. The leopard-like cats of central Asia belong to the species *F. uncia* Schreber, the snow leopard or irbis.

The living leopard is widely distributed and inhabits many kinds of environments, including tropical rain forests, steppes and mountains even above the snow line. No wonder that the species flourished in Europe during the Ice Age, even in the coldest phases of the 4-Würm. A skilful climber, the leopard stalks its prey or lies in ambush in the trees or among the rocks. The prey consists of medium-sized herbivores, of which there were a great number in Europe during the Ice Age, such as the boar and several kinds of deer. Long of body, with short legs, a small head and a very long tail, the leopard is unique among the larger cats in having almost mustelid-like proportions [246].

The Giant Cheetah, *Acinonyx pardinensis* Croizet & Jobert. This species is known mainly from the Villafranchian, beginning with the basal level (Villafranca d'Asti; Etouaires), although it did survive in the early Middle Pleistocene (Saint-Estève; Hundsheim). The Villafranchian was clearly its heyday in Europe; its slender limb bones and typical teeth are quite common in the Villafranchian, at Pardines, Villaroya, Senèze, Olivola and other sites.

Equal in size to a modern lion, it was indeed a giant of a cheetah. The living cheetah is known to be the fastest runner of all animals, galloping easily at 56 mph [103]. In this respect the giant cheetah was probably as advanced as its living relative, judging from the skeletal elements. *A. pardinensis* on the hunt must have been a fabulous sight.

The giant cheetah was gradually reduced in size during the Villafranchian. The late Villafranchian fossils are decidedly smaller than the early ones and in the Middle Pleistocene the size reduction had gone so far that this form has even been regarded as a distinct species (*A. intermedius* Thenius). However, the Saint-Estève form is as large as the Villafranchian one [34].

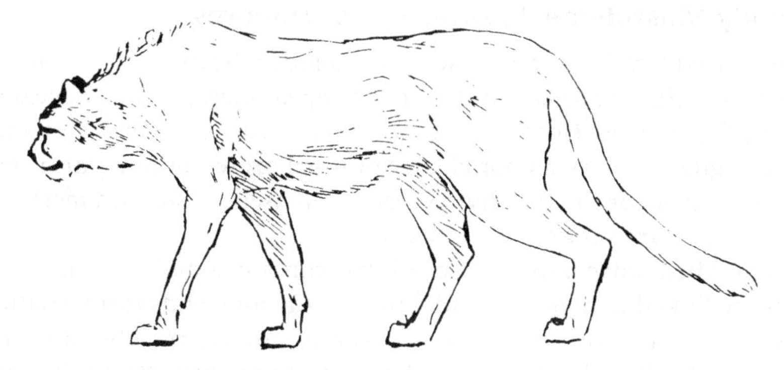

Figure 35. Restoration of Giant Cheetah, *Acinonyx pardinensis*, after a mount in the Museum of Basel (skeletal parts from various Villafranchian sites).

The giant cheetah has also been found in the Villafranchian of India and China. Again, there is a transition to a smaller form during the Middle Pleistocene in China and the Late Pleistocene cheetahs in this area approach the living species closely enough to be classified within it (*Acinonyx jubatus* Schreber), though they are still a little larger on average. It would seem that there was a gradual transition from the great Villafranchian species to the living cheetah. On the other hand the pre-Villafranchian history of the cheetah is unknown to us. As the earliest Villafranchian finds already show all of the cheetah characters, it may be assumed that the genus *Acinonyx* had a long history in the Pliocene.

At present, *A. jubatus* inhabits most of the African steppes and savannas; it also existed up to recent years in southwestern Asia and India, but not in eastern Asia, where it became extinct at the close of the Ice Age. Unfortunately the species has now been exterminated in all of its Asiatic range, where it used to be tamed and kept as a hunting animal. The range in Africa has also shrunk seriously in the north. Cheetahs were present in historical times in Asia Minor; they have, for instance, been found at Troy in post-Homeric (Greek Ilion) association.

Ecologically the cheetah is dependent on open ground and avoids forests, where of course its great speed would be useless. Its presence in southern and central Europe during the Villafranchian, together with *Euryboas*, indicates that extensive grasslands were available even during the forest episodes [243; 302].

Family Mustelidae, Weasel-like Carnivores

The varied family of the Mustelidae includes the majority of the carnivores of Europe today; and so, to all appearances, it must have done in the Pleistocene. But the bones of mustelids are comparatively small and fragile and have a poor chance of fossil preservation. This is particularly true for the tree-living forms – weasels, stoats and martens – all of which are very scarce as fossils.

The Mustelidae is one of the oldest carnivore families. As the viverrids are linked to hyaenids and felids, so the mustelids form a natural group together with the dogs, procyonids and bears. The mustelids have established themselves in the trees, on the ground, in the earth and in the water. Some are exclusive meat-eaters, others are omnivorous with a strong vegetarian bias. They live in environments ranging from the Arctic to the tropics. No other carnivore family exhibits such a diversity of adaptive types.

The family may be divided into a number of subfamilies. Of these the Mustelinae, with the weasels, polecats, martens and gluttons, is especially rich in species. The subfamilies of the ratels (Mellivorinae) and skunks (Mephitinae) are absent in the Pleistocene of Europe, although they were well represented in the Tertiary. The badgers or Melinae on the other hand form an important element and so does the otter subfamily, the Lutrinae.

Schlosser's Glutton, *Gulo schlosseri* Kormos. Glutton-like mustelids are entirely lacking in the Villafranchian of Europe and the earliest Pleistocene evidence of this group comes in the early Middle Pleistocene at Episcopia (B-Waalian). The species has also been identified in strata dating from I-Günz II and C-Cromer (Mosbach I; Forest Bed; Gombasek; Erpfingen; Stránská Skála), but these may be transitional to the true glutton. The size is almost the only key to the identification of Schlosser's glutton and it is actually only at Episcopia that a form decidedly smaller than the living glutton has been found.

Indirectly, however, another character differentiating Schlosser's glutton from the living species is evident. It occurs at Episcopia in a decidedly warm fauna together with macaque and other warmth-loving species. Perhaps this means that Schlosser's glutton did not yet possess the boreal adaptation of its living descendant; it would seem that the glutton embarked upon its modern way of life as late as the 2-Mindel glaciation.

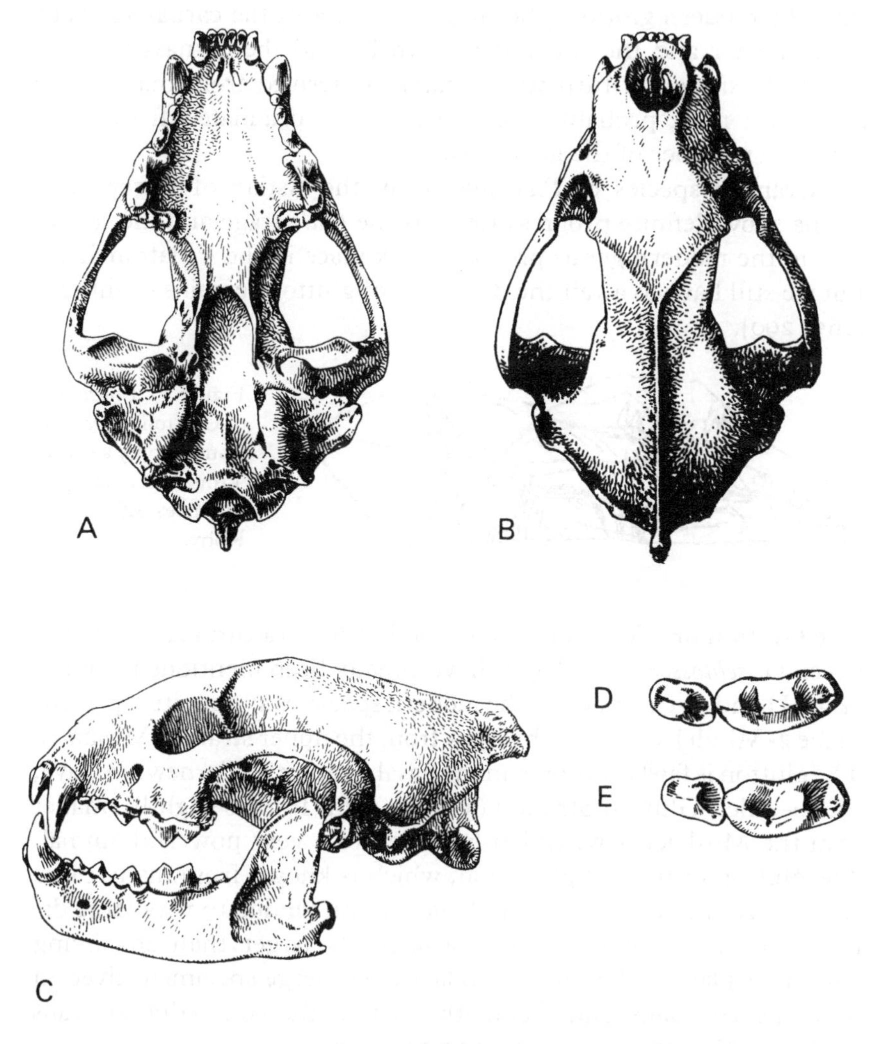

Figure 36. A–C, skull and mandible of Glutton, *Gulo gulo*, Recent; one-fourth natural size. D–E, lower P_4 and carnassial of D, Schlosser's Glutton, *G. schlosseri* and E, Recent Glutton; two-thirds natural size. A–C after Gromova; D–E after Stehlin.

The transition between Schlosser's glutton and true glutton is so gradual that it can hardly be doubted that *G. schlosseri* was the immediate ancestor of the living species. To trace the history of the glutton stock backward from the *schlosseri* stage is more difficult, but at least the general outline is known. There existed during the Pliocene, both in Eurasia and North America, a genus called *Plesiogulo* which is probably ancestral to *Gulo*. It has a less specialized dentition than the modern glutton: the shearing blades of the carnassial teeth were shorter and the tubercular teeth behind the carnassials were larger. *Plesiogulo* in turn was apparently derived from marten-like ancestors; most probably the glutton stock originated from some Miocene member of the genus *Martes*.

Advanced species of *Plesiogulo* from the Astian of Europe and China show definite progress towards the *Gulo* stage and the transition to the modern genus probably took place in the Villafranchian; but we still have to await the discovery of glutton remains of this age [265; 290].

Figure 37. Engraving of Glutton, *Gulo gulo*, transfixed by a spear. Trois-Frères in the Pyrenees. After Koby.

The Glutton or Wolverine, *Gulo gulo* Linné. Transitional forms between *G. schlosseri* and *G. gulo* have already been mentioned from 1-Günz II and C-Cromer and the modern species was definitely present in the 2-Mindel, as proved by a jaw from the upper strata at Mosbach. The glutton is fairly common in the basal stratum of Tornewton Cave, the so-called Glutton Stratum (3-Riss). This form is slightly smaller than the Mosbach jaw, which belonged to a very powerful animal. The glutton of the last glaciation, which is known from many caves and ranged into the Balkans and Italy, again tended to very large dimensions; some of these are considerably larger than any living glutton. As late as the early Postglacial very large specimens lived in northern Germany and Denmark, so that the size reduction falls entirely within the last 8,000 or 10,000 years.

In China, too, true glutton appears in the 2-Mindel (at Choukoutien). The species now has a circumpolar distribution; the living

American form belongs to the same species as the Eurasian. It seems to have immigrated in North America in the 3-Riss, for the earliest specimens come from cave deposits of about this age (Port Kennedy Cave in Pennsylvania; Cumberland Cave in Maryland).

The glutton is highly dependent on bogs, an environment in which it is superior to the wolf, elsewhere its most dangerous rival [151]. Its summer diet consists of eggs, caterpillars, berries and small rodents; in the winter it will kill mammals and birds as available and may prey on animals very much larger than itself, for instance the elk. It also eats carrion, being able to crack bones with its strong premolars almost like a hyena. The prey is usually killed by a bite in the back of the neck. When attacking a large animal, the glutton leaps on to its back and clings to it, while its teeth work into the neck. Perhaps the large Pleistocene variety may on occasion have brought down such large game as the giant deer [28; 164; 290].

The Primitive Marten, *Martes vetus* Kretzoi (*M. intermedius* Heller). The genus *Martes* probably existed in Europe throughout the Pleistocene, but fossils are scarce and only part of its history is known. In the late Astian and perhaps the early Villafranchian a large species, *M. wenzensis* Stach, lived in Europe; it was apparently related to the North American Fisher (*M. pennanti* Erxleben). There is no later evidence of the fisher group in Europe, and the martens make their next appearance in the early Middle Pleistocene (Sackdilling; Schernfeld; Süssenborn; Kövesvarad; Forest Bed) with the species *M. vetus*.

The primitive marten may well be ancestral to both of the present-day European martens, the pine marten and beech marten; it has characters resembling both. Unfortunately the species is not too well known. A suggestion of its ecology may be gained from the fact that *M. vetus* is absent in the Hungaro-Roumanian faunas, which otherwise tend to be richer in small mammals than those of central and western Europe. Its absence in the steppe faunas may suggest that *M. vetus* was arboreal in habits, that is to say an ecological forerunner of *M. martes*.

M. vetus cannot be derived from *M. wenzensis*, so that its origin is unknown at present. The genus *Martes* is present as early as the beginning of the Pliocene in Eurasia [97].

The Pine Marten, *Martes martes* Linné. This carnivore has been definitely identified in deposits from the F-Eemian, both in travertine and cave deposits. One of the most interesting finds has been recorded in the cave of Drachenloch in St Gallen, Switzerland, which lies 1800 ft. higher than the highest present-day record for the species. This is one of many indications of a particularly warm climate in the F-Eemian.

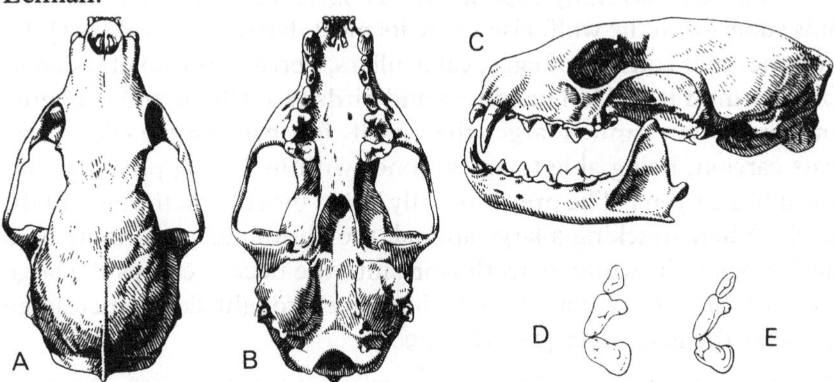

Figure 38. A–C, skull and mandible of Pine Marten, *Martes martes*; one-half natural size. D–E, upper P^3–M^1 of D, *M. martes* and E, Beech Marten, *M. foina*; two-thirds natural size. A–C after Gromova; D–E after Miller.

Scattered remains of martens have been found in earlier deposits, but species determinations are not certain. As regards earlier Middle Pleistocene finds, reference to *M. vetus* would seem probable; that species is likely to be ancestral to *M. martes*. In the case of later finds, confusion with beech marten is a possibility to be kept in mind. Specimens from Tarkö and Breitenberg Cave may represent the D-Holsteinian pine marten.

The pine marten of the 4-Würm and early Postglacial is a large form; the present-day species attains similar size only in the far north. In most areas there seems to have been a gradual decrease in size during the Postglacial.

Externally the pine marten may be separated from the beech marten by its slightly larger size, longer legs, and a yellow throat patch. It usually keeps to pine forests, so that it is a good climate indicator. An accomplished climber, the pine marten does most of its hunting in the trees and leaps from branch to branch when pursuing the squirrel, its staple diet. The marten will also take birds, eggs and insects and eats vegetable food such as berries and fruits. The species inhabits large tracts of Europe and Asia but is replaced to the east by a related species the sable (*M. zibellina* Linné), a somewhat smaller form [52; 265].

The Beech Marten, *Martes foina* Erxleben. This species has been found in Postglacial deposits in Europe. There are also several records that may be Late Pleistocene but it is possible that it has been confused with pine marten. In Palestine and Iraq, however, *M. foina* is common in cave deposits from 4-Würm, while *M. martes* is rare or absent. Probably the beech marten entered Europe from the east at the end of the Ice Age.

Separation of beech marten and pine marten on the basis of teeth and skeletal fragments is rather difficult, so that determinations of some of the Pleistocene material may not be reliable. The differences in the dentition are slight; for instance in the beech marten the carnassials are relatively more powerful, with slightly longer blades; also the skull is somewhat broader. A beech marten in the flesh is easily identified by its white throat patch.

The beech marten is less dependent on woodlands than the pine marten and is well adapted to a steppe environment, including the man-made steppes of the present day; in some countries it is called the house marten. The species is now common in southern and central Europe up to Denmark and in the greater part of Asia. It is less resistant to cold than the pine marten, which makes identification of this species in European deposits from the 4-Würm somewhat suspect. Of course it may have been present in the interglacials.

The species may have been descended from *M. vetus* collaterally with the pine marten; but there is no successional evidence to bear out the relationship, which has only been suggested because of some *foina*-like characters in the primitive marten. In spite of the resemblance and obvious relationship between beech marten and pine marten, the two species are very dissimilar ecologically [179].

The Baranya Polecat, *Baranogale antiqua* Pomel (*B. helbingi* Kormos). This is a representative of the group of banded polecats with the living genera *Zorilla* and *Vormela*. The group was highly varied and numerous in Europe during the Early and Middle Pleistocene and the Baranya polecat is one of the most common carnivores of that time. The oldest finds, from Weze, Podlesice and Csarnóta, date from the Astian, but it ranges to the earliest Middle Pleistocene (Villány; Beremend; Rebielice). A slightly larger form of the same species has been found at Etouaires and Saint-Vallier.

B. antiqua was rather larger than the living banded polecats but otherwise seems to have resembled the Cape polecat, *Zorilla striata*.

Its presence in the faunas of Etouaires and Saint-Vallier may suggest that it was a less exclusive steppe form than its living relatives. Unfortunately we do not know enough of its skeleton to determine its mode of life.

The ancestry of the genus *Baranogale* is uncertain but it has been suggested that the Baranya polecat was an invader from Africa, since its closest living relatives are found there. No living descendants of *Baranogale* are known [302].

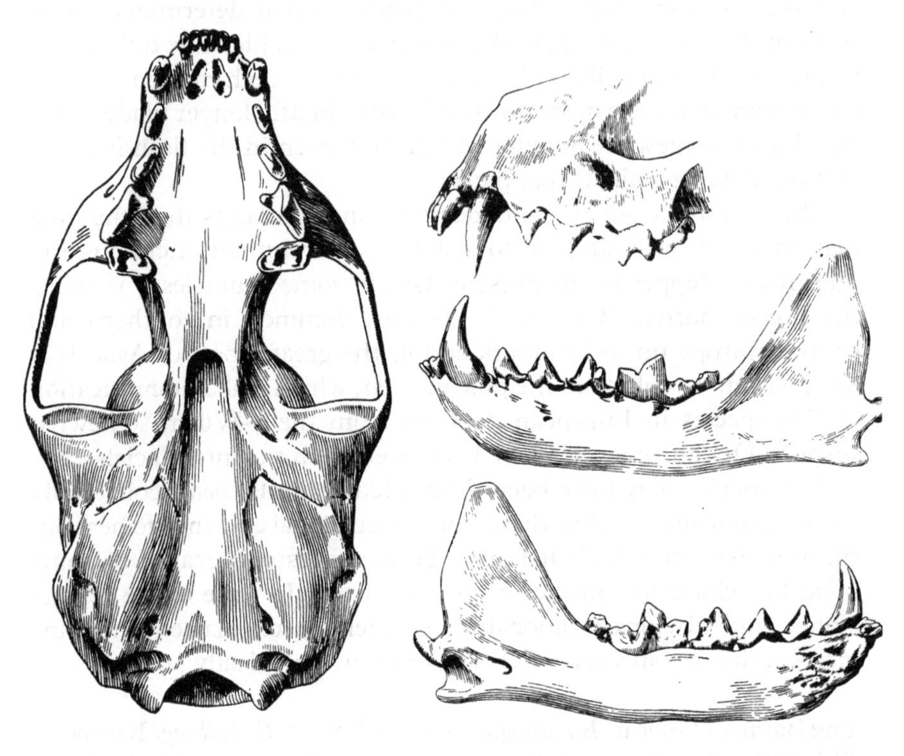

Figure 39. Skull and mandible of Baranya Polecat, *Baranogale antiqua*, Villafranchian, Saint-Vallier. Natural size. After Viret.

The Beremend Polecat, *Vormela beremendensis* Petényi. This species is regarded as the immediate forerunner of the living marbled polecat, which it resembles very closely both as regards size and dental morphology; the head is stated to be somewhat narrower in the fossil form. The species is known from the Astian, Villafranchian and early Middle Pleistocene (Podlesice; Villány; Beremend; Nagyharsányhegy) in Hungary and Poland. Probably the Beremend polecat, like its living descendant, was a steppe animal, preying on burrowing rodents which it hunted in their subterranean nests.

The ancestry of the genus *Vormela* is uncertain, but a Pliocene mustelid from China may have a place here. Most probably the genera *Zorilla*, *Baranogale* and *Vormela* (which some would unite in a single genus) evolved from a common ancestor at a relatively late date in the Tertiary [141].

The Marbled Polecat, *Vormela peregusna* Gueldenstaedt. The living marbled polecat is not known in the fossil state in the Pleistocene of Europe except Russia; in the Caucasus, for instance, it is found in the Binagady fauna perhaps of F-Eemian date. In Palestine there are several finds dating from the 4-Würm; the earliest are interstadial in age (4-Würm I–II).

The marbled polecat has a strikingly multi-coloured fur, brown with yellow patches on the body but black with white patches on the head and neck. Its mode of life has been briefly described above in the section on *V. beremendensis*. Unfortunately the time hiatus between the last known Beremend polecats and the earliest true marbled polecats is so great that it is impossible to give an accurate date for the emergence of the modern species – if the two are indeed distinct.

The species now inhabits steppe and desert areas in southern Russia, on the Balkan Peninsula and eastward through Asia into China [179].

The Ardé Polecat, *Enhydrictis ardea* Bravard (*Pannonictis pilgrimi* Kormos). This species has been described under various names from different sites, causing systematists a lot of trouble. It turns up in the basal Villafranchian (Etouaires) and persists throughout this stage, with a preference for forest faunas (Saint-Vallier; Villaroya; Olivola; Tegelen). In England it has been found in the Red Crag, in Hungary in the late Villafranchian and early Middle Pleistocene up to 1-Günz II. It has also been found in Sardinia. At one time it was thought to be amphibious, somewhat like the minks – hence the genus name – but study of the limb bones has not confirmed this theory. Like the related *Pannonictis*, *E. ardea* was a land animal. Both are now often classified together with the living South American genera *Tayra* and *Grison*. Whether there is a real relationship between these geographically separated groups remains to be seen – the similarity might also result from adaptive response to a similar mode of life. In ecological habits, the grison may well resemble the Ardé polecat. The grison inhabits both steppes and woodlands and this may also have been true for *Enhydrictis* [22; 301].

The Pannonian Polecat, *Pannonictis pliocaenica* Kormos. (The genus *Pannonictis* is doubtfully separate from *Enhydrictis*.) This species is found as early as the Astian (Csarnóta) and ranges through the Villafranchian and into the early Middle Pleistocene in eastern Europe; the last specimens may be of B-Waalian age. It seems to have invaded western Europe in the late Villafranchian (A-Tegelen) and persisted to the C-Cromerian (Forest Bed at West Runton). A closely related form has been described from the early Pleistocene of North America under the name *Trigonictis*.

The Pannonian polecat was a large mustelid. The length of head and body may have exceeded 2½ ft. with an additional 1 ft. 8 in. for the tail. Like the related *Enhydrictis*, the Pannonian polecat probably was not an amphibious form but may have led a life resembling that of the living tayra. This large, marten-like mustelid inhabits grasslands and open forests and preys on small and medium-sized mammals [138].

The European Mink, *Mustela* (*Lutreola*) *lutreola* Linné. Fossil finds of this species are very rare. According to Boule, a specimen from Grotte de l'Observatoire, Monaco (4-Würm) belongs to this species. On the other hand a number of supposed finds from southeast Europe have been revised by Mottl [199], who concludes that they belong to other species.

The genus *Mustela* comprises not only the weasels but also the minks and true polecats, sometimes regarded as distinct genera (*Lutreola* and *Putorius*, here treated as subgenera). The minks are particularly well adapted for an amphibious mode of life and even have webbed toes for swimming. The European mink has now been exterminated in most of its former range in central Europe but is still found in Finland and eastward through Russia and in Siberia. Its closest relative is the American mink, *M. vison* Schreber. It seems likely that the minks originally arose in North America and that *M. lutreola* is a late newcomer in Europe.

Stromer's Polecat, *Mustela* (*Putorius*) *stromeri* Kormos. The earliest known representative of the true polecats appears in the late Villafranchian (Beremend). It has also been found at Episcopia, Erpfingen and Hundsheim, and thus ranges approximately to 1-Günz II. The remains are somewhat incomplete but suggest that Stromer's polecat was closely allied to the living *M. putorius*, although considerably smaller. This may indicate that the subgenus *Putorius* evolved at a relatively late date, perhaps in the Villafranchian, from weasel-like ancestors. In the Astian there existed large weasels (*Mustela*) which exceeded the stoat in size.

The mode of life of *M. stromeri* is a matter of speculation, for the two species that inhabit Europe at the present day have quite dissimilar habits. In any case it seems probable that both *M. putorius* and *M. eversmanni* were derived from Stromer's polecat [141].

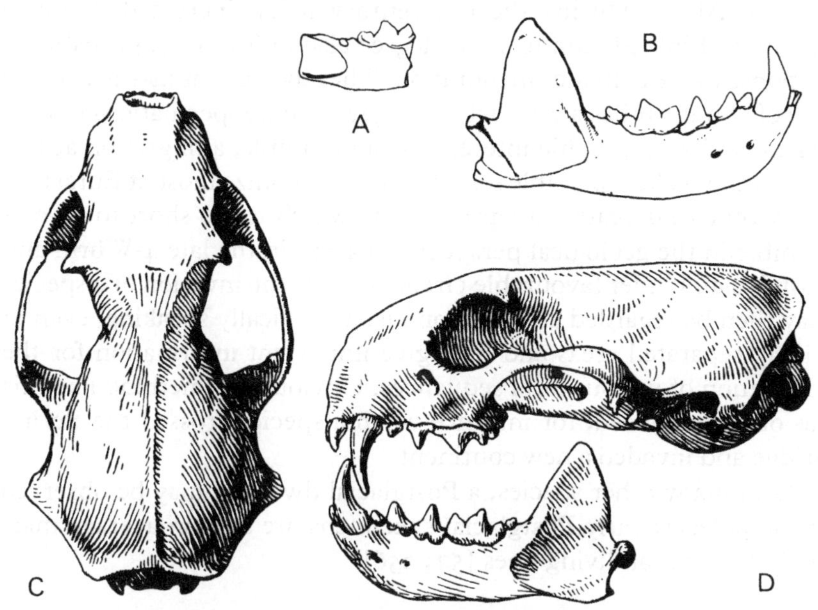

Figure 40. A, mandible fragment of Stromer's Polecat, *Mustela stromeri*, Villafranchian, Beremend. B, mandible of Polecat, *M. putorius*, Recent. C–D, skull and mandible of Steppe Polecat, *M. eversmanni*, Recent. All natural size. A after Kormos; B after Miller; C–D after Gromova.

The Polecat, *Mustela (Putorius) putorius* Linné. This species is mainly distinguished from the closely related steppe polecat by its somewhat smaller size, some details in the dentition (relatively smaller canine and carnassial teeth, larger post-carnassial molars) and the more moderate constriction of the skull behind the eyes. Fragmentary material may often be quite impossible to identify, which means that Middle Pleistocene finds generally cannot be determined as to species.

Polecats of modern type (*M. putorius* or *M. eversmanni*) have been discovered in deposits dating from the C-Cromer and 2-Mindel

(Forest Bed; Mosbach 2), while other finds date from the D-Holsteinian (Lunel-Viel; Breitenberg) and the 3-Riss (Tornewton Cave). F-Eemian and later finds are very common in cave deposits and may be definitely identified as a large form of *M. putorius* (*M. putorius robusta* Newton).

The polecat seeks much of its food in and along streams but is not an aquatic form to the same degree as the mink. It eats amphibians, fish and invertebrates as well as some small mammals. It ranges over most of Europe and the Asiatic taiga belt and is also found in north-western Africa. During the last century it has increased its range greatly in Finland (and to some degree also in Sweden), probably in response to the climatic amelioration. The advance has had an average rate of about 5 miles a year, which suggests some speculations of geological and stratigraphic interest. As an example, a migration at this rate would make it possible for a species to colonize most of Eurasia in a few thousand years – a lapse of time which is too short to be discernible in the geological perspective except in the late 4-Würm and Postglacial. Under favourable circumstances the invasion of a species may then be regarded as synchronous, geologically speaking, even in widely separated areas and may give important information for the correlation of various local sequences. Obviously something like this has often happened, for instance when a species crossed the Bering Bridge and invaded a new continent.

As in many other species, a Postglacial dwarfing may be observed in the polecat; early Postglacial specimens are still on average markedly larger than living ones [52; 256].

The Steppe Polecat or Ferret, *Mustela* (*Putorius*) *eversmanni* Lesson. This species is definitely present in the F-Eemian (Lambrecht Cave; travertines at Weimar) and 4-Würm (various cave finds, especially in central and eastern Europe). Whether Middle Pleistocene finds represent this species or *M. putorius*, or both, is still uncertain. It is possible that the two species were not yet distinct in the Middle Pleistocene; their close similarity may suggest that the splitting up is relatively recent. In the area of overlap between the species ranges in eastern Europe, hybrids are said to be fairly common and many authors regard the steppe polecat merely as a subspecies of *M. putorius*.

Its habits, however, are quite different from those of the true polecat. Its staple diet consists of the rodent *Citellus major* Pallas which it hunts in its subterranean nests, so that its habits resemble those of the marbled polecat. The somewhat larger canine teeth of *M. eversmanni* as compared with *M. putorius* probably evolved in adaptation to this mode of life; other changes in the skull and dentition have been shown by Soergel [256] to be merely correlatives of that primary change.

The steppe polecat ranges from southeastern Europe to western China, but is extinct over most of its Pleistocene range in Europe (Germany, France, perhaps England). The ferret is the domestic form of *M. eversmanni* and is used to hunt rabbits and rats; its domestication probably dates back to a few centuries BC [119; 129].

The Primitive Stoat, *Mustela palermina* Petényi. One of the most common carnivores in central and eastern Europe during the earlier part of the Middle Pleistocene was a forerunner of the stoat. It is only very slightly different from the living species and is obviously directly ancestral. The primitive stoat is found at a number of sites (Beremend; Villány; Episcopia; Gombasek; Sackdilling; Koneprusy, etc.) and ranges at least up to the C-Cromer.

A stoat-like form has also been found in the Astian (Wölfersheim; Weze; Csarnóta) and has been described under the name *M. plioermina* Stach. It is thus evident that there is a continuous phyletic sequence in Europe from the Late Pliocene to the present day. Whether the degree of change that took place in this sequence actually merits the recognition of three distinct successive species-stages is more doubtful.

In the present-day stoat there is a tendency for the average body size to decrease northwards: the Scandinavian forms are markedly smaller than those of central Europe and the Scottish stoat is smaller than the English one. This is, of course, contrary to Bergmann's rule. The weasel, incidentally, behaves in the same way; so did *M. palerminea*, for the interglacial specimens of B-Waalian age are decidedly larger than the 'cold' ones from 1-Günz I and II.

Generally speaking the stoat may be regarded as a woodland form, though it does range marginally into open country and the high mountains [141].

The Stoat or Ermine, *Mustela erminea* Linné. The ermine stock of the later Middle Pleistocene is incompletely known; in the F-Eemian and 4-Würm, on the other hand, the species is common enough. In the Drachenloch at Vättis in Switzerland the ermine has been found in interglacial deposits 2445 m. above sea level; but this is within its present-day range, for it reaches 3000 m. in the Alps. It may also be found in the tundra of the north, but it is mainly a forest animal. Its present-day range covers nearly all of Europe except the Mediterranean peninsulas, but Pleistocene records from Portugal and Italy indicate that the range extended further south in the Ice Age. (But there is always the risk that the large south European races of the weasel may be mistaken for stoat.) The ermine is also found in northern Asia and has been present in North America since the 3-Riss.

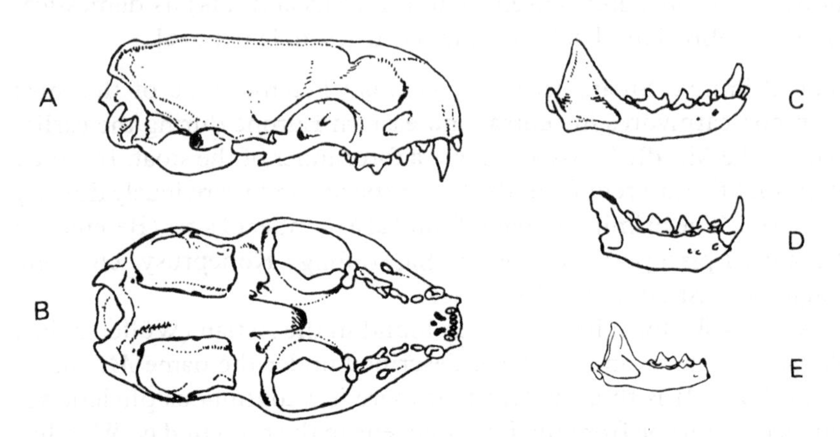

Figure 41. A–B, skull and C, mandible of Recent Stoat, *Mustela erminea*; D, mandible of Primitive Stoat, *M. palerminea*, Middle Pleistocene, Episcopia; E, mandible of Primitive Weasel, *M. praenivalis*, Middle Pleistocene, Nagyharsány. All natural size. A–C after Miller; D–E after Kormos.

The exclusively carnivorous habits and hunting skill of the stoat are well known. Besides eggs, rodents and insectivores it will attack mammals and birds much larger than itself [265].

The Primitive Weasel, *Mustela praenivalis* Kormos. Like the ermine, the weasel has a predecessor in the Middle Pleistocene and Villafranchian. Recorded sites include Villány, Beremend, Schernfeld, Erpfingen, Sackdilling, etc. Finds from the C-Cromerian (Cromer Forest Bed; Abbeville; Kövesvarad) may represent the same species or may be transitional to the living form.

The primitive weasel in turn was probably derived from an Astian form, *M. pliocaenica* Stach, so that there is a continuous weasel succession from the Pliocene to the present day, just as in the case of the stoat. The exact demarcation of the successional species is as uncertain as in the *erminea* sequence [141; 261].

The Weasel, *Mustela nivalis* Linné (*M. vulgaris* Erxleben). The transition between *M. praenivalis* and the modern species seems to be gradual. Finds from the earlier Middle Pleistocene are mostly classified in the extinct species, while the specimens from the 2-Mindel (e.g. Mosbach 2) and the D-Holsteinian (Tarkö, Breitenberg and various other caves) are regarded as *nivalis*. The modern species is quite common in cave deposits from the F-Eemian and the 4-Würm.

The weasel inhabits almost all of Europe and ranges eastward to central Siberia. Like the ermine it tends to grow larger in the south and in the Mediterranean countries the weasel attains almost the size of a Scandinavian ermine. The weasel is even more dependent on forest than the stoat, which it otherwise resembles in habits. Its closest living counterpart is the Least Weasel and the two species have frequently been confused. In *M. nivalis* the boundary between the brown colour on the back and the white on the belly is diffuse; in the least weasel it is sharp. As regards fossil material, size is the only clue. The analysis is not facilitated by the fact that the female in all weasels is smaller than the male and that a female of *M. nivalis* is of the same size as a male of *M. rixosa*. The identification of fossil weasels on the basis of incomplete material, perhaps only a few teeth and jaw fragments, is a tricky matter indeed [117].

The Least Weasel, *Mustela rixosa* Bangs (*M. minuta* Pomel). This species, obviously a very close relative of *M. nivalis*, used to be regarded as nothing but a variety of the latter; but this opinion cannot now be upheld. It is a good species, in some areas a vicar of *M. nivalis*, in others existing alongside with its relative. Its distribution is circumpolar, extending from northern Scandinavia and Finland over Russia and Siberia to North America.

Fossil remains of the least weasel have been found in central and western Europe and Italy, apparently only in the 4-Würm. It is the smallest of all living carnivores and so it must be regarded as a highly specialized form, probably of rather late date. It probably arose as a side branch of *M. nivalis*, perhaps in the east and thence spreading into Europe during the last glaciation [312].

Thoral's Badger, *Meles thorali* Viret. The badgers, which form the subfamily Melinae, have a long fossil record in the Tertiary. It shows gradually increasing emphasis on the tubercular teeth at the back of the jaws, while the shearing carnassial teeth tend to become reduced. This is an adaptation to an omnivorous diet, almost as in the bears.

The badger is a common fossil in the Middle and Late Pleistocene, but in the Villafranchian of Europe it has only been found at one site, Saint-Vallier near Lyon, of mid-Villafranchian date. There is a closely similar form in the Middle Villafranchian of China, while Late Villafranchian badgers in eastern Asia can hardly be distinguished from the modern species. Thoral's badger is also very close to the modern form and evidently ancestral to it.

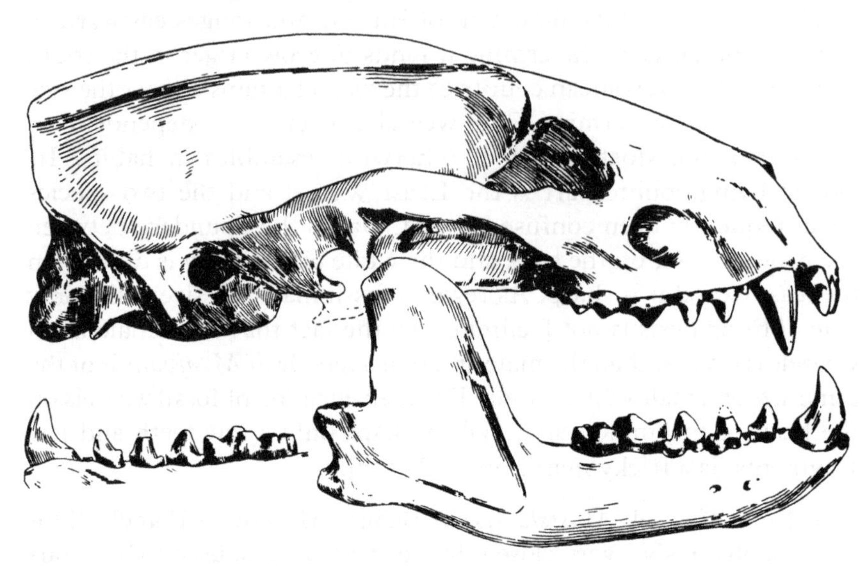

Figure 42. Skull and mandible of Thoral's Badger, *Meles thorali*, Villafranchian, Saint-Vallier. Two-thirds natural size. After Viret.

Somewhat more primitive species of the genus *Meles* are present in the Astian of Europe and Asia; the genus may have originated from the Pliocene genus *Melodon* in China [300; 302].

The Badger, *Meles meles* Linné. In the early Middle Pleistocene Europe was inhabited by badgers similar to the modern species. They were originally described under a specific name of their own, but are now simply held to constitute a subspecies *M. meles atavus* Kormos. Sites include Episcopia, Gombasek, Süssenborn, Hundsheim, Erpfingen, Koneprusy, Mosbach 2, Stránská Skála; thus the oldest records date from 1-Günz I or B-Waalian times. Records from the late Middle Pleistocene (e.g. Heppenloch, Lunel-Viel, Breitenberg, Achenheim, Montmaurin) and Late Pleistocene (numerous caves, travertine and loess deposits) are plentiful. Occasionally the burrowing habits of the badger may lead to its bones being deposited in strata formed at a much earlier date; for instance, the rich badger material from the F-Eem interglacial deposits at Barrington, Cambridge, seems to represent Postglacial intruders throughout. (The fox, which often frequents badger holes, is also found here.) A trained observer can usually spot a fossil badger hole without difficulty.

The badger now inhabits most of Europe except the extreme north. It does not reach the northern end of the Bothnian Gulf; the population in Sweden and Norway has immigrated over Denmark, that in Finland from the southeast. The species ranges eastwards to China and Japan. It has a good fossil record in China during the Middle and Late Pleistocene.

In the summer the badger lives mainly on earthworms, while birds, mammals, eggs and insects are also eaten. Fungi and berries are important vegetarian elements in the diet. The badger spends much of the winter in its nest but does not actually hibernate [52; 179].

Bravard's Otter, *Aonyx bravardi* Pomel. The Pleistocene otters of Europe fall into two quite distinct groups, represented in modern times by the common otter (genus *Lutra*) and the small-clawed otter (genus *Aonyx*). The last-mentioned line is the dominant or only otter in Europe for most of the Pleistocene.

The small-clawed otters have a much broader head than *Lutra* and the teeth are broad with blunt cusps. Such a dental battery indicates that a large proportion of the food consists of shellfish, especially crustaceans, but also molluscs. In the modern sea otter this tendency has been carried to an extreme; molluscs form the staple diet and the teeth are very broad, blunt-cusped, crushing discs. Such a degree of specialization has not been reached in the genus *Aonyx*.

Bravard's otter was originally described on a maxillary fragment from Etouaires, in the basal Villafranchian; a possibly ancestral form is known from the Astian of Montpellier. The later Villafranchian history of the otters is too fragmentary to permit any certain conclusions. A tibia from Saint-Vallier may belong to the same species. A single tooth from the late Villafranchian Norwich Crag (*A. reevei* Newton) might also belong to the same form.

In the Villafranchian of India and China, *Aonyx*-like otters have also been found; the whole group may be derived from a Lower and Middle Pliocene species such as '*Lutra*' *aonychoides* Zdansky in China [218].

The Corsican Otter, *Aonyx antiqua* Blainville. In the Middle Pleistocene the small-clawed otters were represented in Europe by a species ranging from D-Holstein (Lunel-Viel) to 3-Riss (Montsaunès; Tornewton Cave). A specimen from the Roter Berg at Saalfeld in Thuringia may be still more recent, perhaps early 4-Würm. The species has also been described (under the genus name *Cyrnaonyx*) from Grotta del Margine in Corsica.

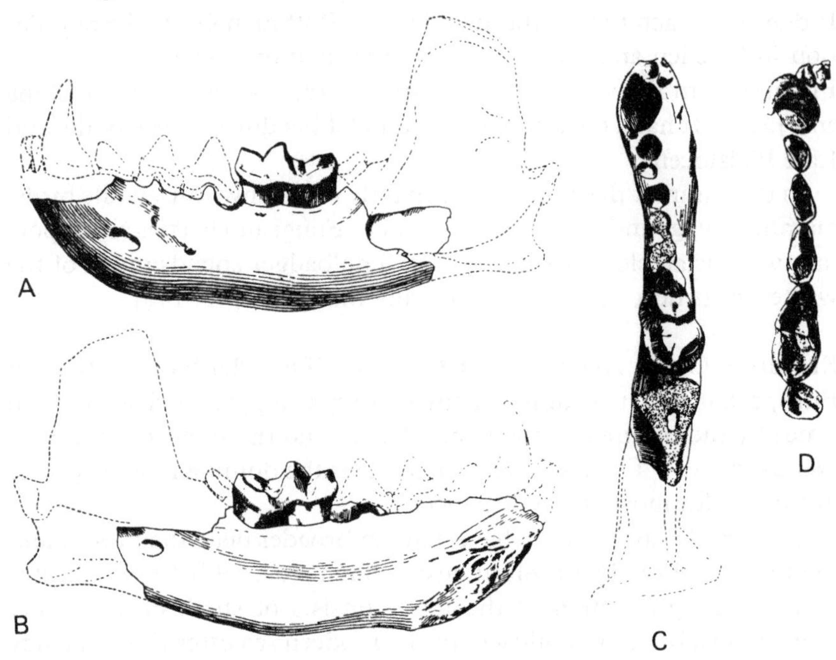

Figure 43. A–C, partially restored mandible of Corsican Otter, *Aonyx antiqua*, Late Pleistocene, Roter Berg at Saalfeld; D, lower teeth of Otter, *Lutra lutra*, Recent. Natural size. A–C after Helbing; D after Miller.

The Corsican otter presents some problems. The sites where this species has been recorded (those mentioned above plus a locality in the Hautes-Pyrénées) have yielded only skull and tooth material, but no limb bones. On the other hand, limb bones of an otter, without skull or teeth, have been discovered in Malta; they have been described as *Nesolutra euxena* Bate. Perhaps *Nesolutra* is simply the limb skeleton of *A. antiqua*; or perhaps it may be identical with Bravard's otter or *A. reevei* (or both). In either case the D-Holsteinian lutrine would form a link between the Villafranchian and later Middle Pleistocene *Aonyx* and suggest that small-clawed otters inhabited

Europe more or less continuously from the Astian to the end of the Middle Pleistocene. All this, however, is speculative and what we really need is a discovery of limb bones and teeth in direct association.

A. antiqua is more advanced than the Villafranchian form, but a direct phyletic connection between the two remains still to be proved.

The living small-clawed otters inhabit Africa and south Asia; in both areas their history goes back to the Pliocene [23; 96].

The Hundsheim Otter, *Lutra simplicidens* Thenius. The first Pleistocene appearance of the genus *Lutra* appears to be at Hundsheim and Voigtstedt (I-Günz II and C-Cromer) where the remains of an otter closely related to the living European form have been found. Its dentition is slightly more primitive than that of *L. lutra*, while the limbs are actually more specialized; this may suggest that the Hundsheim otter is not a direct ancestor to the living form [286].

The Otter, *Lutra lutra* Linné. The modern European otter makes its first appearance in the F-Eemian, for instance in the travertines at Weimar, and is common enough in Late Pleistocene deposits. But it seems to be entirely absent in the Middle Pleistocene, so that apparently it immigrated in the F-Eemian. Where did it come from if it did not evolve from *L. simplicidens*? Neither in China nor India has anything resembling an ancestor of the species been found. Some Pleistocene fossils from Africa have been referred to *L. lutra*, but to conclude that it originated in Africa (where its present-day range is restricted to the northwest corner of the continent) would seem rash. Its nearest living relative is the North American *L. canadensis* Schreber and perhaps a North American origin is the most likely alternative. The main problem is that the Canadian otter, which has a long Pleistocene history in North America, would appear to be somewhat more specialized than *L. lutra* and thus is unlikely to be directly ancestral.

To a far greater degree than minks and polecats the otter is an aquatic animal; it lives almost exclusively on fish. The adaptation to swimming is reflected in the long, cylindrical body, the short legs, the powerful tail and the flattened head – a tendency to flattening of the skull is found in all aquatic carnivores, including the seals. The otter swims with powerful undulating movements of the body and tail, while it keeps the short legs pressed along its sides. It is able to dive for up to five minutes. In spite of its superb swimming adaptation, the otter moves easily and gracefully on land. It may perhaps be said that the otter represents the most elegant solution to the problem of constructing an amphibious carnivore that Nature has so far effected.

The species is widely distributed in Eurasia, including the Indo-Malayan region as far as Java [52].

Family Canidae, Dogs

The history of the dog-like carnivores begins in the Eocene; they have a rich fossil record and during the Tertiary were almost as varied as the mustelids. There were cat-like dogs; hyena-like dogs; bear-like dogs; and dog-like dogs. But all of them except the dog-like dogs became extinct and the Pleistocene canids of Europe constitute a rather uniform cohort. They are common as fossils at all levels and many species have a long, well-documented record. Most belong to the fox or wolf group; the exotic element is represented by the genera *Nyctereutes* (raccoon-dogs), *Cuon* (dholes) and *Lycaon* (hunting dogs), of which the two former are frequently met with in European Pleistocene deposits.

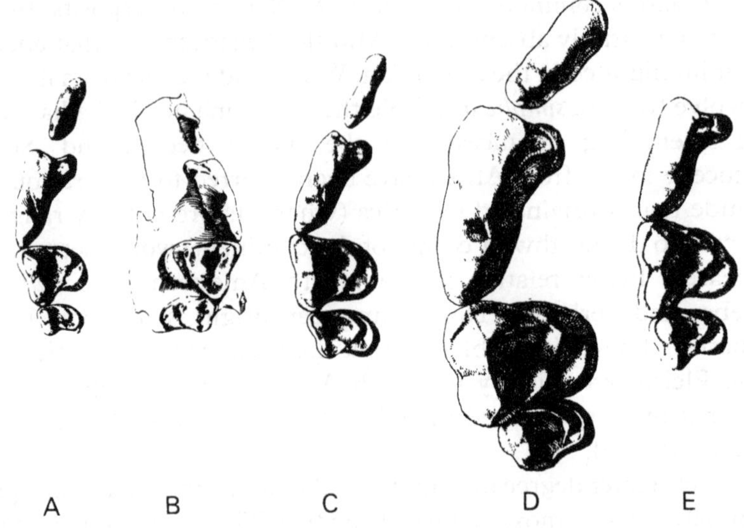

Figure 44. Larger upper cheek teeth of canids. A, Arctic Fox, *Alopex lagopus*, Recent; B, Alopecoid Fox, *Vulpes alopecoides*, Villafranchian, Val d'Arno; C, Red Fox, *V. vulpes*, Recent; D, Wolf, *Canis lupus*, Recent; E, Golden Jackal, *C. aureus*, Recent. All natural size. A, C–E after Miller; B after Stehlin.

The type of dog represented by the Pleistocene and living canids in Europe is a conservative one, close to the earliest known members of the family. The long nose and the big nasal cavities indicate that the dogs rely on their excellent sense of smell. In the dentition the carnassial teeth are well developed in most species, but the tubercular teeth are generally also fairly large; thus the canids may turn to vegetable food when meat is in short supply and can survive under conditions that would spell death to more specialized animals. In the hunt, stress is often laid on tenacious long-range pursuit rather than stalking and this may lead to the development of group action with tactical cooperation. With their moderate level of specialization, great adaptability and high intelligence, as well as their tendency to social cooperation, the canids have some points in common with the higher primates. They have colonized every continent except Antarctica.

Falconer's Dire Wolf, *Canis falconeri* Major. A large wolf, so far definitely identified only in the late Villafranchian (Val d'Arno), where it is rare. It may possibly be related to the big Dire wolf, *C. dirus* Leidy, which lived in North and South America in the Late Pleistocene; it was a large-headed, very powerfully built animal, with a dental development that may suggest hyena-like habits [57].

The Arno Dog, *Canis arnensis* Del Campana. A small species about the size of a modern jackal and perhaps related to it. It has been found in great numbers at Val d'Arno and in the Saint-Estève cave; there are also less certain records from Senèze and Chilhac. A cranial fragment from the Red Crag in England may also belong to this species. Perhaps the dog described as *C. strandi* Kormos from Episcopia, Gombasek and Erpfingen may also be identical with the Val d'Arno species [57].

The Etruscan Wolf, *Canus etruscus* Major. Wolf-like animals of about the size of a sheepdog are found all through the Villafranchian, but we do not know for certain whether they were ancestral to the true wolves of the Middle Pleistocene. The Etruscan wolf is the most common species; it appears in the basal Villafranchian (Etouaires) and seems to favour woodland faunas (Olivola; Val d'Arno; Erpfingen) but is absent in the extreme steppe associations. Perhaps ecologically it resembled the Indian wolf of the present day.

It is probable that one of the species *C. falconeri*, *C. arnensis* and *C. etruscus* is ancestral to the true wolf and *C. etruscus* seems to be the most likely alternative [57].

The Wolf, *Canis lupus* Linné. The true wolf appears for the first time in the 1-Günz II (Hundsheim; Süssenborn) and is abundant in the C-Cromer and 2-Mindel (Forest Bed; Mauer; Mosbach). The early wolf in Europe was comparatively small, about as large as the living wolf of the Near East, and is referred to the subspecies *C. lupus mosbachensis* Soergel. Small wolves, perhaps the same subspecies, are also found in the D-Holsteinian (Tarkö; Lunel-Viel; Heppenloch; Grays Thurrock). The wolf of the 3-Riss (many localities; large sample in Tornewton Cave) is transitional to the large form that lived in Europe during the Late Pleistocene and the remains of which have been found in almost all caves containing a representative macromammalian fauna, as well as numerous open-air sites. There is little if any difference in size between the F-Eemian and 4-Würmian wolf. In Postglacial times the size was again slightly reduced, so that the present-day wolf of northern Europe is on an average somewhat smaller than the Late Pleistocene form.

Figure 45. Engraving of the head of a Wolf, *Canis lupus*. Les Combarelles, Dordogne. After Kühn.

The wolf has a long fossil record in Asia too, starting with relatively small forms in the 2-Mindel (Choukoutien). The species was common in Europe at a fairly recent date and is widely distributed in Asia and North America. It seems to have entered the New World in the 3-Riss, if not earlier; the American timber wolf is not now regarded as a distinct species.

The wolf is a good exponent of the main canid characters. Its climatic tolerance is very great, so that it is able to live in the tundra, the forest, the steppe and the desert. It is gregarious and thanks to its powers of tireless pursuit, strength and tactical skill is able to prey on very large animals such as the elk (moose to Americans); during the Pleistocene, the various kinds of deer probably were important game to the wolf. In summer time the wolf lives mostly on small game, mainly rodents.

The dog is a domestic wolf. It used to be thought that some species of jackal might have contributed to the ancestry of the dog, but this theory is now discredited. The idea that domestic animals have been produced by cross-breeding of two or more wild species, so often found in earlier literature, cannot be accepted any more. It has been suggested that the dog was derived from the small wolf of the orient and fossil finds from the Mesolithic cave deposits in Palestine have been cited in support of this theory. But later studies have shown that the fossils in question are true wolf, not dog, and the same seems to be true for canid remains from the early farming village at Jarmo in Iraq, about 7000–6500 BC. Domestication is, however, suggested by clay figurines of small dogs from Jarmo. At this time domestic dogs were already in existence in Europe, as shown by several finds from the Maglemose culture in Denmark. Recently, Degerbøl [52] has been able to prove the presence of domestic dog at a still older European living site, Starr Carr in Yorkshire, 7500 BC. This is the oldest definitely identified dog known at present [53; 187; 276].

The Golden Jackal, *Canis aureus* Linné. This species is now found in North Africa and a large part of Asia; it reaches Europe only in a small area in the southeast, partly in Russia, partly in the Balkans. Oddly enough the golden jackal is very little known in the fossil state, although its larger relative the wolf jackal, *C. lupaster* Ehrenberg, is rather common in the Late Pleistocene in Palestine and North Africa. There are however some records of *C. aureus* from the Late Pleistocene of Italy and North Africa. In the Middle Pleistocene, small jackals of the type of *C. aureus* occur in North Africa (e.g. Ternifine), while other species of jackal are reported as early as the Villafranchian in Africa [179].

The Primitive Dhole, *Cuon majori* Del Campana (*C. dubius* Teilhard). The dholes, together with some other living canids (hunting dog, bush dog) used to be referred to the subfamily Simocyoninae together with some extinct canids. More recent research has shown that both the dhole and the hunting dog are closely related to *Canis* and probably diverged from that genus as late as the Astian.

In the case of the dhole the evidence is complete enough. Its history begins with the species *C. majori*, which actually resembles *Canis* so much that it would be referred to that genus had not its descendants evolved into full-fledged dholes. In modern *Cuon* the tubercular teeth have been greatly reduced and their few remaining cusps have become sharply trenchant points. In *C. majori* of the late Villafranchian (Val d'Arno; also in China) this trend had barely started: the molar dentition was still complete but somewhat reduced in size and the cusps are more pointed than in *Canis*. By early Middle Pleistocene times another step in this direction had been taken: *C. majori stehlini* Thenius from Rosières, perhaps of B-Waalian or C-Cromerian age, had lost the last lower molar. This subspecies happens to be very large, like a big wolf. From the Rosières dhole the step is not long to the oldest representatives of *C. alpinus* [57; 276; 277].

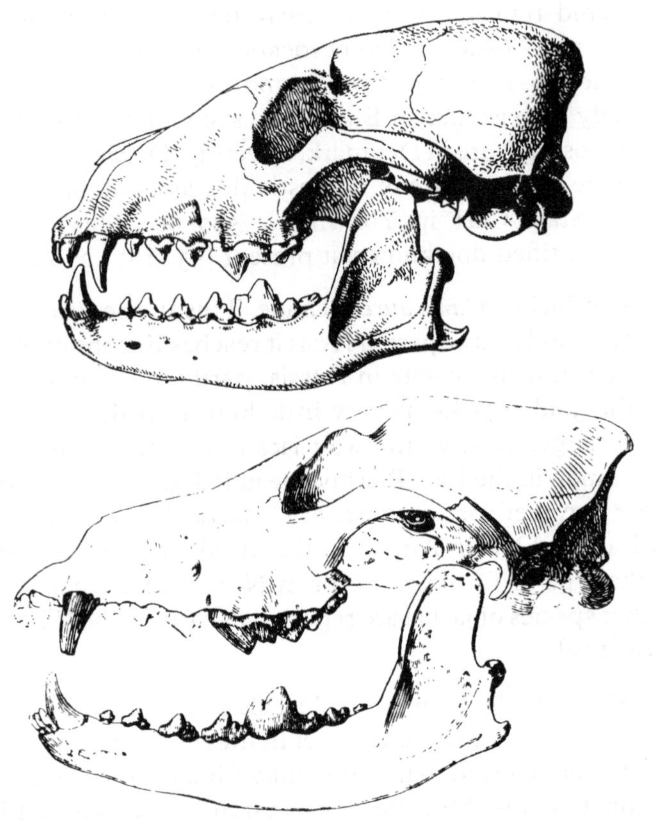

Figure 46. Skull and mandible of Dhole, *Cuon alpinus*, Recent (above) and Wolf, *Canis lupus*, Pleistocene, USSR (below). One-third natural size. After Gromova.

The Dhole, *Cuon alpinus* Pallas. This species makes its first appearance at Hundsheim (1-Günz II) and Mosbach (2-Mindel). In these early variants the remaining lower tubercular molar still carries three cusps but it is more reduced than in *C. majori*; the early Middle Pleistocene dhole is called *C. alpinus priscus* Thenius. The change in the tubercular teeth continued in the later Middle Pleistocene. The D-Holsteinian *C. alpinus fossilis* Nehring from Heppenloch represents an intermediate stage; in the Late Pleistocene *C. alpinus europaeus* Bourguignat, the transformation of the lower tubercular into a single-cusped, sharply trenchant tooth has been completed. This dhole is quite modern looking except for its large size which is almost comparable to the wolf.

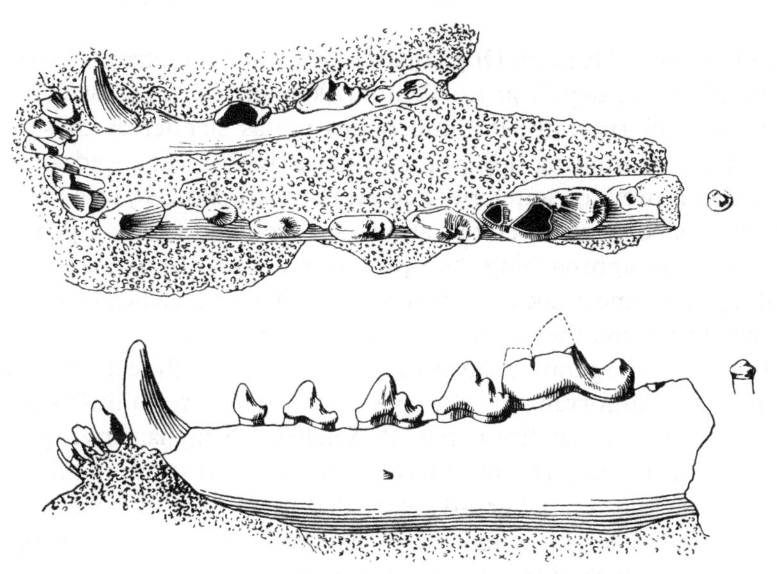

Figure 47. Mandible of Sardinian Dhole, *Cuon alpinus sardous* Studiati, Pleistocene, Tramarigli, Corsica. A small subspecies found on Corsica and Sardinia. Two-thirds natural size. After Stehlin.

Late Pleistocene dhole has been found at eight sites in Czechoslovakia, Hungary, Austria, Switzerland and on the Riviera (for instance in Monaco). The species became extinct in Europe at the end of the 4-Würm; it still inhabits a large territory in Asia.

The story is somewhat complicated by the presence of *Cuon*-like forms retaining three lower molars like the early *C. majori* as late as the 2-Mindel (Mosbach 2, a large form resembling *C. majori stehlini*) and D-Holsteinian (Grays). Whether these represent atavistic individual variants or a separate species has not yet been clarified. A name that occasionally crops up in connection with medium sized canids like the Grays dog is *Canis neschersensis* Blainville; it goes with a mandible from Mt. Perrier that apparently does not come from a Villafranchian deposit and its status is uncertain.

As reflected in its dentition, the dhole is more exclusively carnivorous in habits than canids in general. It hunts in large packs which may overpower and destroy the most powerful adversary; Kipling's dramatic description in the Jungle Book of the hunting of the dhole is very apt.

The record of the true dhole in China goes back to the 2-Mindel (Choukoutien); this form resembles *C. alpinus priscus* [5].

The European Hunting Dog, *Lycaon lycaonoides* Kretzoi. The living African hunting dog, *L. pictus* Temminck, might be called a wolf modified for swift running; its limb bones are as slender as those of a cheetah. The hunting dog hunts in packs with great speed and endurance; it preys on antelopes and other medium-sized, fast-running herbivores. Eyewitnesses tell amusing accounts of how a group of hunting dogs approaching their prey trot along assuming a pose that will suggest some innocent herbivorous animal until, when recognized for what they are, they explode into furious pursuit.

In Europe the genus makes an appearance during the earlier Middle Pleistocene; unfortunately the species *L. lycaonoides* is little known. It has been found at Episcopia (B-Waalian), Nagyharsányhegy and Gombasek (1-Günz II) and thus has a rather short stratigraphic span. The size of this animal greatly exceeded that of the modern species, which attains a length of about 3½ ft. excluding the tail. The European hunting dog reached the size of a large wolf. Like the African species it was probably a steppe and savanna animal. Perhaps lack of a suitable biotope kept the species out of western Europe.

Fossil *Lycaon* have been found in the Pleistocene of North Africa as well as within the present-day range of the hunting dogs south of the Sahara. According to Thenius [254] the genus was derived from *Canis* and probably evolved at about the same time as *Cuon*, so that the three genera really form a quite closely-knit systematic group. The dhole and hunting dog resemble each other ecologically so much that it may be suspected there would be no room for both in the same area; perhaps this is the main reason why the genus *Lycaon* failed to establish itself for a longer time in Europe [160; 276].

The Alopecoid Fox, *Vulpes alopecoides* Major (*Alopex praeglacialis* Kormos). The species appears in the middle Villafranchian and survives until 1-Günz II; localities include Villaroya, Saint-Vallier, Senèze, Val d'Arno, Beremend, Villány, Episcopia, Brassó and perhaps Hundsheim. The eastern form has been regarded as a distinct species belonging to the same genus as the arctic fox (*Alopex*), but it does not have the dental specialization of the modern form.

The alopecoid fox is of about the same size as the living arctic fox but resembles more the red fox in dental characters. It may be ancestral to the arctic fox or the red fox or both; information on fossil foxes postdating 1-Günz II and antedating 3-Riss is so incomplete that we simply cannot tell.

The Pliocene ancestry of the foxes is also uncertain. In the Villafranchian of China there is a species that is closely allied to the alopecoid fox [140; 265].

The Red Fox, *Vulpes vulpes* Linné. The common fox has been found in the D-Holsteinian (Grays Thurrock; Heppenloch; Tarkö; Montoussé; Lunel-Viel) but records from earlier deposits are doubtful as to species. It is also found in the 3-Riss and E-Ilford and is very abundant in the Glutton Stratum of Tornewton Cave (3-Riss II). Late Pleistocene records, both from F-Eemian and 4-Würmian deposits, are very common both in caves and open-air sites. Outside Europe the history of the species extends back to the 2-Mindel (Choukoutien).

The red fox was probably derived from Villafranchian populations of *V. alopecoides* or perhaps from the related Chinese *V. chikushanensis* Young. Of course it may also be descended from some ancestor at present unknown to us. However this may be, the ancestral form is likely to have been small judging from the size of the red fox in the D-Holsteinian and 3-Riss. The size of the living fox varies approximately in accordance with Bergmann's rule: the northern subspecies *V. vulpes vulpes* is large, the central European *V. vulpes crucigera* Bechstein is distinctly smaller and the North African *V. vulpes barbara* Shaw is quite small. A somewhat analogous variation in time may be observed, for instance in Palestine, where the red fox increased in size all through the 4-Würm to a maximum in the early Postglacial (in Mesolithic times, ca. 8000 BC); but in the next few millennia its size was rapidly reduced and the present-day form in Palestine is very small.

The range of the living red fox may exceed that of all other wild carnivores and covers Europe, North Africa, most of Asia and North America (introduced red fox may also be found in Australia). The colonization of North America probably dates from 3-Riss, for the oldest American fossils of this species have been found in fissure deposits apparently of that age (Conard Fissure, Arkansas). The North American red fox has been regarded as a distinct species (*V. fulva* Desmarest) but this has been proved wrong.

The environmental tolerance of the red fox is very great; although it is not found on the actual tundra, it inhabits forests, steppes and deserts. In contrast with the *Canis* group the fox is not gregarious and its hunting method depends on stalking. The staple diet is formed by rodents but the fox also eats a variety of other foods, including birds, medium-sized mammals (hares, rabbits, roe deer fawns), reptiles, amphibians and fish, as well as vegetables [265; 272].

The Primitive Corsac, *Vulpes praecorsac* Kormos. The small present-day steppe fox had a forerunner in the Villafranchian and Middle Pleistocene only slightly different from the living species. It has been found at Villány, Episcopia, Nagyharsányhegy and Gundersheim, which suggests a stratigraphic range from the late Astian to 1-Günz II. Whether all are the same species and whether it is really distinct from *V. corsac* is uncertain. The corsacs may have branched from an early alopecoid fox or *vice versa*; study of the early history of *Vulpes* is still in its beginning [140].

The Corsac or Steppe Fox, *Vulpes corsac* Linné. This species is almost unknown in the fossil state in Europe although it has been reported, somewhat uncertainly, from Late Pleistocene cave deposits in Switzerland and Bohemia. In China on the other hand it has a good record going back to the 2-Mindel, thus linking it chronologically with *V. praecorsac*.

The steppe fox, which is slightly smaller than the arctic fox, inhabits the south Russian and Asiatic steppes. Pikas and small rodents are thought to be its main food [214; 265].

The Arctic Fox, *Alopex lagopus* Linné. This species is probably descended from the Middle Pleistocene alopecoid fox but it has not been certainly identified in deposits antedating the 3-Riss (for instance Tornewton Cave, Glutton Stratum). There are a few records from the F-Eemian (Lambrecht Cave; Fontéchevade), but it is only in the 4-Würm that the arctic fox became really common; it then ranged over most of Europe down to the Riviera and even into Spain and southern Russia, a messenger of the Ice Age. As early as 1933 Stehlin mentioned that the arctic fox had been found at more than 80 sites ranging from Kiev in the east to Ireland in the west.

The arctic fox differs from the red fox in being slightly smaller and in a number of dental traits, reflecting a tendency to more purely carnivorous habits. The tubercular teeth have a simplified pattern compared to *Vulpes*, while the canine teeth are relatively larger. The arctic fox mainly preys on rodents and its population is greatly boosted during the so-called lemming years [9; 187; 265].

The Great Raccoon-Dog, *Nyctereutes megamastoides* Pomel. A typical Villafranchian species in Europe, the great raccoon-dog is found at Etouaires, Roccaneyra, Pardines, Villaroya, Saint-Vallier, Senèze and in the Villafranchian of Hungary. Towards the end of the Villafranchian it apparently became extinct in Europe, since it is absent at Val d'Arno and other sites from the terminal part of the stage.

N. megamastoides resembled the living raccoon-dog (*N. procyonides* Gray) but was considerably larger. The present-day form attains a head-and-body length of up to 23 in., which is about the same length as an arctic fox but its short legs make it seem much smaller. The Villafranchian form may have been up to 25 per cent larger. Another large species (*N. sinensis* Schlosser) has been found in the Villafranchian of China; it may be identical with the European form. In China, however, the species survived in the Middle Pleistocene and its body size was gradually reduced. Later Pleistocene Chinese forms carry through a complete transition to the modern species, so that the evolution of the raccoon-dog is very completely documented. Elsewhere in Asia large raccoon-dogs may have survived to a later date; there is a form in the F-Eemian of Palestine that may belong to this genus (*N. vinetorum* Bate).

A direct forerunner of the raccoon-dog is known from the Astian, *N. donnezani* Depéret; it was almost as large as the Villafranchian species.

Like the modern raccoon-dog, the Villafranchian species may have lived on fruit and other vegetables as well as meat. The tubercular molars are well developed in both species, while the carnassial teeth are small. The modern form lives mainly on frogs in the summer but it also takes molluscs, fish and some small rodents.

Though the raccoon-dog is mainly an inhabitant of the plains, it is also found in open woods. Its modern distribution is primarily confined to eastern Asia but after introduction into European Russia it has spread rapidly from there, for instance into Finland and northern Sweden [214; 302].

Family Procyonidae, Raccoon-like Carnivores

Procyonids were not uncommon in Europe during the Tertiary but in the Pleistocene only one species has been found and it belongs to the panda group. Most of the living procyonids inhabit the New World. The family appears to be somewhat heterogeneous and the relationships between the raccoons and coatis on one hand and the pandas on the other has been questioned.

The English Panda, *Parailurus anglicus* Boyd Dawkins. The species was originally described on the basis of fragmentary material from the Red Crag in East Anglia, evidently of early Villafranchian date. Later on, more complete material has been found in the Astian forest faunas of Europe, so that the Crag record seems to represent a relict population.

The fossil form is closely related to the living Panda, *Ailurus fulgens* F. Cuvier, a forest animal living on the slopes of the Himalayas. The panda has a curiously inverted colour pattern: the belly, which is light-coloured in most mammals, is black; while the face, usually dark in multi-coloured carnivores, is white; the back is a magnificent rusty red. The diet is predominantly vegetarian (bamboo shoots, fruit, fungi, etc.) but does include some animals like small mammals and insects. The choice of food is reflected in the well developed tubercular teeth. The Giant Panda, *Ailuropoda melanoleuca* Milne-Edwards, is even more specialized in this respect: bamboo forms its staple diet.

The English panda is classified in a separate genus because the reduction of the dentition has shifted one step further back in this species than in *A. fulgens*. However, the habits of this animal probably resembled those of the living panda, so that we may assume that much of East Anglia was forest-clad in Red Crag times.

The ancestry of the *Ailurus* group is unknown. The modern species has been found in the Middle Pleistocene of China (Yunnan) but it is rare as a fossil; the giant panda, on the other hand, is quite common in the Pleistocene of China [144].

Family Ursidae, Bears

Like the hyenas, the bears are a 'young' family; they date from the early Miocene. The transition from the dogs is well recorded. The earliest bears were small animals (genus *Ursavus*) but most forms tended to increase in size. At the same time the tubercular back teeth tended to become larger at the cost of the 'carnivorous' part of the dentition, so that the bears developed a marked adaptation to an omnivorous mode of life. At first sight it may even be difficult to see any difference between the back teeth of a bear and a pig.

The primitive bears, in which the teeth had not yet become strongly specialized, are referred to the subfamily Agriotheriinae, of which one member survived at the beginning of the Pleistocene. All the other Pleistocene bears of Europe belong to the subfamily Ursinae, which also includes all the living bears of the world except the South American spectacled bear.

The bear family contains relatively few species, which makes it easy to survey. Its fossil record is excellent; indeed, as regards the Pleistocene bears of Europe, almost incomparable. This holds especially for the Late Pleistocene, a time when the cave bear on the continent and the brown bear in England used caves as winter quarters. No bear caves of this type have yet been found from the Early and Middle Pleistocene but open-air sites from these stages may yield great numbers of bear fossils. Presumably the large size of bear bones increases the chances for their preservation and recovery.

The Hyena Bear, *Agriotherium insigne* Gervais (*Hyaenarctos insignis*). This archaic form, which was common in the Astian, survived in the earliest Villafranchian forest faunas (Etouaires; Vialette). It was a large animal even for a bear, but its dentition retains some dog-like characters and the upper back teeth are not elongated as in ursine bears. The name *Hyaenarctos* presumably alludes to the robustness of the cheek teeth; there is no reason to assume that the habits of this animal resembled that of the hyena in the least. On the other hand it is possible that the hyena bear was less vegetivorous than most living bears.

The hyena bear was probably a forest animal, like most modern species of bears. Its extinction in the Villafranchian may be related to the evolution of large bears of the genus *Ursus*, which may have proved superior in competition. The Etruscan bear was apparently the ecological successor of *A. insigne*.

The genus *Agriotherium* has also been found in Asia and North America but seems to have become extinct in most areas before the end of the Pliocene [76].

The Auvergne Bear, *Ursus minimus* Devèze & Bouillet (*U. arvernensis* Croizet & Jobert; *U. ruscinensis* Depéret, etc.). This small and rather primitive species initiates the long series of members of the genus *Ursus* that lived in Europe during the Quaternary period; in fact all the living and fossil bears of this genus seem to have descended ultimately from *U. minimus*. The species is typical of the Astian (Montpellier; Perpignan; Wölfersheim; also in eastern Europe) but survived in the early Villafranchian (Etouaires; basal levels of Val d'Arno). It is apparently the immediate ancestor of the Etruscan bear. The tendency to size increase within *U. minimus* (the Etouaires form is larger than the Astian) continued in the daughter species, *U. etruscus*. Again, the Auvergne bear was ancestral to the black bears either directly or through early Etruscan bears.

The ancestors of *U. minimus* may be sought among the Early and Middle Pliocene species of the genus *Ursavus*. The Auvergne bear was of about the same size as the small Malay bear (*Helarctos malayanus* Raffles), but anatomically it resembled the black bears [76].

The Etruscan Bear, *Ursus etruscus* Cuvier. The typical bear of the Villafranchian made its first appearance in the early middle part of the stage (Villaroya; Saint-Vallier) and ranged to the end of it (Senèze; Val d'Arno; Tegelen). The first Etruscan bears were small, about the size of the modern Asiatic black bear; but they tended to increase in size and the terminal forms were as large as a brown bear. At the same time the dentition became more specialized: the back teeth became more elongate, forming large, grinding tubercular surfaces, while the carnassial teeth were reduced and the anterior premolars dwindled to small, useless pegs. All the premolars were however retained in the Etruscan bear and it was not until the Middle Pleistocene that the *Ursus* line began to lose these vestigial teeth altogether.

Closely related forms of late Villafranchian age in China are known, but it has not been definitely settled whether they belong to this species or whether they are early members of the Tibetan black bear line; in any case they are probably ancestral to that species.

The Etruscan bear, itself a direct descendant of the Auvergne bear, gave rise to two distinct evolutionary lines. One of them, localized in Asia at the outset, gave rise to the brown bears. The other, a uniquely European line, is that of the cave bears, of which two successive species have been recognized [76; 283; 302].

Deninger's Bear, *Ursus deningeri* Reichenau. The systematics of the early Middle Pleistocene bears in Europe are somewhat confused, but it is clear that all of them are on the line leading to the cave bear and most students separate the populations up to and including the 2-Mindel as a distinct species, *U. deningeri*. As a matter of fact it seems that the bears of C-Cromer and 2-Mindel could just as well be regarded as early true *U. spelaeus*. This would leave only the smaller and somewhat more primitive forms that lived during the B-Waalian interglacial and 1-Günz II (Bacton Forest Bed, Hundsheim and other sites) which had advanced too far beyond the *etruscus* stage to be referable to that species; for these forms the name *U. savini* Andrews is available.

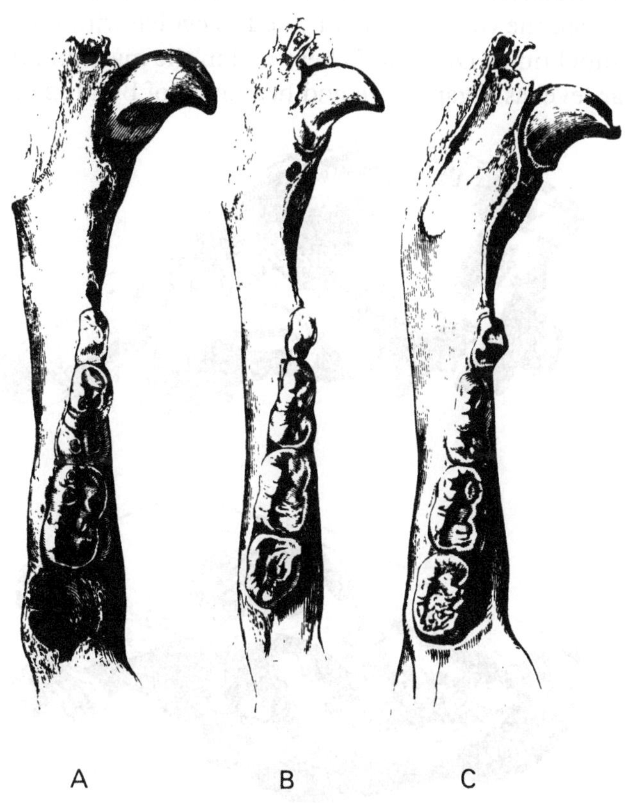

Figure 48. Mandibular dentitions of A, Brown Bear, *Ursus arctos*, Late Pleistocene or Holocene, Cambridgeshire Fens; B, Brown Bear, Late Pleistocene, Kent's Cavern, Torquay; C, Deninger's Bear, *U. deningeri*, Middle Pleistocene, Forest Bed, Norfolk. Not to scale. After Owen.

Savin's bear was well advanced in the *spelaeus* direction as regards the partial loss of the premolars and the increase of the nasal sinuses, resulting in a peculiar, deceptively intellectual-looking doming of the forehead. This form was, however, smaller and less heavily built than the true cave bear, especially as regards the limb bones. It represents a transitional stage in the evolution from the last Etruscan bears (of A-Tegelen date) to the Deninger's bears in the C-Cromerian. The record of the cave bear line in the early Middle Pleistocene is one of the most remarkable examples of evolution in action known to us.

True Deninger's bear is found in the deposits from the C-Cromerian and 2-Mindel (Mauer; Forest Bed; Mosbach 2); it is doubtless a continuation of the line of Savin's bear without any sharp boundary between the two and it continued to evolve within its own span. By 2-Mindel times Deninger's bear had in the main attained the true cave bear level as regards size and heaviness of build [223; 313].

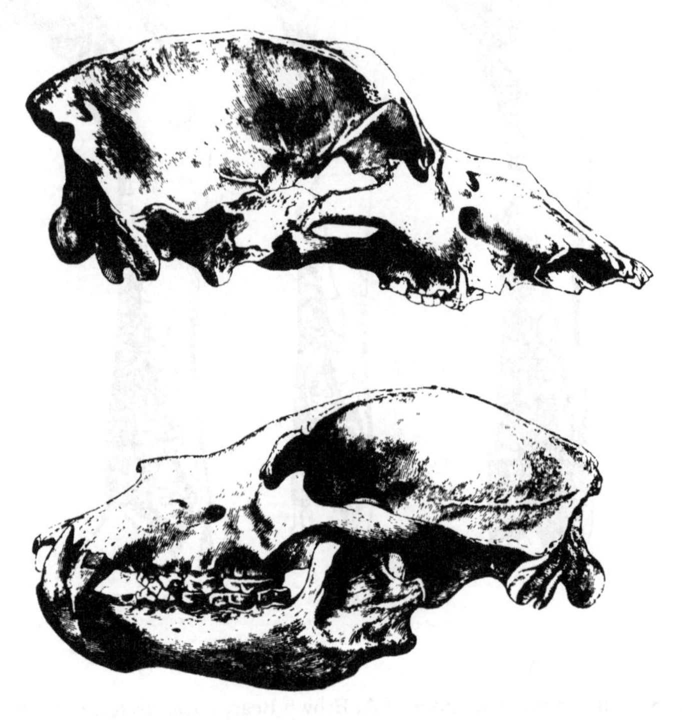

Figure 49. Skull of Cave Bear, *Ursus spelaeus*, Late Pleistocene, Gailenreuth Cavern (above), one-sixth natural size. Skull and mandible of Brown Bear, *U. arctos*, Late Pleistocene of Holocene, Cambridgeshire Fens (below), one-fourth natural size. After Owen.

The Cave Bear, *Ursus spelaeus* Rosenmüller & Heinroth. We now come to one of the best-known Ice Age mammals, whose remains have been unearthed by the hundred thousand in the European bear caves. The mass occurrences date from the Late Pleistocene but the species was in existence far earlier. The Deninger's bear of C-Cromer and 2-Mindel times could very well be included in the species and at any rate the form that lived in the D-Holsteinian was definitely *U. spelaeus* (Swanscombe; Heppenloch; Lunel-Viel). In the early part of the D-Holstein interglacial the species ranged well into England but later on the cave bear lost its foothold in the north, apparently as a result of the invasion of the brown bear (*U. arctos*), so that it is absent at the other D-Holsteinian sites in England. Later on, only marginal populations occasionally ranged into southern England [171].

In the Late Pleistocene the cave bear ranged northward into Belgium, southern Holland, Harz in Germany and the Kraków region in southern Poland. It ranged eastward to Odessa and the Caucasus but is lacking in the greater part of Russia. (The cave bear line inhabited the Caucasus as early as the *deningeri* stage.) In Spain only the area down to Barcelona was inhabited by the cave bear; in Italy the species ranged to Cassino. The southern boundary in the Balkans is uncertain but the species does not seem to have reached Greece. One or two uncertain finds are reported from North Africa and Palestine.

The cave bear was thus an endemic European species and it has not been found in the caves of Asia. Although its geographic range was very small, the species had a strong tendency to split into local races, which suggests limited migration movements between populations; probably individuals of this species were quite localized in habits, unlike most large Carnivora which tend to trek widely [165]. As a result local populations may have been isolated from each other to some extent. In high Alpine caves (the species has been found at an elevation of 2445 m.) and also in the Harz mountains, local dwarf races evolved. This has no connection with sexual dimorphism: the males were much larger than the females, so that the latter have occasionally been misinterpreted as a dwarf race or species of cave bear. In the true dwarf races the males are about as large as normal females, while the females are still smaller.

A few scattered finds in southern England (Wookey Hole south of Bristol and Kent's Cavern in Torquay, etc.) show that the cave bear occasionally ranged here; at both sites, however, it is greatly outnumbered by brown bear. Oddly enough some of the English specimens are extremely small dwarfs, while others represent truly gigantic individuals.

F-Eemian forms are apparently on average somewhat smaller than the full-fledged cave bears of the Last Glaciation; some of the

F-Eemian populations even retain ancient characters such as the occasional presence of anterior premolars [201].

The cave bear was occasionally hunted by man, but the great accumulation of bones in the caves represents animals that died in hibernation [257]. Death in winter sleep was apparently the normal end for the cave bear and would mainly befall those individuals that had failed ecologically during the summer season – from inexperience, illness or old age. As a result the remains found are mostly of juvenile, old or diseased animals.

Figure 50. Restoration of Cave Bear, *Ursus spelaeus*. After a life-size model by Koby and Schaefer (Natural History Museum, Basel).

Since caves form an ideal environment for fossilization, a large percentage of the remains have been preserved; but most specimens have been trampled and crushed and the bones scattered around when bears and other animals moved about in the caves (*charriage à sec*; see Koby [130]). A complete articulated skeleton is a very unusual find [72]. Other evidence of the presence of cave bears may be seen in narrow spaces where the walls have been polished to a high sheen by the passage of innumerable bears during thousands of years. Occasionally claw marks show where the bear stood up to scratch the wall.

The cave bear differed from the living brown bear by its unusually large size, rivalling that in the giant brown bears of Alaska. The head was very large, the forehead strongly vaulted; the legs were short but very robust [70]. The dentition is characteristic too: of the premolars only the hindmost remain but the tubercular teeth are greatly enlarged. The cave bear presumably was almost exclusively vegetarian in habits.

The age distribution of the fossil remains [168] indicates that the annual mortality of the cave bear may have been about 20 per cent but most of this comprises the young, of which nearly 70 per cent died before sexual maturity at four or five years. The number of young in a litter appears to have been one or two. The oldest cave bears probably did not live to more than 20 years or so, for the teeth were worn down rapidly as a result of vegetivorous habits. In the oldest individuals the tooth crowns are gone and even the roots are wearing down.

Pathological deformations are not uncommon: dental caries, rheumatism, inflammations and other maladies have been diagnosed, as well as various kinds of mechanical damages [1]. Some of these may be due to rocks dislodged from the cave ceiling, which could maim or kill a bear. In several instances healed fractures of the penis bone have been observed; it has been suggested that they might be due to fighting between males in the breeding season.

Curiously enough there may be a strong disproportion between the sexes in some caves. In the famous Drachenhöhle at Mixnitz in Styria, where it has been estimated that some 30,000 individuals of the cave bear were represented, the males outnumber the females three to one at some levels; but in other caves, for instance in Switzerland, female individuals are in the majority. Generally speaking, males are more common in large caves that served as winter quarters for numerous individuals, while females accompanied by their young seem to have preferred small, undisturbed caves.

The extinction of the cave bear may be followed in some detail. It was still quite common throughout its range in the interstadial 4-Würm I–II, about 30,000 years ago. But after the invasion of *Homo sapiens* in Europe at about this time the number of cave bears seems to have declined; and in Magdalenian times, near the close of 4-Würm II, only a few caves in central Europe show evidence of occupation by cave bears. There seem to have been small relict pockets in the Swabian Alps, in Westphalia and in Switzerland, with broad empty tracts in between. By the end of Magdalenian times even these straggling populations had become extinct.

It does not seem probable that man hastened the extinction of the species by actual hunting, for evidence of human interference with the bear is scarce in comparison with the mass of naturally

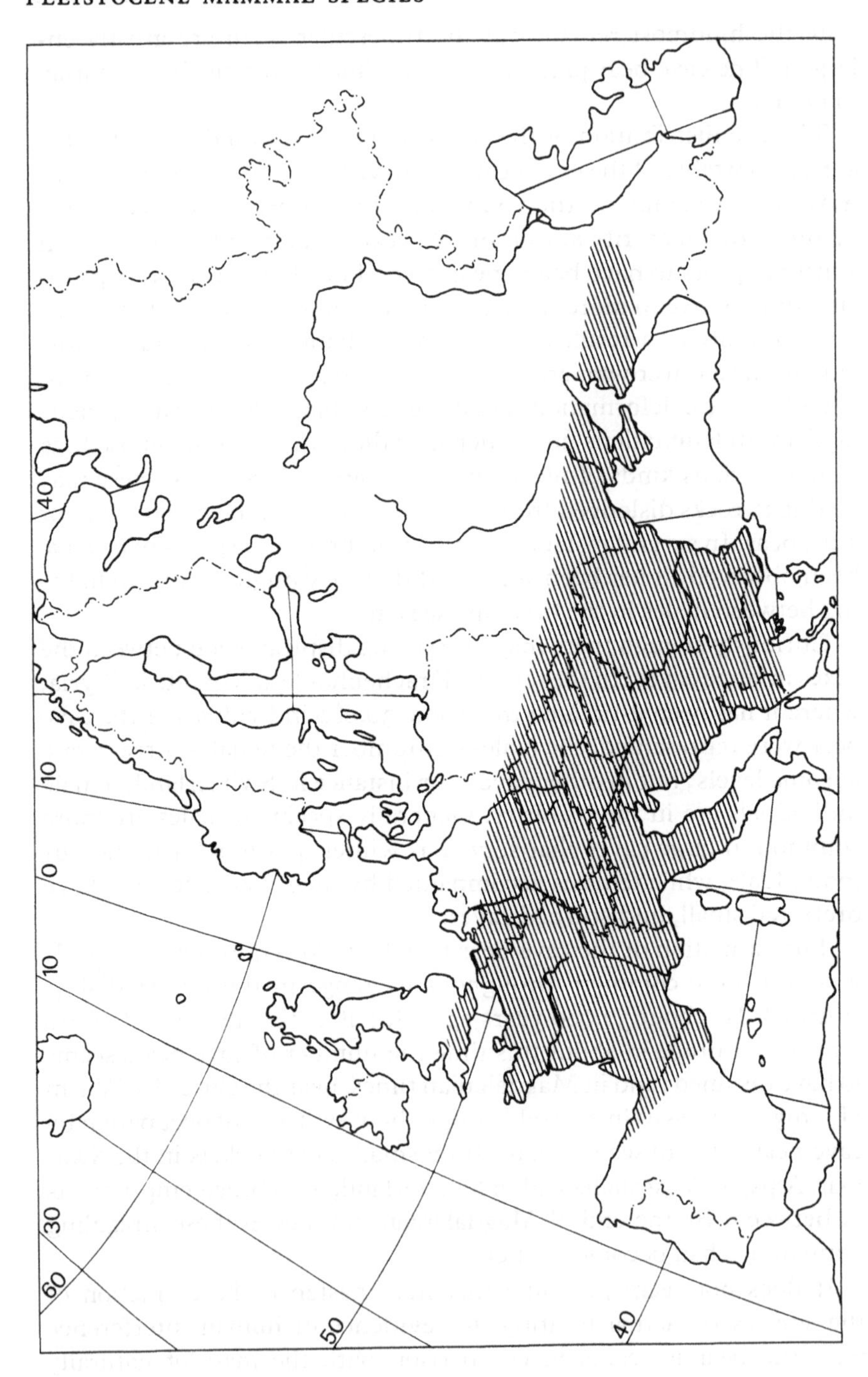
40
10
0
10
30
60
50
40

accumulated remains, although some examples of ritualistic behaviour involving cave bear bones have been reported [1; 71; 76; 134; 165; 168; 201; 204].

The Brown Bear, *Ursus arctos* Linné. The earliest brown bears are known from the 2-Mindel in China (Choukoutien); these are large animals, some even big enough to have been described incorrectly as cave bear. The species has a continuous record in East Asia from the early Middle Pleistocene to the present day. Doubtless it is descended directly from *U. etruscus*; anatomically the A-Tegelen bear is very close to the Choukoutien specimens. Unfortunately the intervening populations of 1-Günz and C-Cromer times are little known.

The species entered Europe slightly later, by mid-D-Holsteinian times; early in this interglacial only cave bear is found in Europe. In central Europe the two species existed together in the D-Holsteinian (Heppenloch; Lunel-Viel) but in England the brown bear apparently ousted its relative completely, for only *U. arctos* is found at Grays Thurrock. From that time to the end of the Ice Age, Britain remained a stronghold of the brown bear: it has been found at E-Ilford and later on in deposits from the 3-Riss (Tornewton Cave) and the Late Pleistocene (numerous caves). Mass occurrences are known in many British caves, beginning with Tornewton Cave; this shows that the species was an ecological vicar of the cave bear.

The presence of the species in central Europe is attested to by a number of finds, though the cave bear was entirely predominant as a cave-dweller. Rich material of *U. arctos* has been found in the F-Eemian travertines near Weimar and even in *spelaeus* caves there are occasional specimens of *arctos* in the 4-Würm. When the cave bear became extinct, several of the caves were inhabited by brown bear.

The brown bear is a forest animal although it has now been driven into the tundra in some regions. As is well known, the bear is omnivorous and eats almost anything that is digestible, from ants, berries, tender grass and succulent herbs to large mammals – reindeer, elk and horse. It also eats carcasses but is unable to crush large bones.

The present-day distribution of the species is very wide. It was common throughout Europe in historical times. It is found in Asia except for the northern tundra and the southern peninsulas and in the western part of North America down to northern Mexico. The invasion in North America dates back to the 4-Würm, but the great continental ice-sheet extending from the Pacific to the Atlantic confined the species to Alaska during most of the glaciation; in Postglacial times it spread southward. (All the pre-Wisconsin *Ursus* finds in the US that I have checked turn out to be black bear.) At one time the local American forms (great brown bear and grizzly bear) were regarded as distinct species but this cannot now be upheld, especially in view of the fact that both races have an evident Old World origin: the great brown bear in the broad-skulled Kamtchatka form, the grizzly in the narrow-skulled bears of northeastern Siberia.

Figure 52. Engraving of Brown Bear, *Ursus arctos*. Teyjat, Dordogne. After Koby.

The largest living forms of the species are found in Alaska and Mongolia, while the European race is rather small. The plasticity in size of the brown bear is extraordinary. In Europe the species varied according to Bergmann's rule: the specimens from 3-Riss and 4-Würm are veritable giants, exceeding even the largest living forms, but those in D-Holsteinian and F-Eemian deposits are much smaller. The size reduction in Postglacial times can be followed step by step thanks to the excellent fossil record and forms a classical example of rapid evolutionary change [52; 76; 169; 171; 177; 280].

The Polar Bear, *Ursus maritimus* Phipps. Surprisingly enough, this species is almost unknown in the fossil state. There are a few Postglacial finds in Yoldia Clay from Denmark and south-west Sweden. Pleistocene finds are uncertain, but at least one lower arm bone (ulna) from Kew, London, appears definitely to be polar bear. It was found in association with numerous reindeer, evidently dating from the beginning of the 4-Würm.

The indication is that the polar bear is a very young species, geologically speaking. It probably evolved in the later part of the Middle Pleistocene; it has some special resemblance with brown bear of 2-Mindel and 3-Riss date. Perhaps the polar bear is descended from a coastal brown bear population which specialized in seal hunting.

In accordance with its entirely carnivorous habits the dentition of the polar bear has been modified and the cusps of the tubercular teeth have grown high and sharp, but the dentition as a whole still bears the stamp of its brown bear ancestry. The polar bear lives almost entirely on seals, which are hunted on land or on the ice but not in the water. The polar bear uses its great swimming powers only to trail the seals when they move from place to place. The distribution of the species is circumpolar [178].

The Asiatic Black Bear, *Ursus thibetanus* Cuvier. This species might almost be regarded as a surviving but slightly modified Etruscan bear, resembling especially the early variety of the middle Villafranchian. Fossils from the end of the Villafranchian in China have already been mentioned as transitional between *U. etruscus* and *U. thibetanus*. In the Middle Pleistocene the species ranged into Europe. At first it inhabited only eastern Europe, but in the C-Cromer it extended its range into central Europe and in the 3-Riss into France. According to a recent review by Thenius [259], the European localities are: Vrhovlje and Podumci in Yugoslavia; Elba and Sistiniana in Italy; Episcopia in Roumania; Villány and Beremend in Hungary; Laaerberg in Austria; Mauer, Mosbach and Bammental in Germany; Montmaurin and Achenheim in France. The European black bear is sometimes regarded as a distinct species, *U. mediterraneus* Major.

This species was evidently always predominantly Asiatic in distribution and it has a well-documented fossil record in China throughout the Middle and Late Pleistocene. The black bears apparently entered North America in 2-Mindel, since the earliest known American specimens appear to date from the D-Holsteinian. This early form, known from a cave near Port Kennedy in Pennsylvania, is still very similar to its Asiatic mother species.

Today the Asiatic black bear inhabits a great tract of Asia from the east coast, including Japan and Formosa, westward into Baluchistan and Afghanistan [282; 283].

Chapter 11

Order Proboscidea

THE Order Proboscidea comprises the elephants and their extinct relatives. This group of animals passed its acme long ago and today only two species remain, both having a limited distribution in the Old World tropics. But in the Tertiary and the Pleistocene the proboscideans were a dominant group, inhabiting the entire world continent – that is to say, all the continents except the island continents of Australia, Antarctica and (in Tertiary times) South America.

The earliest proboscideans appeared in the Late Eocene. Up to the end of the Tertiary, the mastodonts were the dominant group and it was not until the beginning of the Pleistocene that the true elephants, with their lamellar cheek teeth, made their appearance. Mastodonts survived for some time along with elephants, in America into Post-glacial time. There were also the stegodonts, which are transitional between mastodonts and elephants. The stegodonts did not reach Europe, but mastodonts were fairly common here in the Early Pleistocene.

But it is the elephants that dominate the scene in the Ice Age. They evolved on two lines out of a common ancestor, the southern elephant. The mammoth line evolved in adaptation to a steppe environment, feeding on tough, siliceous grass; in this stock the lamellae of the cheek teeth became gradually more tightly appressed. In the other line, that of the straight-tuskers which were forest forms, the cheek teeth remained little modified.

The fossil history of the proboscideans is excellent in the sense that one can almost certainly count on finding a record if they lived in a given area where fossiliferous deposits were laid down. Usually it is the massively built cheek teeth that will be preserved. These teeth are very useful for classification purposes and the rapid evolution of the elephants in the Quaternary makes them important guide fossils. If, for instance, one finds a transitional form between steppe mammoth and woolly mammoth, its date is likely to be early 3-Riss. Advanced true mammoth with tightly packed lamellae in the cheek teeth will suggest 4-Würm. The elephants are also important climatic indicators. The mammoths are steppe and tundra forms, *Palaeoloxodon antiquus* a woodland animal.

Unfortunately it is less common to find well-preserved complete skeletons, to say nothing of the practical and financial difficulties involved in their collection. Only the advanced woolly mammoth of the late Pleistocene is really well known anatomically; we even know the soft parts, skin and hair, thanks to the discovery of frozen specimens in the permafrost area of Siberia.

From the biological point of view, the Pleistocene elephant record is of the greatest importance. It is one of the finest examples of the gradual evolution of new species known to us. To a systematist bent on pigeonholing his finds this is a difficult situation, for there is no way to draw sharp boundaries between the species succeeding each other in time. Evolution consists of a continuous, gradual shift in the averages of various characters.

Family Gomphotheriidae, Pig-toothed Mastodonts

The Auvergne Mastodont, *Anancus arvernensis* Croizet & Jobert. This species dates from the Pliocene but survived during the entire Villafranchian. It was common in the earlier part of this stage before the

Figure 53. Restoration of Auvergne Mastodon, *Anancus arvernensis*, based on skeletons from lower Val d'Arno and Dusino, Piemont. After Osborn & Flinsch.

entrance of true elephants. Later on it became rare and is missing in steppe faunas like that of Senèze. It reappears, however, at Val d'Arno and some discoveries indicate that it straggled on at least into 1-Günz

II. There is one specimen from the basal layers at Mosbach and another from Middle Pleistocene gravels at Gmunden in Austria.

By that time the species had reached a venerable age, for it was in existence at the beginning of the Astian; finds from the Middle Pliocene show it to have evolved from Lower Pliocene ancestors. Like other mastodonts it has tubercular cheek teeth, useful for the chopping of leaves and succulent plants but not grass. The presence of the browsing mastodonts thus indicates an environment of woodlands.

Externally the mastodonts resembled elephants, but they tended to have shorter legs and a relatively large, elongate head. The Auvergne mastodont was peculiarly equipped with a pair of enormously long tusks which reached a length of up to 10 ft. Contrary to the norm in proboscideans, the tusks were almost straight (*Anancus*, 'without curve') and would seem to have been tremendous weapons of defense. Oddly, this character was repeated at a far later date by the ecological successor of the Auvergne mastodon in Europe, the straight-tusker, though there was no direct relationship between the two species.

Two nearly complete skeletons of this species have been discovered (they are now in Turin and Bologna). Related forms have been found in Asia [211; 279].

Family Mammutidae, True Mastodonts

Borson's Mastodon, *Zygolophodon borsoni* Hays. Like the Auvergne mastodon this species is a survivor from the Astian, but it has only been found in the early Villafranchian (Etouaires; Vialette; Chagny). Related species occur in the Pliocene of Asia and in China the genus *Zygolophodon* survives in the early Villafranchian.

Borson's mastodon had much shorter tusks than *Anancus* and they were curved upward; in addition it carried vestigial tusks in the front of the lower jaw. It is closely related to the well-known American mastodon, which lived in North America into Postglacial times and had shaggy, brownish hair [211].

Family Elephantidae, Elephants and Mammoths

The Southern Elephant, *Archidiskodon meridionalis* Nesti. As a symbol for a new epoch in the history of the earth, the great southern elephant marched on to the European stage in the middle Villafranchian.

From the Saint-Vallier level on, into the C-Cromer, it was the dominant animal in Europe and its remains have been unearthed both in Britain (the Crags) and on the continent.

The southern elephant is a direct descendant of the earliest known elephant, *A. planifrons* Falconer & Cautley. The planifrons elephant probably evolved in Asia or Africa and gradually migrated northward. The earliest immigrants in Europe (at Chagny, for instance) are still very close to the mother species, but in general the European branch is regarded as a species distinct from *A. planifrons*.

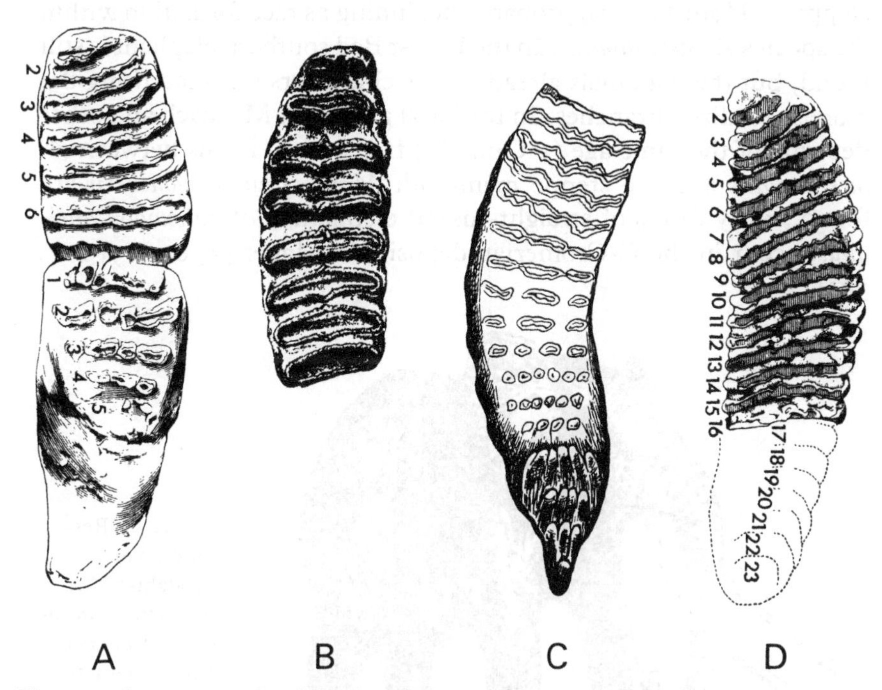

Figure 54. Lower second and third molars of Elephantidae. A, Southern Elephant, *Archidiskodon meridionalis*, early Middle Pleistocene, Durfort, M_2–M_3; B, Straight-tusked Elephant, *Palaeoloxodon antiquus*, late Middle Pleistocene, Grays, M_2; C, Steppe Mammoth, *Mammuthus trogontherii*, early Middle Pleistocene, Süssenborn, M_3; D, Woolly Mammoth, *Mammuthus primigenius*, Late Pleistocene, Siberia. All one-sixth natural size. A, D after Osborn; B after Falconer & Cautley; C after Pohlig.

The body build of the southern elephant may have resembled the modern Indian elephant more than the African species; the back was horizontal or slightly arched but did not slope down from the shoulders as in the African species. The tusks, however, were much larger than in the Indian species and were somewhat curved upward. The cheek teeth were moderately high with short, broad mastication surfaces, traversed by relatively few, wide-spaced lamellae. Evidently the species was not highly specialized but inhabited both savannas, bush steppes and woodlands; it was probably not successful in extreme steppe environments. There is a complete skeleton in Paris (from Durfort) and others less complete in various museums. The Paris specimen indicates a shoulder height in life of 3·7 m. (12 ft.).

In the C-Cromerian there was already a tendency to a splitting into steppe and forest forms, probably beginning as race formation within the species *A. meridionalis*. In the Forest Bed southern elephant is still found, but these animals already show characters foreshadowing the more specialized branches. In the basal gravels at Mosbach and other deposits of the same age (1-Günz II) transitional forms between *A. meridionalis* and the steppe mammoth *Mammuthus trogontherii* are found. Early forms of straight-tusked elephant, *Palaeoloxodon antiquus*, occur in the C-Cromerian deposits of Mauer [3; 64; 65; 211; 302].

Figure 55. Restoration of Straight-tusked Elephant, *Palaeoloxodon antiquus*. After a statuette by Schweizer, based on a skeleton from Steinheim.

The Straight-tusked Elephant, *Palaeoloxodon antiquus* Falconer & Cautley. This species is characterized by its great, almost straight tusks, which in the old bulls protrude like a pair of javelins. In stature and body build the straight-tusker was rather like the southern elephant but it was longer-legged and the head was smaller in relation to the body. The cheek teeth were narrower and more elongate than in the southern elephant but the lamellae were about equally spaced. The straight-tusker was adapted to a temperate climate and a parkland or forest environment and it is typical of the interglacials in Europe, beginning with the C-Cromerian. The earliest forms, at Mauer and in the Forest Bed, still retain many *meridionalis* characters. The immense size of this animal is reflected in partial skeletons from England, Italy and Poland, indicating a shoulder height of about 4 m. (13 ft.); no complete skeleton is known.

The species retired southward during glaciations and the Mediterranean peninsulas served as refuges. The straight-tusker survived in Spain in the 4-Würm at a time when it had already become extinct north of the Pyrenees.

The history of the straight-tusked elephant in the Mediterranean area has a bizarre element in the evolution of dwarf races on certain islands. A rather uniform dwarf type, *P. antiquus melitensis* Falconer, existed on Malta, Sicily and Sardinia, probably in the later Middle Pleistocene; the islands were then in direct land connection with each other. From the size of the upper bone of the fore-leg, the shoulder height of this form may be estimated at 140 cm., little more than one-third of that in the continental form. Later on in the Pleistocene Malta became isolated and as a result of the restriction of the available area (and perhaps adverse environmental conditions) there evolved a local form, *P. falconeri* Busk, which measured only 90 cm. (3 ft.) at the shoulders! This almost incredible pygmy elephant was no larger than a pony. The Cyprus form (*P. cypriotes* Bate) was almost as small as the Maltese pygmy elephant, while another species of about the same size as *melitensis* has been found in Crete. All the Mediterranean dwarf elephants belong to the straight-tusker group; whether they should be regarded as separate species or local subspecies is uncertain.

Dwarf elephants have also evolved elsewhere in the world on islands: for instance Java and Sumatra and the Santa Barbara Islands off the Californian coast. In these instances, however, we deal with other species than *P. antiquus*.

Forms related to the European straight-tusker are common in Asia (*P. namadicus* Falconer & Cautley). Their systematic relationship needs further study [211; 254; 304].

The Steppe Mammoth, *Mammuthus trogontherii* Pohlig. This form initiates the line of steppe and tundra elephants that culminated in the well-known Late Pleistocene woolly mammoth. The transition from southern elephant to steppe mammoth is revealed by finds in southern Germany [6]. For instance in the sands at Aalen, which contain a steppe fauna from 1-Günz II, there occur advanced southern elephants with many traits resembling *trogontherii*. In the slightly younger but still 1-Günz II deposits at Jockgrim, evolution has proceeded to a point where the species has to be called *M. trogontherii* although retaining many *meridionalis* characters. So it is to some extent a matter of taste where the boundary between the species is drawn; what is important of course is that we have evidence of the direct emergence of steppe mammoth by gradual transformation from a steppe race of the southern elephant.

During 2-Mindel *M. trogontherii* was a characteristic element in the fauna and the upper layers at Mosbach contain typical representatives of the species. In size this typical steppe mammoth exceeded all other European elephants, reaching the almost incredible shoulder height of 4·5 m. (more than 14½ ft.) in the largest recorded specimens.

The tusks were moderately curved and in some instances reached great length (e.g. one specimen from Süssenborn probably 5 m. long).

In the upper gravels at Mosbach, from the maximum cold phase of the 2-Mindel, variants of the species somewhat resembling the woolly mammoth are already found; evolution had been going on continuously all through the *trogontherii* stage in the history of the mammoths. Early in the 3-Riss the transformation reached a stage where we have to identify the form as a *trogontherii*-like *primigenius* rather than a *primigenius*-like *trogontherii*. The actual transition is localized in the early part of the 3-Riss [4].

During the long but rather cool D-Holstein interglacial the steppe mammoth was occasionally present in England along with the straight-tusker, but it did not range into central Europe [2; 6; 211; 254].

The Woolly Mammoth, *Mammuthus primigenius* Blumenbach. This species was present in Europe from the 3-Riss to the end of the 4-Würm; the earliest specimens still resemble the ancestral *M. trogontherii* and are often regarded as a distinct subspecies *M. primigenius fraasi*. In later forms the lamellae of the cheek teeth became still more numerous and closely appressed and the size and curvature of the tusks tended to become more and more extravagant, while the body size was reduced. The Late Pleistocene mammoth reached a shoulder height of about 10 ft. or less, so that it was no larger than a modern elephant.

The mammoth did not inhabit central Europe in the F-Eemian, but finds in Scandinavia date from this interglacial and show that the north was a refuge for the European glacial fauna in the warm phases. In the 4-Würm the mammoth returned to central Europe.

From discoveries of frozen carcasses in Siberia something is known about the soft anatomy of the mammoth. The carcasses, of which about 25 have been found, are remains of individuals that were accidentally trapped and buried in the frozen earth. As there have been some odd statements about the deep-frozen mammoths, it may be useful to discuss briefly the best-preserved specimen that has been studied scientifically, the mammoth discovered in 1900 at the Beresovka River. The animal had been buried by a landslide at the river bank, with various resulting injuries; for instance the hipbone was crushed. It had then gradually frozen; not, however, before much of the flesh and all the entrails had rotted away. The stomach contents have been pollen analysed and show that the Beresovka mammoth lived in an Arctic environment and fed on tundra vegetation. A radiocarbon determination shows that this individual lived about 39,000 years ago.

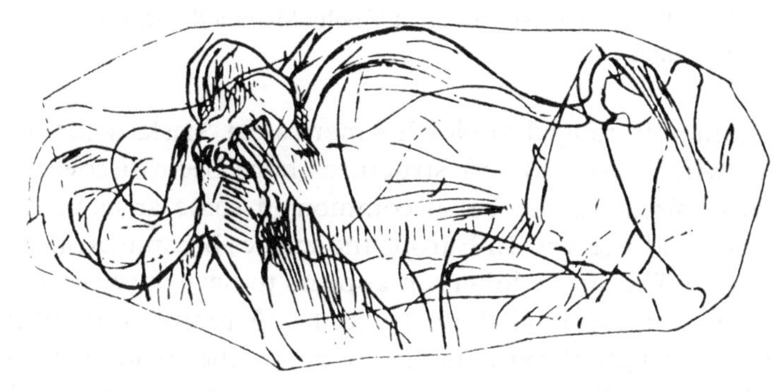

Figure 56. Woolly Mammoth, *Mammuthus primigenius*. Engraving in ivory from La Madeleine, Dordogne. After Breuil.

The exterior of the mammoth was very striking with the long, hanging hair, the immense curved tusks, the high, peaked head and the humped shoulders, behind which the back sloped steeply downward. The hump, which is seen in contemporary pictures, used to be regarded as a fat hump but may actually be due in part at least to the curvature of the spine. The ears were small to reduce heat-loss; insulation was also provided by the hair and by a blubber layer under the skin about 8½ cm. thick. The trunk was rather shorter than in modern elephants.

The mammoth was often portrayed by Ice Age artists and the pictures they have given us of this remarkable animal in their paintings and engravings are fully confirmed by the fossil carcasses. The woolly mammoth was of great importance in the economy of some Ice Age tribes as a game animal; the tusks and bones were used for palisades and other structures. Even in historical times, long after the extinction of the species, fossil ivory has been collected in enormous amounts in Siberia and traded especially to China. It is estimated that remains of about 25,000 mammoths have been unearthed in Siberia.

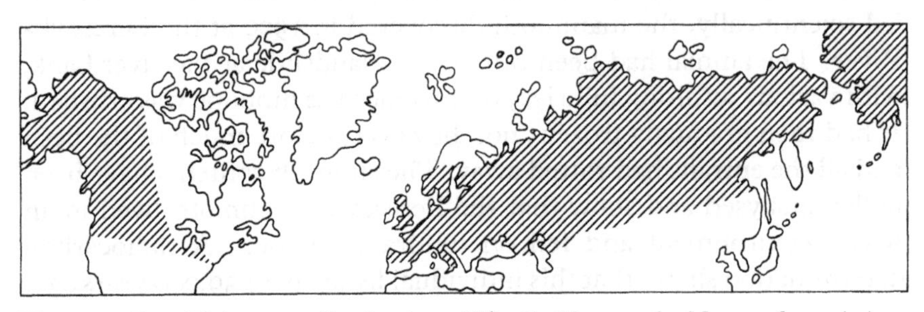

Figure 57. Late Pleistocene distribution of Woolly Mammoth, *Mammuthus primigenius*. After Trofimov.

The mammoth ranged to North America. In Alaska some parts of carcasses with preserved soft structures have been discovered and mammoth hair is said to be so common in some areas as to be a nuisance to gold-diggers. In Europe the species became extinct at the close of the 4-Würm; in Siberia it appears to have survived into the warm Allerød interstadial about 10,000 BC. A recent study of mammoth carcases [94] throws further light on the extinction. It was shown that the radiocarbon dated remains indicate a size oscillation of the type contrary to Bergmann's rule. Specimens dating from 40–50,000 years ago, or 4-Würm I, were decidedly smaller than specimens with a radiocarbon age of 30–40,000 years of 4-Würm I–II interstadial age. Still younger specimens (11–12,000 years old) again show a reduction in size. The authors suggest that the rapid climatic changes at the end of the 4-Würm made it impossible for the mammoth herds to react by sufficiently rapid migrations, thus repeatedly decimating the population until finally the sudden climatic deterioration of the Younger Dryas killed the remainder [85; 93; 211; 292].

Chapter 12

Order Perissodactyla

THE Perissodactyla or odd-toed ungulates comprise the horse, tapir and rhinoceros families. The order passed its apogee long ago; several other perissodactyl families existed in the Tertiary but later became extinct. In the European Pleistocene only the modern families were represented.

The name odd-toed ungulates does not necessarily mean that the number of toes is odd, though this is the rule in most groups. It means that the axis of the foot always coincides with the third or central toe, which is the strongest. In the even-toed ungulates the axis passes between the third and fourth toes, which are equally strong.

The earliest Perissodactyla appeared in the Eocene.

Family Tapiridae, Tapirs

The Tapir family was widely distributed in North America and the Old World in the Tertiary. It is very close to the original perissodactyls of the Early Eocene and so is a primitive group, unspecialized except for the development of a short proboscis. In the Pleistocene tapirs invaded South America; then the northern tapirs became extinct, so that only South America and the East Indies are now inhabited by tapirs. In Europe the family became extinct at the close of the Villafranchian.

The Auvergne Tapir, *Tapirus arvernensis* Croizet & Jobert. This small species of tapir was common in the Astian but has also been found in the beginning of the Villafranchian (Etouaires; Vialette; also in Germany). It was absent in the middle Villafranchian but returned toward the end of the stage and is recorded at Val d'Arno and A-Tegelen. The species does not differ much from living tapirs.

Tapirs are forest animals and partly aquatic; they live on water vegetation but also browse on forest trees, pulling down branches and twigs with the help of the short trunk. In fleeing from a predator the tapir will go for the densest thicket available, which it easily penetrates protected by its tough skin but which will stop most carnivores. Tapirs are usually solitary in habits [83].

Family Rhinocerotidae, Rhinoceroses

The rhinoceroses form an important element in the Pleistocene fauna of Europe. Large, robust bones and teeth are more likely to be preserved than small ones; as a result, the rhinos have a good fossil record. Many species are known and show an interesting range of adaptations to various environments from tundra to forest.

Almost all the Ice Age rhinos of Europe belong to the group of the present-day Sumatra rhinoceros, the genus *Dicerorhinus*. These animals have two nose horns, the one in front being the longest. They also have a somewhat more hairy skin than other living rhinos, which have nearly no hair at all. The dicerorhine rhinos were common in Eurasia in the late Tertiary and the Pleistocene forms were obviously direct derivatives of the Pliocene ones.

The only exception is the Giant 'Unicorn', which however was mostly distributed in eastern Europe. It belongs to a group of its own among the rhinoceroses, with a long, separate record.

Christol's Rhinoceros, *Dicerorhinus megarhinus* Christol. This is the typical rhinoceros of the Astian in Europe but it survived into the earliest Villafranchian (Etouaires; Vialette) in a progressive form, trending towards the daughter species *D. etruscus*. In the Astian form the nasal bone does not connect with the upper jaw in front, or in other words there was no ossification of the septum between the nares, which was entirely cartilaginous. As the nasal bones carry the horns, they would be liable to breakage with the impact of a heavy charge. Accordingly there is a tendency to ossification of the nasal cartilages in the early Villafranchian specimens, especially in the males with their heavier horns; such an arrangement would reduce the risk of breakage by transferring part of the strain on to the bones of the upper jaw. This trend continued in the Etruscan rhinoceros, in which a firm base was formed for the horn-bearing nasal bones.

Like its successor, Christol's rhinoceros was a comparatively small, gracefully built form [278].

The Etruscan Rhinoceros, *Dicerorhinus etruscus* Falconer. This species was the sole representative of the rhinoceros family during most of the Villafranchian in Europe; it has been recorded, for instance, at Villaroya, Pardines, Saint-Vallier, Senèze, Olivola, Val d'Arno, Erpfingen and A-Tegelen. It remained common in the early Middle Pleistocene, where it occurs both in forest faunas (Mauer; Forest Bed) and steppe faunas (Süssenborn; Mosbach) up to the 2-Mindel. At that time, however, a more highly specialized competitor entered the stage in the shape of Merck's rhinoceros. Shortly after the immigration of this species and of the steppe rhinoceros slightly later, the Etruscan rhinoceros died out.

Figure 58. Restoration of Etruscan Rhinoceros, *Dicerorhinus etruscus,* based on a skeleton from Senèze.

The reasons for the extinction may probably be sought in the incomplete specialization of the Etruscan rhinoceros. It was a relatively small form, about the size of the smallest living species, and judging from the angle between the occipital plane and the skull base, the head was carried in an almost horizontal position. It has been shown that the leaf-eating rhinos of the present day habitually carry their heads slightly tilted upward and in their skulls the occipital plane is inclined forward in relation to the skull base. The African black rhinoceros, which eats grass as well as leaves, carries its head horizontally, while the white rhinoceros, an exclusively grazing form, habitually hangs its head. The Etruscan rhinoceros in this respect most closely resembles the black rhinoceros and is likely to have had a similarly unspecialized diet including leaves as well as grass. Savanna and bush steppe would seem to have been the most congenial environments of this form, but it is also found in faunas dominated by woodland species.

There are complete skeletons of this species, for instance from Senèze (Basel). The head-and-body length was somewhat less than 2·5 m., the shoulder height about 1·5 m.; the animal was unusually long-legged for a rhinoceros and probably rather fleet of foot [6; 127; 242; 278].

Merck's Rhinoceros, *Dicerorhinus kirchbergensis* Jäger (*Rhinoceros mercki* Jäger). This species is the constant companion of the straight-tusked elephant in most interglacial faunas from the D-Holsteinian to the F-Eemian. Its first appearance, however, comes somewhat unexpectedly in the 2-Mindel (Mosbach 2), where it is found in a steppe association; it is, however, rare. In England Merck's rhinoceros was particularly abundant during the D-Holsteinian (Swanscombe; Clacton), while the steppe rhinoceros was scarce. At E-Ilford, on the other hand, the steppe rhinoceros was common and Merck's rhinoceros rare. Finally, in the F-Eemian the species was completely absent in England, while *D. hemitoechus* held sway.

In central Europe on the other hand Merck's rhinoceros was common in the F-Eemian and became extinct only at the end of the interglacial. In Spain the species persisted during the first part of 4-Würm I.

The ecological successor of the Etruscan rhinoceros, *D. kirchbergensis* resembles its forerunner in many respects; the skull has the same general shape. But Merck's rhinoceros was considerably larger, which may have given it an advantage in the competition. Like the Villafranchian species, Merck's rhinoceros probably inhabited woodland, parkland and occasionally savanna environments but it does not appear in extreme steppe milieus apart from the very beginning of its history; such environments were taken over by the steppe rhinoceros and the woolly rhinoceros. Unfortunately, no complete skeleton of this species has been found to date.

D. kirchbergensis may have evolved from extra-European representatives of the Etruscan stock. A Chinese Villafranchian member of the genus *Dicerorhinus* is closely allied to Merck's rhinoceros and may well be its ancestor. Merck's rhinoceros had a great range in Asia (southern Siberia), which may also suggest an Asiatic origin [6; 262; 316].

The Steppe Rhinoceros, *Dicerorhinus hemitoechus* Falconer. This species appears to be represented as early as at Val d'Arno but is not found again until the D-Holsteinian, when it invaded Europe; from that time on it occurs in the interglacials with a frequency inversely related to that of Merck's rhinoceros. The steppe rhinoceros carried its head in a hanging position like the present-day white rhinoceros of Africa and its high tooth crowns are adapted to deal with the abrasive grasses of the steppe. It may therefore be taken as an indicator of open grasslands, but it does not seem to have developed any special adaptation to cold so that it was unable to colonize the tundra biotope; its main habitat lay in temperate areas. However, a head with preserved skin and hair of a rhinoceros found in frozen earth at a tributary of the Jana River in Siberia in 1877 is stated to belong to *D. hemitoechus*. Presumably this indicates survival well into 4-Würm.

The steppe rhinoceros is closely related to the Etruscan rhinoceros and may represent an offshoot from early representatives of that group [16; 262; 316].

The Woolly Rhinoceros, *Coelodonta antiquitatis* Blumenbach (*Rhinoceros tichorhinus* Cuvier). The woolly rhinoceros was a highly specialized derivative of the *Dicerorhinus* stock, which invaded Europe in the 3-Riss and remained a typical member of the cold fauna there until the end of the 4-Würm. It apparently evolved in northeast Asia, where predecessors of this species dating from the Early Pleistocene have been found. Discovery of frozen remains in Siberia, and especially the find of a remarkably well preserved cadaver in deposits impregnated with salt and petroleum at Starunia in Galicia, have given detailed information on the appearance in the flesh of this animal; there is a complete stuffed specimen in the Museum of Kraków. The woolly rhinoceros also figures in Stone Age paintings and engravings.

This rhino had a thick, woolly coat as a protection against the Arctic cold. A fairly well-developed pelage may well have been a common character of all the dicerorhine rhinoceroses but *Coelodonta* was probably unusually endowed in this respect. The ossification of the nasal septum was extreme also; and in addition the nasals grew down in front to fuse with the upper jaws – the resulting odd appearance of the rostrum has led French palaeontologists to name it 'the rhino of the closed nares'. The strengthening of the horn base corresponded to the development of two powerful, elongated nasal horns. The cheek teeth were very high-crowned, even more so than in the steppe rhinoceros. Not only the head but also the neck was carried low in this species, thus showing extreme grass-eating specialization.

Figure 59. Restoration of Woolly Rhinoceros, *Coelodonta antiquitatis*, based on stuffed specimen from Starunia; original in the Zoological Museum, Kraków.

The woolly rhinoceros is perhaps mostly thought of as an extreme tundra form, but unmistakable specimens of this species have also been discovered in completely different surroundings. In Spain, for instance, *C. antiquitatis* occurs at Cueva del Toll, north of Barcelona, in deposits from the interstadial 4-Würm I-II; pollen analysis of the same deposit indicates a dry, temperate climate with extensive grasslands and a few broad-leaved trees; there is no possibility whatever that this could be a 'cold' steppe [259].

The Giant 'Unicorn', *Elasmotherium sibiricum* Fischer. The homeland of this species was the steppes of southern Russia, but a few stray finds indicate that it made a temporary incursion in central Europe in the Middle Pleistocene. This species is the terminal form of an evolutionary line that has its roots far down in the Tertiary; we have glimpses of it in the Miocene of Spain, the Pliocene of China and the Pleistocene of Russia.

It was a truly gigantic animal, far larger than any living rhinoceros. It had no nose horn, but instead an immense horn on the forehead: it grew to a length of two metres. This animal was thus a veritable unicorn. Its great, prismatic cheek teeth with their complicated enamel pattern are as highly specialized as those of the horses and show a wonderful adaptation to grazing habits. Knowledge of the postcranial skeleton is incomplete [42].

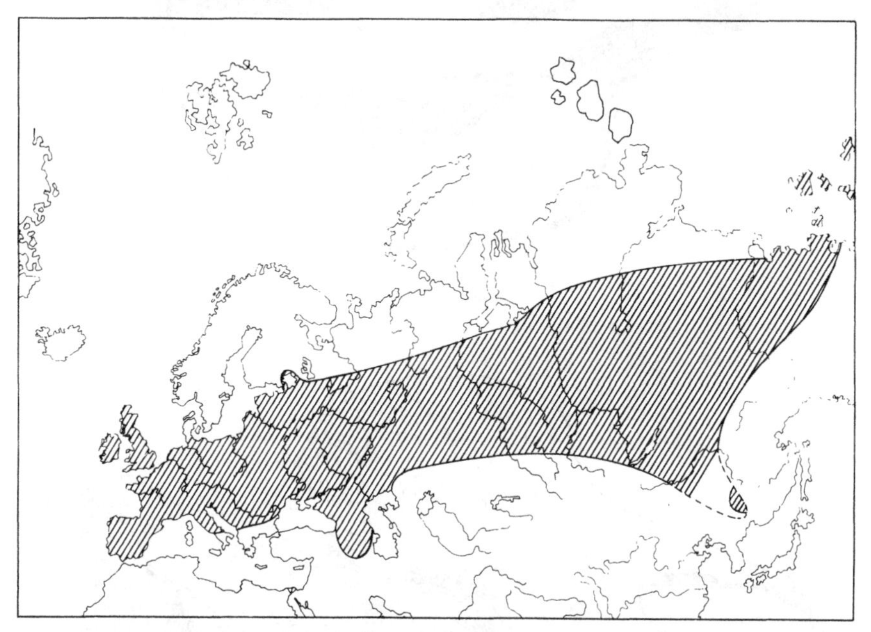

Figure 60. Distribution of Woolly Rhinoceros, *Coelodonta antiquitatis*, in the Pleistocene. After Trofimov.

Family Equidae, Horses

The history of the horse family in the Tertiary is often cited as a classical example of evolution demonstrated by palaeontology, though in fact several other mammalian families have as detailed a record. The earliest equid, the Eohippus or *Hyracotherium* of the Eocene epoch, was a small animal with four-toed fore-feet and three-toed hind feet and a very primitive dentition. These animals, which of course did not in the least resemble a modern horse, are linked to the latter by an unbroken evolutionary series spanning almost 50 million years. In addition various side branches of different types arose; some of them were the dominant horses of their day and could have made a case for

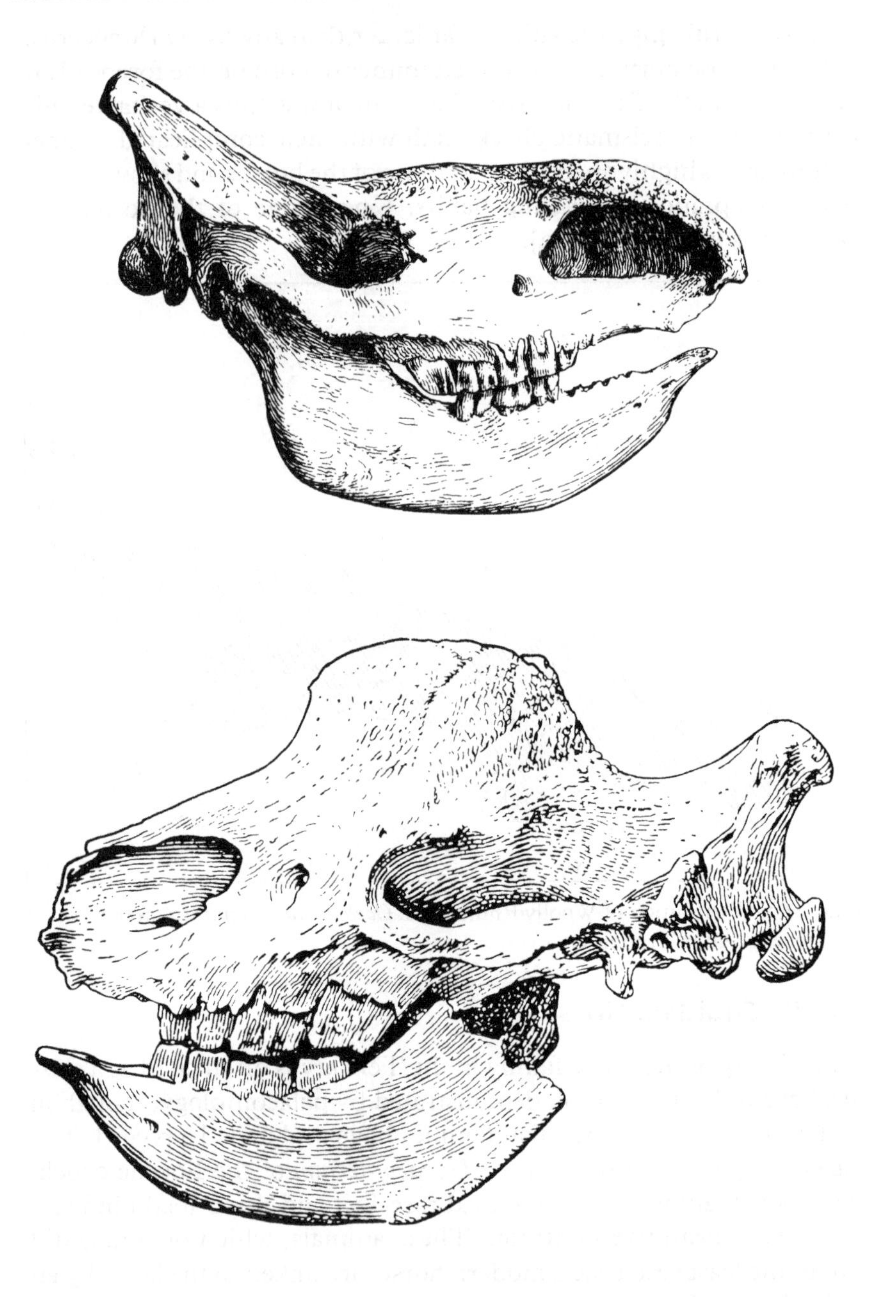

Figure 61. Skull and mandible of Woolly Rhinoceros, *Coelodonta antiquitatis* (above) and Giant 'Unicorn', *Elasmotherium sibiricum* (below), both one-fifteenth natural size. Pleistocene, USSR.

regarding our one-toed horse as an evolutionary side-branch. One of these earlier types of dominant horses, the three-toed *Hipparion*, survived in the Early Pleistocene of Europe. Otherwise one-toed horses of modern type are predominant in the Pleistocene and their migration from North America across the Bering Bridge constitutes one of the main faunal markers inaugurating the Pleistocene.

Crusafont's hipparion, *Hipparion crusafonti* Villalta. With this species the line of three-toed horses expired in Europe after having been a dominant faunal element for some 10 million years. The genus *Hipparion* originated in North America, which was the centre of equine evolution throughout the Tertiary. The invasion of *Hipparion* in the Old World is usually taken as the event marking the beginning of the Pliocene, so that this genus is one of the most important index fossils of its time. The Pliocene hipparions were usually small, about the size of a pony but lighter in build, and there were even one or two dwarf species of gazelle-like proportions in Spain. One line tended to gradual size increase and this trend culminated in the Villafranchian *H. crusafonti* which was almost as large as a modern horse. Its immediate ancestor seems to be the Pliocene *H. rocinantis* Hernandez-Pacheco, a fairly large form that lived in Spain. In China, too, a large hipparion was present in Villafranchian times.

The hipparions differ from *Equus* in that all four feet still have three functional toes instead of one. In the modern horses the middle toe is very strong and linked to the lower part of the leg by a powerful system of ligaments which give it a spring-like action. The middle toe had the same function in the hipparions but was more slenderly built; when galloping or trotting, the side toes would touch the ground and assist the main toe in taking up the impact.

This system apparently functioned well enough, especially in small, lightly built forms such as most of the hipparions. In a large form it must have been less efficient than the one-toed arrangement, which may be one of the reasons why the invading *Equus* rapidly eclipsed its three-toed competitors. In Europe the competition between one-toed and three-toed horses was enacted in the early Villafranchian; in Africa, however, hipparions survived to a much later date.

The external appearance of the hipparions probably would have been that of a slim, delicately built horse. *Hipparion crusafonti* occurs in great numbers at Villaroya. A somewhat smaller form of *Hipparion* is present at Perrier (Roccaneyra, Pardines) but very rare. In

Pardines times *Equus* invaded Europe, and *Hipparion* apparently became extinct almost immediately [108; 298].

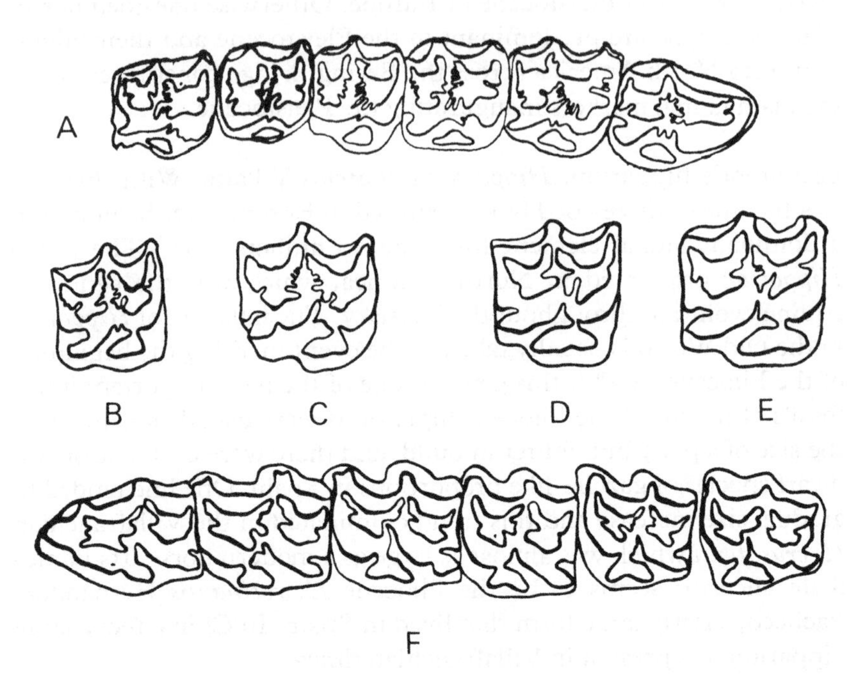

Figure 62. Upper cheek teeth of Equidae. A, tooth series in Crusafont's Hipparion, *Hipparion crusafonti*, Villafranchian, Villaroya; B–C, isolated teeth of European Zebra, *Equus stenonis*, Villafranchian; D–E, isolated teeth of Wild Horse, *E. przewalskii*, Recent; F, tooth series in European Wild Ass, *E. hydruntinus*, Late Pleistocene, Grotta Romanelli. About one-half natural size. A after Villalta; B–F after Gromova.

Zebrine Horses, *Equus stenonis* Cocchi and *E. süssenbornensis* Wüst. The earliest *Equus* to enter the Old World seem to have been zebra-like as regards their dentition, at any rate; whether they were striped like modern zebras is, of course, unknown. But in Europe the zebra-like *Equus stenonis* appears at a later date, in the middle Villafranchian of Saint-Vallier, at a time when the 'caballine' robustus horse (*E. bressanus*) was already present. *Equus stenonis* was abundant in the later half of the Villafranchian (Senèze; Olivola; Val d'Arno; Doveholes; Tegelen) and finally was replaced by, or evolved into, the Süssenborn zebra. Exactly where in the time scale the boundary between the two species should be drawn is not clear at present. At any rate it may be concluded that the zebras in Europe ranged in time from the middle Villafranchian to 1-Günz II.

The European zebras were relatively small, lightly built forms. A complete skeleton of *E. stenonis* (at Basel) indicates a shoulder height of about 140 cm.; the specimen comes from Senèze. The earliest immigrants of this species in Europe were considerably larger, like the Villafranchian zebrine horses found in India (*E. sivalensis* Falconer & Cautley) and China (*E. sanmeniensis* Teilhard & Piveteau) are probably closely related to the European zebra; they are about the same size or slightly larger. This group may be directly derived from the North American Pliocene genus *Plesippus*. However, the early Villafranchian zebras soon gave rise to caballine and asinine horses, which later on were to crowd out the zebras throughout Eurasia; zebras now survive only in Africa [17; 87; 302].

Caballine Horses, *Equus bressanus* Viret (*E. robustus* Pomel), *E. mosbachensis* Reichenau, *E. germanicus* Nehring, *E. przewalskii* Poliakoff and perhaps other species. The systematics of the 'true' or caballine horses in Europe during the Ice Age are somewhat chaotic. A great number of species have been proposed, but it is most improbable that all or even a majority of them will turn out to be valid. However, as Stehlin [265] says, it is '*plus facile de reconnaître la faute que d'y remédier*'. In brief, the following seems to be the main outline of the history of the caballine horse in Europe.

The first invaders in Europe of the genus *Equus* were of caballine type and belonged to the species *E. bressanus*, the so-called robustus horse, which appeared at Pardines; later on it lived side by side with the zebrine *E. stenonis* (Senèze). Other localities include Erpfingen, Chagny, Norwich Crag, A-Tegelen. This great horse is easy to distinguish from the zebrine horse because of its enormous size, rivalling that of the biggest living carthorses. The caballine horse type must have evolved at an early stage from the first zebrine invaders in the Old World.

A slightly smaller form, which has been called *E. mosbachensis*, occurs in the C-Cromerian and 2-Mindel (Mosbach; Mauer; Forest Bed; Koneprusy, etc.). Material from 1-Günz times (Episcopia; Nagyharsányhegy; Hundsheim; Gombasek; Jockgrim, etc.) is sometimes referred to the robustus horse and sometimes to the Mosbach horse (or species of their own).

The tendency to size reduction continues later on in the Pleistocene.

Horses from the D-Holsteinian, 3-Riss, F-Eemian and 4-Würm I seem however to vary only moderately in average size and may perhaps be regarded as a single species, although several names have been proposed (besides *E. germanicus*, which has already been mentioned, *E. steinheimensis* Reichenau and *E. taubachensis* Freudenberg belong here). These horses were inferior to the Mosbach horse in size but were still quite powerful.

Figure 63. Wild Horse, *Equus przewalskii*. A group of five horses and one bison. Wall painting, Lascaux, Dordogne. The small head, erect mane, short legs and hanging belly are typical of this form. After Laming.

Finally, in 4-Würm II the modern species *E. przewalskii* makes its appearance. It seems to be this rather small form that was depicted by the Stone Age artists, for instance in the cave paintings of Lascaux. This was also the form hunted at the famous site of Solutré.

The record suggests that the change in size was comparatively sudden in some instances, which would indicate replacement by immigration; this is the case in the *germanicus–przewalskii* shift. On the other hand a gradual transition *in situ* may also be possible, for instance in the *bressanus–mosbachensis* sequence.

Przewalski's horse is the stock from which the domestic horse has been derived and local races both in the east and in the west were probably tamed. The horse does not appear to be among the earliest domestic animals. In Europe the wild form became extinct long ago, but a small population survives in the Gobi desert in Asia. Before its recent decimation the population formed large herds led by an experienced stallion; the same holds for the European tarpan or wild horse of Russia and Poland, which became extinct in 1918.

The extinction of the tarpan is due to human activity, but the history of the equid family contains several examples of inexplicable extinctions. As late as the end of the Ice Age the plains of North and South America swarmed with horses and yet they all died out in the Postglacial and had passed completely from living memory among the American Indians. Reintroduced by Europeans, horses thrived and soon multiplied into enormous feral herds on the prairie and the pampa. Why then did their forerunners become extinct? In England, horse is missing in the F-Eemian interglacial faunas; perhaps it became extinct when the island had become separated from the continent by the high stand of the interglacial sea. The problem is made still more intriguing by the absence of man in England at the same time [18; 87; 187; 192; 265; 271; 302].

The European Wild Ass, *Equus hydruntinus* Regalia. This species is somewhat intermediate between the Asiatic wild asses (*E. hemionus*) and the true donkeys; for while its slender limb bones resemble those of the Asiatic asses, the lightly built 'microdont' teeth are of the same type as in the African form. The earliest evidence of this group in Europe comes from the late Villafranchian; Val d'Arno has yielded a jaw of the same type and an incomplete find (part of a shoulder blade) from Senèze may belong to the same form. These are now regarded as a distinct, Villafranchian form, *E. stehlini* Azzaroli. Finds from the earlier Middle Pleistocene are scarce (a member of the subgenus *Asinus* is reported, e.g. from Bad Frankenhausen and Koneprusy, 2-Mindel), but in the D-Holsteinian and 3-Riss finds are common enough (Lunel-Viel; Châtillon-Saint-Jean; Achenheim, etc.) and in the Late Pleistocene there are numerous records from caves and open-air sites in western, central, southern and eastern Europe. The species became extinct at the end of the 4-Würm.

Reports of donkey (*E. asinus* Linné) from Pleistocene deposits in Europe may be due to misidentification of *E. hydruntinus*, which as we have seen had donkey-like teeth. However, a donkey-like form has recently been described from the Late Pleistocene of Val di Chiana, Italy, under the name *E. graziosii* [18]. *E. asinus* has been recorded from the Pleistocene of Africa; domestic forms appear in the Neolithic of the Near East [17].

The Asiatic Wild Ass or Kulan, *Equus hemionus* Pallas. This species has been reported at several localities in Europe including Achenheim (3-Riss) and various cave deposits from 4-Würm. However, records based on limb bones may represent *E. hydruntinus*, while some records based on teeth may represent the small Przewalski horse [243]. Finally, however, Dietrich [66] found both teeth and limb bones in the 4-Würm I–II interstadial Rixdorf Horizon of Berlin and was able to substantiate the presence of kulan.

The present-day kulan has a rather wide distribution in northern

Asia, where it inhabits the steppes preferably in the neighbourhood of rivers and lakes. Whether the herds are normally led by a stallion (as in the case of wild horses) or an old mare (as with donkeys) is not clear; information is conflicting.

A related but smaller species, the Onager (*E. onager* Pallas), which originally inhabited the steppes and deserts from Palestine to northern India but is now very rare, appears never to have ranged into Europe.

Chapter 13

Order Artiodactyla

IN the even-toed ungulates the axis of the foot passes between the third and fourth toes. The first toe, corresponding to our thumb and big toe, is missing in all except the most primitive artiodactyls. The side toes – the second and fifth – are more or less reduced.

The artiodactyls are an old group, dating from the Eocene, but their definitive ascendance began in the Pliocene and continues today. Since the Pliocene the even-toed ungulates have gradually been crowding out the odd-toed ungulates and the great majority of those still living belong to the order Artiodactyla. Among the factors that may have contributed to this success may be mentioned the mechanical superiority of the artiodactyl tarsal joint and the ability to ruminate (chew the cud), found in all higher artiodactyls.

The fossil record of this order is excellent. In the Pleistocene of the area discussed here, four of the six living Old World families are well represented: pigs, hippopotami, deer and bovids. The two remaining families, camels and giraffes, are both represented in the Pleistocene of eastern Europe. The camels evolved in North America and entered the Old World at the same time as the one-toed horses, while the Giraffidae have a long history in the Old World and were plentiful in Europe as late as Pliocene times.

Family Suidae, Pigs

The suids are the most primitive of living Artiodactyla. Highly varied during the Tertiary, they have since lost most of their variety, but the few surviving stocks are vigorously holding their own. Pigs were probably present throughout the Pleistocene in Europe, although only one species existed at a time. Unfortunately their fossil record is rather unsatisfactory except in the Late Pleistocene.

The pigs are omnivorous and their diet, like their dentition, resembles that of the bears. The pig menu will, for instance, include acorns, herbs, ferns, tubers, grubs, small rodents and other small mammals, reptiles, amphibians and carrion. Because pigs root a great deal of their food out of the ground, strong winter cold will limit their geographic range and their short legs make them nearly helpless in deep snow. They are thus characteristic members of the interglacial and interstadial forest faunas.

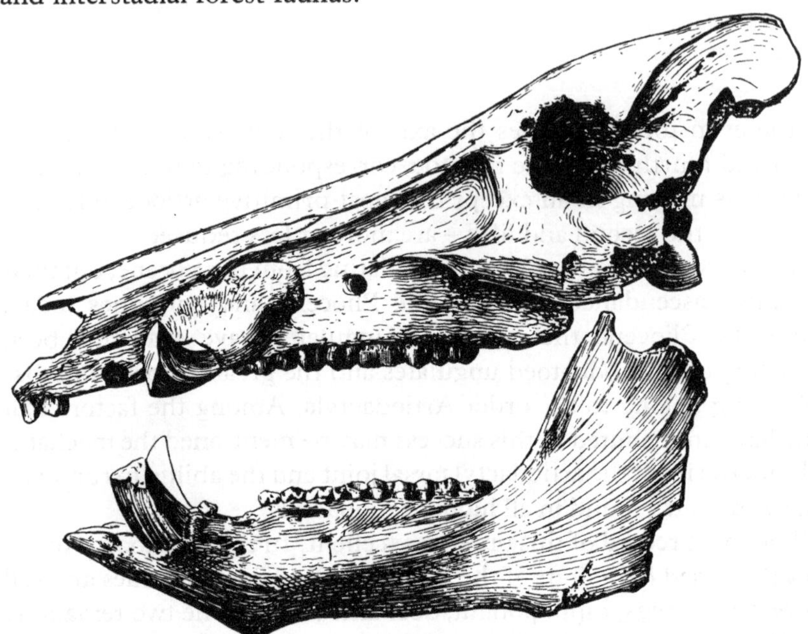

Figure 64. Skull and mandible of Strozzi's Pig, *Sus strozzii*, Villafranchian, Val d'Arno. One-sixth natural size. After Azzaroli.

Strozzi's Pig, *Sus strozzii* Meneghini. This species is typical of the later Villafranchian of Europe (Senèze; Olivola; Val d'Arno; A-Tegelen); some incomplete remains from Villány may also belong to this species. However, it was in existence in the early Villafranchian too, for it has been found at Bethlehem in a fauna of the Villaroya type with *Hipparion* instead of *Equus*.

S. strozzii was closely allied to the living *S. verrucosus* Müller & Schlegel of the East Indies. The *verrucosus* group differs from other members of *Sus* by the presence of a number of tufted warts on the face (it is not, however, related to the true wart hog of Africa); odontologically it may be identified by the development of the enamel facets of the lower canine teeth. The Villafranchian species was somewhat larger than a modern wild hog. A skeleton of a young animal from Senèze (at Basel) indicates a head-and-body length about 150 cm., corresponding to a full-grown boar, while incomplete adult specimens suggest lengths approaching 6 ft.

A Pliocene species from Roussillon, *S. minor* Depéret, is now regarded as the immediate ancestor of *S. strozzii*; it is somewhat smaller and slightly more primitive. A small form at Etouaires has been known for a long time under the name *S. arvernensis* Croizet & Jobert, but the material is not well enough preserved to show whether this is a suid of the *verrucosus* or the *scrofa* group. If, as is most likely, it represents the surviving *S. minor* stock, it may be ancestral to *S. strozzii*.

Generally speaking, the fossil record indicates that the wart-bearing *verrucosus* swine were widely distributed in Eurasia up to the end of the Villafranchian. Since then, however, they have been gradually driven out by competition with *S. scrofa*, finally losing their entire foothold on the Eurasian continent and surviving only as insular relicts.

The habits of *S. strozzii* probably were rather similar to those of the living Eurasian wild hog, though its slightly broader feet and shorter legs might suggest adaptation to swampy ground and possibly to a partly aquatic mode of life, paralleled in some living populations of *S. scrofa* [14; 108].

The Eurasian Wild Hog, *Sus scrofa* Linné. The first appearance of the modern species comes in the early Middle Pleistocene (Süssenborn; Gombasek; Mauer; Mosbach). It is also found in the D-Holsteinian (Grays; Lunel-Viel; Steinheim) and is very common in later deposits, especially of interglacial date. The Pleistocene form was markedly larger than the living one and is regarded as a distinct subspecies, *S. scrofa priscus* De Serres; it grew to even larger size than *S. strozzii*.

The species also appeared in the early Middle Pleistocene of China, at Choukoutien. So far, its precise origin has not been determined. The history of the genus *Sus* as a whole goes back to the Early Pliocene.

The Recent geographic range of the wild boar originally included most of Europe except the north, but it has since been extirpated in many areas. Domestication probably took place independently in several different regions. Very early material of domestic pig comes from Jarmo in Kurdistan, Iraq; it dates from about 6500 BC. Domestic specimens are easily identified by the much smaller size of their back teeth compared with those of the wild form [184; 222].

Family Hippopotamidae, Hippopotami

This family has only two living species, the hippopotamus (*Hippopotamus amphibius*) and the pygmy hippo (*Choeropsis liberiensis* Morton), both in Africa. The fossil species in Africa and Eurasia are also very few. The history of the family goes back to the Pliocene. It was formerly thought to be derived from the Suidae, but more recent work indicates that the Hippopotamidae probably emerged from a group of animals called anthracotheres – a primitive artiodactyl group resembling the pigs in some respects.

The Hippopotamus, *Hippopotamus amphibius* Linné. The European form is sometimes regarded as a distinct species (*H. antiquus* Desmarest) but it is more likely to represent a subspecies of the living species. The hippopotamus appeared in Europe at the end of the Villafranchian (Val d'Arno) and made repeated incursions in later interglacial faunas; in Italy it seems to have been present throughout the Pleistocene. In interglacial times it probably migrated northward following the river valleys and coasts. The Rhône, Danube and Rhine seem to have formed important highways for this species. In addition the modern hippopotamus is known to make long cross-country treks which may aid its dispersion and help it to cross some water divides.

The species was fairly common in the early Middle Pleistocene, where it appears both in the forest faunas (Mauer; Abbeville; Forest Bed) and the cool steppe fauna at Mosbach. The latter find indicates that the climate was not cold enough for the Rhine and its tributaries to freeze over in winter. With the increasing cold of the 2-Mindel the species became extinct.

In the D-Holsteinian there is a record from Lunel-Viel, but after that the species is absent to return only in the F-Eemian, when it seems to have been very numerous in England. Particularly rich material, including a complete skeleton, has been found at Barrington (Cambridge). Several Late Pleistocene records from France indicate the route along which the hippopotamus migrated to England, but finds in central Europe are scarce (there is a record from the Balver Höhle in Westphalia).

The European form differed from the African one mainly by its larger size, which is remarkable enough considering that the largest modern hippopotami may reach a length of 4 m. and weigh more than 4 tons. An archaic character found in some individuals of the European form was the presence of three incisors in each lower jaw half instead of two, the normal number in *H. amphibius*. Only a few individuals have this atavistic trait ('hexaprotodonty') in which they resemble the Villafranchian hippopotami of Asia.

Though the fossil record shows that the species could survive for some time in a cool-temperate climate, it is probably significant that the hippopotamus did not reach England in the relatively cool D-Holsteinian interglacial, whereas the climate of the C-Cromerian and F-Eemian, when the species was common there, was warmer than that of the present day.

Dwarf races of hippopotamus evolved in several Mediterranean islands. They have received various specific names – *H. minor* Desmarest in Cyprus, *H. pentlandi* Meyer in Crete, Sicily and Malta, and *H. melitensis* Major in Crete and Malta. Their relationship to each other and to the European form remain to be elaborated. As in the case of the insular dwarf elephants the reduction in size may have given the species a chance to keep the number of individuals over a certain critical level (ensuring, for instance, the need of genetic variability) in spite of a limited food supply.

The hippopotamus feeds on water plants like reeds and lotus as well as grass. Its only dangerous enemy is man, but fights between males during the breeding season may occasionally have fatal results [62; 107; 183].

Family Cervidae, Deer

The deer have evolved from a group of Miocene ruminants, the Palaeomerycidae; the same group apparently also gave rise to other higher artiodactyls including the giraffes, which are the closest living relatives of the deer. The earliest real deer in the Miocene and Pliocene lacked antlers or had simple, bifurcated antlers (*Dicrocerus*) or a peculiar arrangement with a saucer-like, tine-rimmed plate at the end of a short beam (*Stephanocemas*). In the Pliocene, forms representing the three major living groups in Europe (red deer, roe deer and elk) had already appeared, but the majority were so-called pliocervines (*Damacerus*, etc.), of which several forms survived in the early Villafranchian.

The deer have a rather primitive dental battery with low-crowned teeth; they are more or less confined to woodlands. The reindeer is an important exception; this is the only member of the deer family that has been able to colonize the Arctic tundra biotope.

The deer family is abundantly represented throughout the Pleistocene in Europe, but the fossil deer pose numerous problems to the palaeontologist. The dentition is rather stereotype, so that the systematics are mainly based on the antlers. Unfortunately skulls with attached antlers are relatively rare, while shed antlers are fairly common. It may then be difficult to combine the antlers and the remainder of the skeleton as the fauna may contain more than one species of about the same size. Individual and age variation in the antlers is great and doubtless many species have been based on such variants. Much remains to be done to unravel the history of the deer in the Pleistocene, especially in the Villafranchian and Middle Pleistocene.

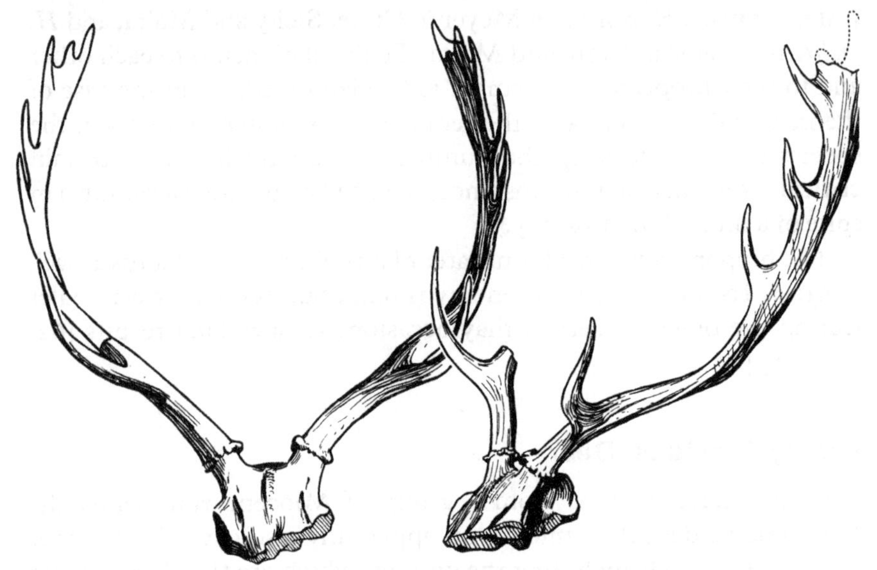

Figure 65. Frontlet with antlers of Ramosus Deer, *Anoglochis ramosus*, Villafranchian, Saint-Vallier; two views. One-twelfth natural size. After Viret.

Etouaires Deer. A number of primitive deer species have been described from Mt Perrier. '*Cervus*' *ardei* Croizet & Jobert, '*C*'. *issiodorensis* Croizet & Jobert and '*C*.' *cladocerus* Pomel apparently became extinct in the early Villafranchian, whereas '*C*.' *cusanus* Croizet & Jobert has been identified also in middle and late Villafranchian strata (e.g. Chilhac) and '*C*.' *perrieri* Croizet & Jobert has been tentatively reported in the late Villafranchian as well [39].

The Ramosus Deer, *Anoglochis ramosus* Croizet & Jobert. This is a fairly large form which ranges from the early Villafranchian (Perrier;

Villaroya) throughout the stage (Saint-Vallier; Erpfingen; Saint-Estève). The somewhat flattened, evenly curved beam carried a series of tines pointing forward and subequal in size; each antler carries up to seven tines. The terminal part of the beam tends to be lobated. A subspecies of the same species (*A. ramosus pyrenaicus* Depéret) has been reported from the Astian of Roussillon. It is probably a direct descendant of the typical earlier Pliocene 'pliocervines' of the *Damacerus* group [13].

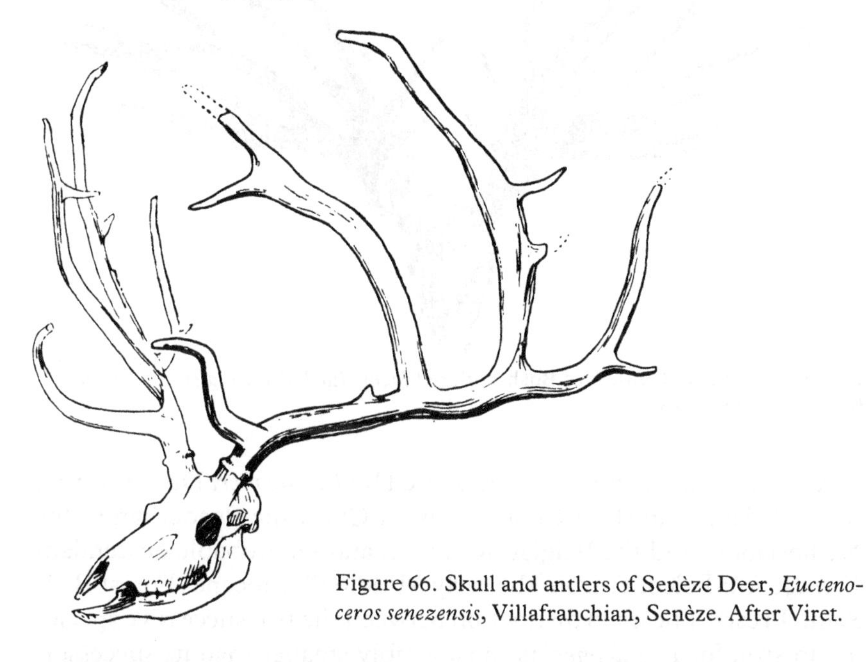

Figure 66. Skull and antlers of Senèze Deer, *Euctenoceros senezensis*, Villafranchian, Senèze. After Viret.

The Tegelen Deer, *Euctenoceros ctenoides* Nesti (*E. tegulensis* Dubois). The genus *Euctenoceros* has medium-sized to large antlers with relatively few long, straight or slightly curved tines, openly arranged. Several species related to the Tegelen deer have been described, for instance *E. senezensis* Depéret and *E. tetraceros* Dawkins, but it remains to work out the species taxonomy. All are large, powerful animals. The Senèze form has been tentatively identified as early as Roccaneyra and Pardines. Other localities include Saint-Vallier, Val d'Arno, Chilhac, Peyrolles and the Cromer Forest Bed [13].

The Bush-antlered Deer, *Eucladoceros dicranios* Nesti. In this genus the antlers reach a stage of almost fantastic complication, as seen in the magnificent skull from Val d'Arno (at Florence). The highly digitate antlers carry a total of 12 tines each and their great length and dense arrangement give this head adornment a tangled, bush-like appearance. This animal reached the size of an elk.

Figure 67. Skull and antlers of Bush-antlered Deer, *Eucladoceros dicranios*, Villafranchian, Val d'Arno.

A closely related group is represented by *E. falconeri* Dawkins from the Villafranchian (Red Crag; Norwich Crag; the Poederlian in the Netherlands; and the Belgian Kempen) and its probable descendant *E. sedgwicki* Falconer from the early Middle Pleistocene (Forest Bed; Saint-Prest). The main distinction between the two successive species lies in size, for *E. falconeri* is considerably smaller than its successor. The antlers resemble those of *E. dicranios* but have a greater tendency to flattening and the brow tine is higher in position. A Chinese form (*E. boulei* Teilhard & Piveteau) also belongs to this group.

Dentitions of *Euctenoceros* and *Eucladoceros* are difficult to separate and there are various records both from the Villafranchian (Erpfingen; Villány) and even the Astian (Ivanovce; Hajnacka) which cannot be definitely determined [10; 13; 189].

The Philis Deer, *Cervus etuerarium* Croizet & Jobert. A small deer with three-tined antlers has been described under various names from different localities (*C. philisi* Schaub, *C. rhenanus* Dubois). Their relationships to each other and to modern species is not well understood. Localities include Etouaires, Saint-Vallier, Chagny, Senèze,

Erpfingen, Doveholes, Tegelen. In the 2-Mindel remains of a form of this type have also been recovered (*C. elaphoides* Kahlke from Mosbach). These animals may have resembled the axis deer or sambar but whether they should really be classified accordingly (in the genus *Axis* or the subgenus *Rusa* of the genus *Cervus*) is still disputed. The species is very common at Senèze and a complete skeleton is to be seen at Basel; the shoulder height is about 100 cm. The comparatively low crowned teeth indicate a browsing mode of life, perhaps in shrubs and spinneys [123; 239].

Figure 68. Restoration of European Sambar or Philis Deer, *Cervus etuerarium*, based on skeleton from Senèze.

The Red Deer, *Cervus elaphus* Linné. We now come to the genus *Cervus* proper. The earliest known specimens of this species are sometimes separated as distinct species, for instance *C. acoronatus* Beninde. The acoronate (or 'crownless') deer differs from typical European stags in that the terminal part of the antler carries a simple fork; the cup or 'crown' formed by three or more tines in typical red deer is absent. But although this type is frowned upon by sportsmen, it still occurs as an individual variant in the European stag and several races of the living species normally have antlers of this type, for instance the Asiatic red deer and the American elk.

Figure 69. Restoration of 'acoronate' type of Red Deer *Cervus elaphus*, from Süssenborn and Mosbach, early Middle Pleistocene.

According to Dietrich [63], the acoronate deer may have evolved from an early Villafranchian species at Perrier, *Metacervocerus pardinensis* Croizet & Jobert. That form also carried a simple fork, but more primitive than in the acoronate deer, for one of the tines was simply formed by the straight termination of the beam. This animal had a total of three or four tines on each antler, including the terminal fork. The acoronate deer carried up to five tines.

While *M. pardinensis* was approximately as large as a modern European stag, the Middle Pleistocene red deer tended to be as large as the big Asiatic forms. Acoronate deer have been found in deposits from 1-Günz II (Süssenborn) and 2-Mindel (Mosbach). 'Coronate' forms with a cup formed by three or four tines also appeared in 1-Günz II at Hundsheim at a time when acoronate forms occurred further north. In the C-Cromerian (Mauer) the red deer of central Europe carried a 'crown' of modern type; in the 2-Mindel the acoronate form returned and coronate forms are again seen in the D-Holsteinian (Steinheim). This alternation is evidently caused by environmental changes: the acoronate forms are found in cold, continental phases, the coronate ones in warmer and more oceanic phases. Each new cold phase tended to send into Europe a wave of the acoronate deer that still populate the Asiatic taiga today. It would seem that no valid specific distinction can be made between the two forms.

The red deer is very common in the Late Pleistocene all over Europe. Cave art from the Palaeolithic often features a stag motif; a famous example is the beautiful frieze at Lascaux showing five swimming red deer.

Dwarfed forms of this species have evolved in various islands. The modern British subspecies is comparatively small, but still smaller forms existed on the islands of Jersey (in the Eemian: *C. elaphus jerseyensis* Zeuner), Capri (*C. tyrrhenicus* Azzaroli), Sicily (*C. siciliae* Pohlig), Malta and Pianosa; the last-mentioned was the smallest, with a shoulder height less than two-thirds of that of the living Scottish red deer.

The red deer feeds mainly on leaves and twigs, herbs and fruits, as well as spruce bark and hanging lichens in the winter. Its modern geographic range includes the temperate forests of Eurasia, North Africa and North America. The American form, the wapiti or American elk, is very large; the Asiatic race, too, is larger than the European. The entrance of the species in North America may possibly date from the 3-Riss, for the earliest American fossils occur in deposits of that age or from the F-Eemian (e.g. the Conard Fissure in Arkansas) [15; 122; 123; 126].

The Verticornis Deer, *Praemegaceros verticornis* Dawkins. This enigmatic form, the first of the giant deer, flourished in the earlier Middle Pleistocene approximately from the B-Waalian to the 2-Mindel (Rosières; Solilhac; Süssenborn; Mosbach 1–2; Forest Bed; Koneprusy, etc.). Whether it inhabited Europe in the Villafranchian is not quite clear; specimens from Leffe, the Red Crag, etc., have been tentatively referred to this group. It became extinct on the continent in the 2-Mindel but survived in dwarfed forms in the Mediterranean islands up to the Late Pleistocene.

This species, an important guide fossil, used to be referred to the giant deer genus (*Megaloceros*) but has such peculiar and aberrant antlers that a separate genus has been created for it. The beam jutted out laterally from the burr; the brow tine was reduced or absent, the other tines very small. The outer part of the antler extended backward and upward at right angles to the beam and was digitate or slightly palmate, terminating in a number of small tines. The details of these antlers are highly variable and a great number of species have been based on individual variants, but all the types seem to be linked together by intermediate forms.

The whole arrangement (the right-angle bend of the beam and the terminal palmations) is faintly reminiscent of that in some reindeer but the details are wholly different. The size of the verticornis deer was about the same as that of a wapiti.

Dwarf types of verticornis deer have been found in Corsica (*P. cazioti* Depéret), Sardinia, Sicily (*P. messinae* Pohlig) and the island of Candia (*P. cretensis* Simonelli). Some of these are evidently Late Pleistocene. The last-mentioned is an example of extreme dwarfing, with a shoulder height of only 60 or 65 cm. [15; 120; 123].

Savin's Giant Deer, *Megaloceros savini* Dawkins (*Praedama süssenbornensis* Kahlke). This species is a forerunner of the true giant deer; some authors consider it directly ancestral to the latter, while others regard it as an early offshoot of the line. If the latter alternative is correct, *Praedama* may be valid as a subgenus. Contrary to later giant deer, this species does not have palmate antlers; the tines are long and cylindrical, but relatively few in number and openly arranged. It has been suggested that this type is derived from *Euctenoceros*. But the brow tine has a very peculiar spoon-like form, which is typical of *Megaloceros*. Evidently this is the beginning of the trend towards palmation of the antlers which is so evident in later giant deer.

Figure 70. Giant Deer, restorations. Top left, Verticornis Deer, *Praemegaceros verticornis*; bottom left, Savin's Giant Deer, *Megaloceros savini*; top centre, early form of Irish Elk, *M. giganteus antecedens*; right, Late Pleistocene Irish Elk, *M. g. giganteus*. After Knight & Thenius.

The species occurs at Süssenborn; Forest Bed; Mosbach 2 and in southern Russia [126], but continued study will probably show it to be present at many other sites [13; 15; 122].

The Giant Deer or Irish 'Elk', *Megaloceros giganteus* Blumenbach. This species is very common in the Late Pleistocene; it has, for instance, been found in great numbers in the Irish bogs, so that many skeletons are in existence. But the species dates back to the D-Holsteinian (Steinheim, etc.). The early forms are regarded as a distinct subspecies, *M. giganteus antecedens* Berckhemer, which is slightly smaller and in which the antlers are deflected backward.

In the typical Late Pleistocene form, the antlers extended laterally, sometimes with a total spread of four metres. The main part of the antler was formed by the great palmation, the rim of which carried a series of tines. Unlike the true elk antler, which lies in the same plane as the forehead and muzzle, the palmation of the giant deer lay in a plane almost at right angles to the long axis of the head, so that the tines along its front pointed upward. The Irish 'elk' grew to the same size as the great American moose but in posture probably rather resembled the red deer, the head being held in a raised position. The western and northwestern forms of Ireland and Denmark surpassed those from central Europe in size.

The species was widely distributed in Europe and northern Asia. A related form, perhaps a local race of the same species, lived in China; the earliest Chinese specimens date from the 2-Mindel. The giant deer is not found in the steppe faunas of the Carpathian basin. In the plains of eastern Europe the species became extinct at the close of 4-Würm I, but in the Crimea it survived to the end of 4-Würm II. In western Europe it existed in the Allerød interstadial, 10,000 BC, and certain finds from Styria and the Black Sea area may suggest that it survived until 700 or 500 BC; some authors have tried to identify it with the '*schelch*' of the Nibelungen Lied.

A very peculiar character found in giant deer and a few other cervids was a thickening of the lower jaw, so that the jawbone may be almost circular in cross section. The evolution of this character may be studied in the *antecedens* stage at Steinheim. The long series of deposits at Choukoutien in China also permits us to trace this character, the specimens from the earliest deposits have a flat jawbone, but in younger strata it gradually grows thicker [30; 119; 224].

Nesti's Fallow Deer, *Dama nestii* Major. This species occurs in the late Villafranchian (Val d'Arno; Olivola; Erpfingen) and ranges into the early Middle Pleistocene (Forest Bed). According to Azzaroli [13] it represents an early type of fallow deer with four tines but no terminal palmation. Its relationship with *Dama* has been questioned by other authors.

The Clacton Fallow Deer, *Dama clactoniana* Falconer. This species was very common in England during the D-Holsteinian interglacial (Clacton; Grays Thurrock; Swanscombe). However, some fragmentary finds from the Forest Bed (Upper Freshwater Bed) and Abbeville may also represent this species, which would then be present in the C-Cromer.

The Clactonian species was about 20 per cent larger than the living one and carried one more tine on its beam. In the living fallow deer the beam carries two tines, the brow tine and the bez tine, and terminates with the characteristic triangular palmation with its lobate rear edge. But some individuals do carry three tines just like the Middle Pleistocene form.

Dama clactoniana was evidently directly ancestral to the living fallow deer. It may be questioned whether it merits specific distinction at all [9; 210; 248].

The Fallow Deer, *Dama dama* Linné. This species appears in the F-Eemian and has been found in many caves in England, France, Spain, Germany, Poland and especially in Italy, as well as in deposits of other types (river gravels at Barrington, last interglacial terrace of the Thames, travertines at Weimar). The F-Eemian form exceeded the modern in size but was not as large as *D. clactoniana* (the material from Fontéchevade, however, was referred to that species). Evidently there was a gradual transition from *D. clactoniana* to *D. dama*.

In the 4-Würm fallow deer were only present in the south and even here they became extinct later on. In Palestine they are abundant throughout 4-Würm. The modern distribution of the fallow deer was originally limited to Mesopotamia and Asia Minor, but it has been successfully introduced in various European countries. This species, which is highly gregarious in habits, lives on grass, twigs and bark [271].

The Roe Deer, *Capreolus capreolus* Linné. The earliest roe deer in Europe differ from their living descendants mainly by their larger size; the antlers, like those of the present-day European form, usually carried three tines. The living Asiatic subspecies *C. capreolus pygargus* Pallas is as large as the fossil form.

The earliest records come from the 1-Günz (Süssenborn; Hundsheim; Voigtstedt); other early Middle Pleistocene finds date from the C-Cromer (Forest Bed; Abbeville; Mauer) and 2-Mindel (Mosbach). It is also found in the D-Holsteinian (Bruges; Tarkö; Steinheim; Heppenloch), 3-Riss (Achenheim; Montmaurin, etc.) and is very common in the Late Pleistocene.

The distribution of the roe deer today encompasses the greater part of Europe and the forests of southern Siberia as far as northern China and Korea. The roe deer prefers woodlands and dense bush but occasionally will seek open ground, especially by night. In Scandinavia the roe deer has spread in a remarkable way during the last 100 years. About 1850 there was only a small local herd in Scania in southernmost Sweden; by now, however, the species has invaded the greater part of Sweden up to southern Lapland, a massive advance of at least 1000 miles. The spread was evidently favoured by the extirpation of the wolf in most of the area, as well as the climatic amelioration. This is an interesting example of the rate at which a species is able to colonize a new area as soon as an obstacle to its dispersal is removed. It throws new light on the migration of species in geological time.

The roe deer feeds on leaves, twigs, fruits, fungi, and tender grass. The ancestry of *Capreolus* is little known [9; 122].

The Gallic Elk, *Alces gallicus* Azzaroli. Here is the beginning of the history of the true elk (moose to Americans) in Europe. This early species is known from the later part of the Villafranchian (the Crags; Senèze, Erpfingen), but a possible forerunner has been found at Perrier (Roccaneyra). A fine skeleton from Senèze is mounted in Lyon.

Figure 71. The Pleistocene Elk sequence. Left to right: Gallic Elk, *Alces gallicus* Villafranchian; Broad-fronted Elk, *A. latifrons* Middle Pleistocene; modern Elk, *A. alces*, Late Pleistocene and Holocene.

The Gallic elk was about the same size as its living descendant but the shape of its antlers was quite different. Each antler was formed somewhat like a spoon with a long handle. The small palmation was cupped upward and rimmed with small tines; the beam was enormously long. The antlers were spread horizontally and reached the enormous span of three metres or more. Obviously these structures would have been a real encumbrance in any forest and it seems quite evident that the Gallic elk must have kept to open ground of the parkland or savanna type.

Together with *A. latifrons*, this species may be classified in the subgenus *Praealces*. Ancestral elk (*Alces*) are found as early as the Early Pliocene, but details in their history are as yet unknown. In North America the true elk group was represented in the Late Pleistocene by the remarkable genus *Cervalces* which may be descended from early representatives of the *Praealces* group [12].

The Broad-fronted Elk, *Alces latifrons* Johnson. The transition from Gallic elk to broad-fronted elk is shown by some interesting specimens from the earliest Middle Pleistocene (Mosbach 1; B-Waalian fauna of the Forest Bed). In these specimens, which are still regarded as *A. gallicus* by some authors, the 'spoon bowls' are still relatively small and the beams longer than in typical *A. latifrons*. In the slightly later and better-known populations from the C-Cromerian and 2-Mindel stages (Forest Bed; Mauer; Mosbach 2, etc.) typical broad-fronted elk is found, with the beam and the bowl each making up about one-half of the total length of the antler. The reduction in length of the beam was approximately made up for by the increase in size of the bowl or palmation, so that the total span of the antlers remained very great. Apparently this animal, like its predecessor, was foreign to the deep forest.

There is great size variation in the early Middle Pleistocene *Praealces* elks and some specimens reach truly gigantic proportions, exceeding even the great American *Cervalces*; whether these are individual variants or representatives of, for instance, a temporal subspecies (*P. latifrons reynoldsi* Azzaroli) remains to be seen. This animal stood 190–200 cm. at the withers. Its skull was very heavily built, but the limbs are long and slender as in other elks [13; 126].

The Elk or Moose, *Alces alces* Linné. In the modern elk the beam of the antler is short and the palmation tends to extend fore-and-aft rather than laterally so that the total spread of the antlers has been somewhat reduced, evidently in adaptation to the forest environment. In connection with the change in shape, the scoop-like palmation tends to divide into two parts, a larger in the rear and a smaller in the front; the latter has sometimes been regarded as an equivalent of the brow tine of the stag, but this interpretation is negated by the evolutionary history of the antlers.

Apart from palmate or shovel-like antlers, cervine or digitate antlers may occur. This is something quite new in the history of the elk, though it may perhaps be viewed as an extreme result of the tendency to lobation. The cervine elk is not admired by game managers, but there may be some selection in favour of this type. During pairing fights a cervine elk may in exceptional instances kill its rival by thrusts in the body or neck, a feat which is hardly possible for a palmate elk; in the long run this may contribute to the increase of cervine types at the cost of the palmates, though selective hunting probably has been an even more important factor. So far, however, this has only happened in southern Scandinavia and especially in Finland, where the cervine type is predominant.

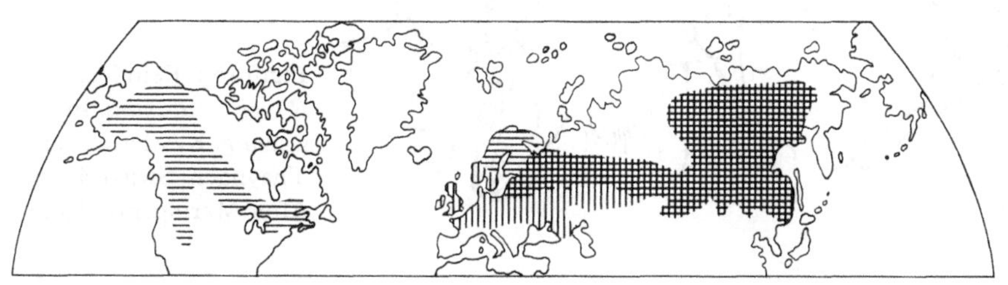

Figure 72. Geographic distribution of the Elk (*Alces alces*) in Pleistocene times (vertical hatching) and in the Recent (horizontal hatching). After Trofimov.

The elk appears in the 3-Riss (Achenheim) and is common in the temperate faunas of the Late Pleistocene (F-Eemian and 4-Würm interstadial). It may probably be derived from *A. latifrons* but this has not yet been directly demonstrated. Many representations of the elk are known in cave art. The modern range of the elk comprises northern Eurasia and North America; as with the red deer, the American form, called moose, reaches exceptional size. There are no definite data on the immigration in North America, although the 4-Würm may perhaps be accepted as a tentative date; it seems that the endemic *Cervalces* was the sole representative of the elks in North America during most of the last glaciation.

The Pleistocene elk in Europe was larger on average than its modern descendant (150–210 cm. at the withers in living European elk) and Postglacial finds in Scandinavia also tend to have very large dimensions.

The main habitat of the living elk in Europe is the coniferous forest and bogs of the north. It browses on willow, ash, birch and pine and also eats various herbs, grass, lichens, fungi and berries [120].

The Reindeer, *Rangifer tarandus* Linné. The oldest fossil of a reindeer known at present is an antler of tundra reindeer type from the sands of Süssenborn, thus evidently of 1-Günz II date. That the species is so old may seem surprising as it is a highly specialized form, but of course there is no definite proof that the reindeer had acquired all of its modern adaptive characters at that early date.

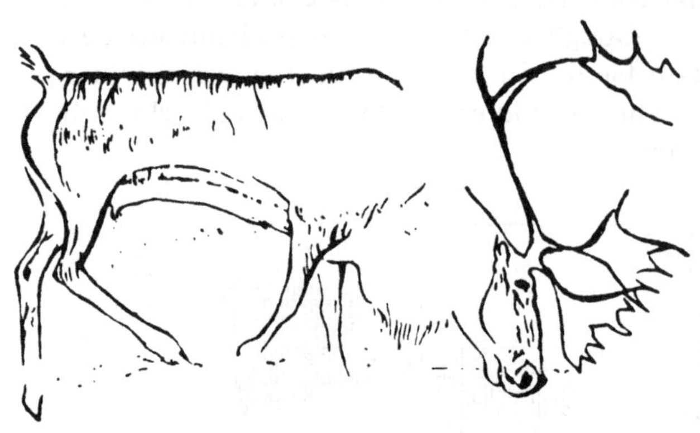

Figure 73. Reindeer, *Rangifer tarandus*. Engraving, Kesslerloch at Thayngen, Switzerland. After Maringer & Bandi.

Reindeer are next met with in the 2-Mindel (Mosbach 2, Bad Frankenhausen; Steinheim basal gravels, etc.); there does not appear to be any certain record from the Forest Bed. The species also occurs in the 3-Riss (for instance in the terraces of the river Saale). Its range in Europe in the 4-Würm is very extensive; it is found in Spain, Italy and southern Russia. Reindeer is particularly abundant in the Magdalenian deposits from the late part of the 4-Würm II just before the end of the Ice Age: at that time and in the early Mesolithic it was the main game animal for many tribes. The supply began to get low during the Mesolithic, when reindeer definitely retired to the north.

In contrast with conditions in other deer, both sexes usually carry antlers (though a proportion of woodland reindeer females may be antlerless). The shape of the antlers is highly variable but there is some superficial resemblance to *Praemegaceros*, for instance in the angulation of the beam, the formation of terminal palmations and the shortness of the tines; the resemblance may have ecological significance. The tundra form usually has cylindrical, digitate antlers, while those of the woodland form tend to be more flattened and more highly palmate. In the American form, often called the caribou – especially the woodland form – the beam forms a smooth curve rather than an angle, and the Pleistocene reindeer of Europe are in this respect more nearly similar to the caribou. In spite of the great variation, all the Pleistocene and living reindeer belong to a single species.

The present distribution of *R. tarandus* is circumpolar. The earliest American specimens date from the 4-Würm and have been found in Alaska.

An important feature in the ecology of the reindeer is the seasonal migration between forest and tundra; an engraving found at a Magdalenian site (Teyjat) gives a striking picture of an advancing herd of reindeer. The reindeer feeds on grass, herbs and leaves during summer and mainly on lichens in the winter [20; 122; 124].

Family Bovidae, Bovids

The bovids evolved in the Miocene; their closest relatives are the American pronghorns. In the Pliocene, the bovids increased enormously in numbers and variety so that now they are the dominant ungulate family of the world. Bovids may be found in almost any environment inhabitable to terrestrial mammals – Arctic tundras, high mountains, tropical rain forest, savanna, steppe, desert and so on. We even know a cave-dwelling bovid, the Pleistocene *Myotragus*.

The Middle Pliocene steppes of Europe swarmed with gazelles and antelopes of every kind, but the forest episode at the end of the Pliocene drove most of them out of Europe. Only a fraction of the antelopes returned to people the Villafranchian steppes and in Europe the dominance passed to other groups of bovids – goats, musk oxen and cattle.

The Bourbon Gazelle, *Gazella borbonica* Depéret. The true gazelles are among the most primitive of bovids. The Early Pliocene gazelles still had an archaic low-crowned dentition and were apparently browsing woodland animals. In the Middle Pliocene typical steppe gazelles with high-crowned cheek teeth evolved. The Bourbon gazelle is of this type but is not an extreme desert form like some living species. Its horns were rather long and moderately divergent, slightly recurved as in the living Indian or Mountain gazelle (*G. gazella* Pallas), except that in the latter the points curve forward. The Bourbon gazelle was of about the same size as most living species, for instance the Dorcas gazelle which has a shoulder height of about 60 cm.

The first Bourbon gazelles are known from Etouaires, but there is an allied species in the Astian of Roussillon which may be ancestral.

G. borbonica has also been found in the younger deposits at Perrier and at Villaroya, Saint-Vallier and Val d'Arno. Other finds of gazelle (for which other species have been erected) come from the Crags in East Anglia and the Villafranchian of the Netherlands.

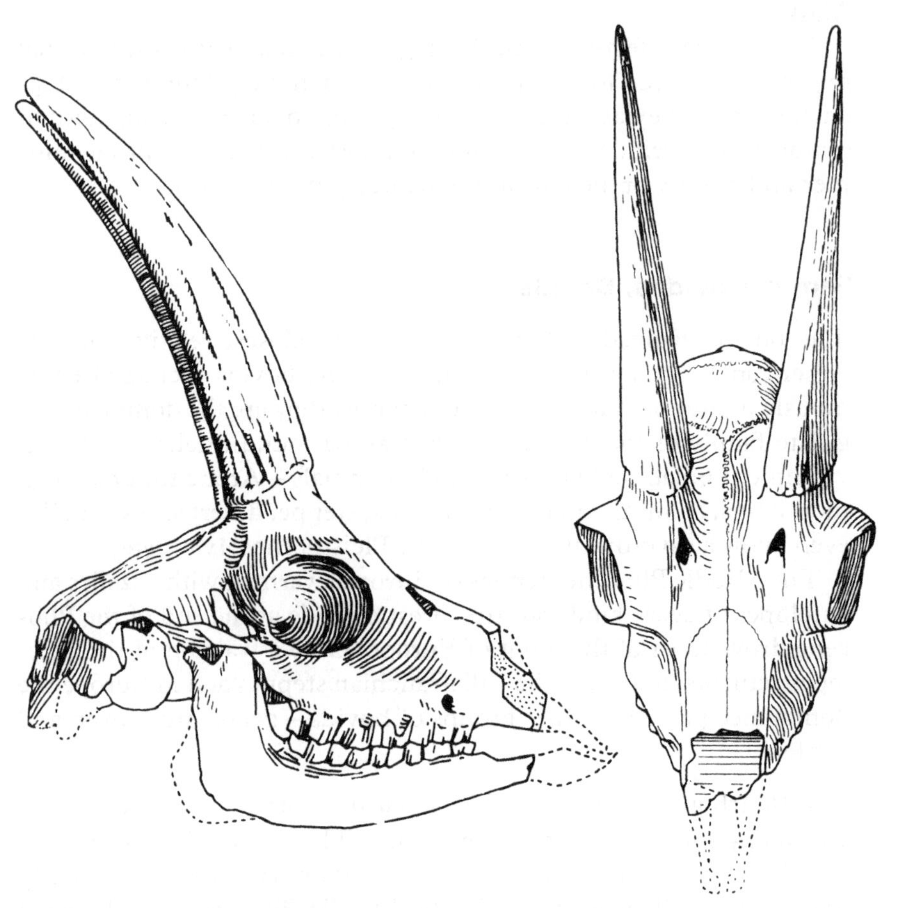

Figure 74. Bourbon Gazelle, *Gazella borbonica*, Villafranchian, Saint-Vallier; side and front views of skull. One-quarter natural size. After Viret.

The species became extinct before the A-Tiglian and since then gazelle has not been found in western Europe [106; 302].

The European Gazelle-Antelope, *Gazellospira torticornis* Aymard. A representative of the gazelle-antelopes with spiral horns, this species is a guide fossil of the Villafranchian. It is absent in the Etouaires fauna but appears at Roccaneyra, Pardines, Villaroya, Senèze, Olivola,

Val d'Arno, Villány and Erpfingen; the species has also been found in the Near East (Bethlehem).

Probably its external appearance was rather similar to the blackbuck (*Antilope cervicapra* Linné) with which it was closely related, but the horns diverged more than in the living species. The European form stood about one metre at the withers; its limb bones are slim and the teeth high-crowned. Related forms, mostly with less high-crowned teeth, are known from the Early and Middle Pliocene, a time when the eastern shores of the Mediterranean seem to have been a veritable centre for the evolution of spiral-horned antelopes [108; 217].

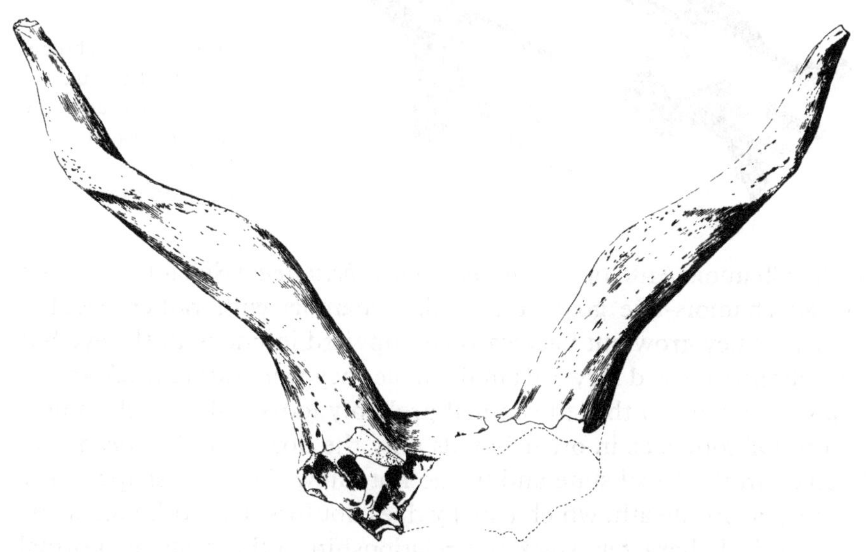

Figure 75. European Gazelle-Antelope, *Gazellospira torticornis*, Villafranchian, Senèze; front view of horn-cores. One-quarter natural size. After Pilgrim & Schaub.

The Saiga Antelope, *Saiga tatarica* Linné. This peculiar, short-legged antelope with its oddly swollen muzzle existed in Europe during the 4-Würm and ranged into England, Belgium and France. It has been gradually forced eastward during the Postglacial and is now distributed from the steppes of southern Russia to Mongolia; as late as the eighteenth century it still ranged to the Carpathians. The saiga also crossed the Bering Bridge and inhabited Alaska during the last glaciation.

The saiga, which stands 75–80 cm. at the withers, forms large herds that seem to thrive well not only on grass but also on the sparse vegetation of the saline barrens in Siberia. The earlier history of the saiga group is obscure. A closely related form, the chiru antelope (*Pantholops hodgsoni* Abel), lives in Tibet and has never been found in Europe, but Dorothea Bate [26] suggested that the enigmatic 'fabulous beast' in one of the wall-paintings of Lascaux shows great resemblance to this species.

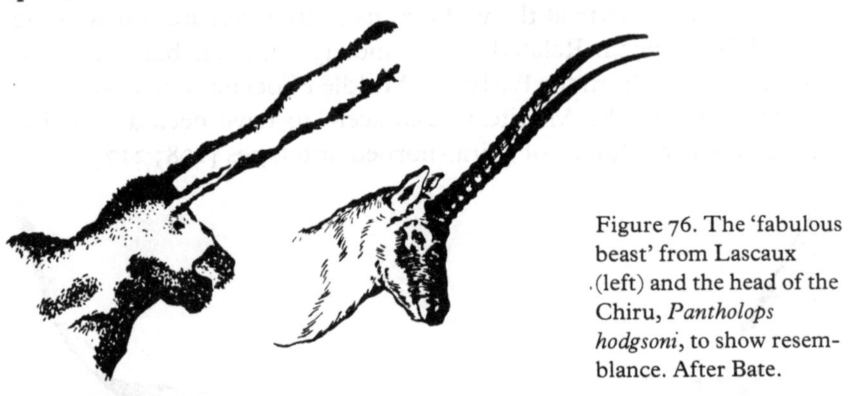

Figure 76. The 'fabulous beast' from Lascaux (left) and the head of the Chiru, *Pantholops hodgsoni*, to show resemblance. After Bate.

The Chamois Antelope, *Procamptoceras brivatense* Schaub. This is a small chamois-like goat or goat-like chamois with rather peculiar horns. They grow out backward and upward in line with the eye but are slightly curved forward; in the male they were rather thick and so closely appressed that the animal probably looked almost like a unicorn. Of course as in other bovids only the horn-core has been preserved in the fossil state and we do not know the exact shape of the outer, horny sheath, which usually does not fossilize. Although some anatomical characters suggest a relationship to the goats, its normal attitude seems to have more nearly like that of the chamois, with head held well forward rather than raised. What is known of the slim limb bones also indicates relationship with the chamois. The size of this animal is intermediate between the gazelle and gazelle-antelope of the Villafranchian fauna.

The chamois antelope has been found in the Villafranchian, ranging from Roccaneyra and Pardines to Olivola and Senèze; it has also been found at Csarnóta, Villány and Beremend [233; 237].

The European Goral, *Gallogoral meneghinii* Rütimeyer. This is also a member of the Villafranchian fauna of Europe. It occurs at Etouaires, Senèze, Chilhac, Olivola, and Val d'Arno, or almost throughout the Age. The material from Senèze is particularly good; two skeletons may be seen in the Basel collection.

The Villafranchian form is related to the living Goral, *Nemorhaedus goral* Hardwicke, and the Serow, *Capricornis sumatrensis* Bechstein; probably all these species could be fitted into a single genus (*Nemorhaedus*). The European form was however larger than the living forms; the head-and-body length was about 2 m., or 1½ times that in the living Goral. It also had longer horns; as in *N. goral* they are directed backwards and lie close together, but they are long and recurved. The neck is somewhat longer but the limbs relatively shorter than in the living species.

The modern goral, which inhabits mountainous country in Asia, is an agile and fearless climber and has been observed at altitudes up to 12,000 ft. It lives in small herds up to a dozen individuals, feeding on grass and shrubs; when resting and chewing the cud it seeks wildly inaccessible positions like isolated rock pillars, trees growing at an angle over an abyss and the like. The Villafranchian goral probably lived in the same way, for all the sites that have produced remains of this species are in mountainous areas. Probably both the living and the fossil goral may be regarded as vicars of the chamois, to which they are related. Like the chamois, gorals have small horns lying fairly close together [90; 235].

The Sardinian Goral, ?*Nemorhaedus melonii* Dehaut. The exact systematic position of this form is still in doubt. It is found in Sardinia from the Middle Pleistocene to the Postglacial together with *Prolagus*, *Nesiotites* and other members of the Tyrrhenis fauna.

The Chamois, *Rupicapra rupicapra* Linné. The earliest finds of this species are no older than the F-Eemian; afterwards it occurred in Alpine caves like Wildkirchli and the Drachenloch. During the 4-Würm it was widely distributed in the mountainous areas of Europe; north of its present range it then inhabited the French Massif Central, the Vosges, Black Forest and even the Ardennes; it ranged southward to Elba and Palmaria. Towards the end of the 4-Würm II, however, the species became extinct in these areas and it now survives in the Pyrenees, the Cantabrian Mountains, the Alps, Apennines, Abruzzi, Carpathians, and some of the Balkan mountains; it also ranges eastward to the Caucasus. The origin and evolutionary history of the Chamois remain a mystery; though the species today is confined to Europe, there is no evidence that it evolved in this continent.

The chamois inhabits mountain slopes except the highest. It lives mainly on clover and various other herbs, or mosses and lichens in the winter. The high-crowned cheek teeth of the chamois wear out very slowly and the potential length of life in this animal is unusually great. The longevity of most mammals is strictly limited by the durability of their teeth, so that senility may be said to consist in the wearing out of their dentition. In chamois under natural conditions the teeth are excellent even in animals enfeebled with old age. The horns, which are small and moderately divergent, rise vertically from the crown of the head and curve backward at the end. The shoulder height is about 75 cm.

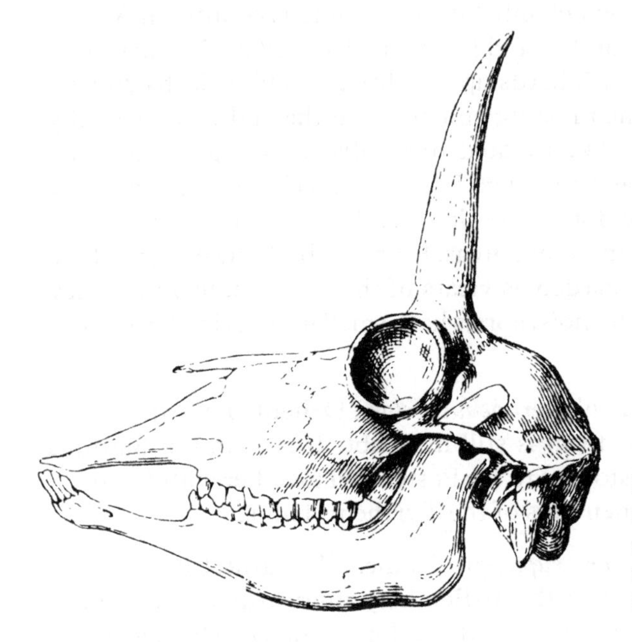

Figure 77. Skull of Chamois, *Rupicapra rupicapra*, Recent. One-third natural size. After Gromova.

The chamois is highly gregarious; summer groups are comparatively small but in winter, when the chamois migrate to the lower valleys, they unite to form large herds [49].

The Cave Goat, *Myotragus balearicus* Bate. This very aberrant species, which may have some relationship with the chamois-like bovids, is only known from the Baleares. It is very common in the caves of Mallorca and Menorca, occurring both in Pleistocene and Postglacial deposits. A very large material has been found in association with Neolithic pottery, human remains, and bones of Domestic Goat. Most other samples are evidently Late Pleistocene in date, but at one locality, Cala Morlanda, the relationship to interglacial shore lines permits the dating of cave goat as D-Holsteinian. The stratigraphic range at present established thus extends from the D-Holstein to the Neolithic.

The cave goat was a relatively small animal, only about 50 cm. high at the shoulders. It had very short cannon bones (the long bones of the middle fore- and hind feet) which must have given it an odd appearance when moving. Curiously, the same shortening of the metapodials is also found in the cave bear and cave hyena. But the most remarkable trait in the cave goat is the development of its incisors. As in other bovids the upper jaw lacks teeth in front, but the middle pair of incisors in the lower jaw grew into huge, chisel-like teeth with open roots; they grew constantly throughout life, as in rodents. The other incisors were usually missing, though a vestigial

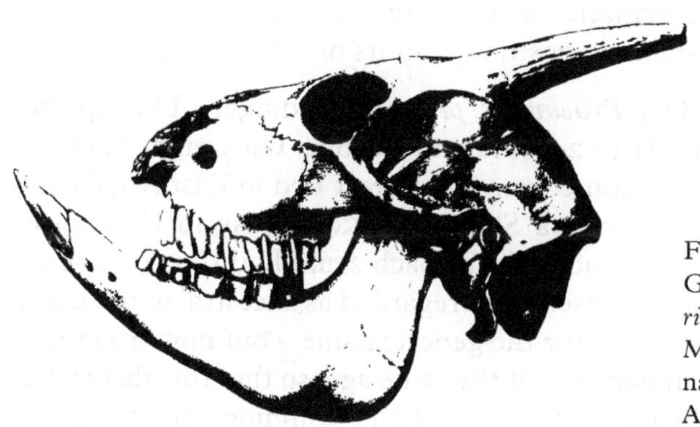

Figure 78. Skull of Cave Goat, *Myotragus balearicus*, Pleistocene, Mallorca. One-half natural size. After Andrews.

second pair has been observed in one or two cases. Just what the cave goat used to gnaw on we do not know, but suggestions range from lichens and mosses on rock-faces to bark of bushes and trees and tough wood fibre. That the food was highly abrasive is also shown by the strong wearing down of the very high-crowned cheek teeth [8].

An ancestral form has recently been discovered at Génova near Palma de Mallorca, apparently dating from the Villafranchian. This form, *M. bateae* Crusafont, was less advanced in many respects, although it shows an incipient stage of the typical incisor specialization. However, it still retained the full number of three incisor pairs, and the premolar series is less reduced than in true cave goat.

Somewhat similarly specialized antelopes occur in the late Tertiary of Italy, but they are not related to *Myotragus* [8].

The Ardé Antelope, *Deperetia ardea* Depéret. (It may be mentioned that the genus *Deperetia* was named by Schaub; Depéret described the species as a member of the genus *Oryx*.) With this species we return to the Villafranchian. It has been found at Etouaires, and ranged up to Senèze and Erpfingen; its geographic range extends from France to Roumania.

The horns of the Ardé antelope differ from those of all other bovids: they are rather short and implanted immediately above the eye-socket, from where they jut out in a broad V. There is a suggestion of slight spiraling. The head was very broad, the jaws powerful. A few remains of the postcranial skeleton indicate a very large animal: the radius measures 50 cm. in length and the metacarpal cannonbone 40 cm., figures in the size class of the elk (*Alces*). Neck vertebrae from Olténie in Roumania indicate that the neck was relatively short. The Ardé antelope is enigmatic in many respects, both as regards its mode of life, its systematic position, and its origin [33; 233].

The Giant Musk Ox, *Praeovibos priscus* Staudinger. This species ranges from 1-Günz II to 2-Mindel in Europe. Only eight finds are known, of which three come from the Forest Bed in East Anglia and the remainder from Stránská Skála and Koneprusy in Czechoslovakia, Bielszowice in Poland and Mosbach 2 and Bad Frankenhausen in Germany. The species used to be regarded as ancestral to the living musk ox – which accounts for the generic name – but now true musk ox has been found in deposits of the same age, so that this theory has to be modified. Probably both species had a common ancestor in the Villafranchian and the giant musk ox represents a conservative hold-over of this primitive type.

P. priscus was notably less specialized than the modern species. Though larger in size, it had much longer and less massive limb bones in proportion and probably it was more fleet of foot. Perhaps this species never developed the tendency to massive herd co-operation which is so typical of the modern musk ox.

Musk ox-like animals are known as early as the Early Pliocene, but to what extent they were really allied to *Ovibos* and *Praeovibos* is not known. The forms that have hitherto been found seem too specialized in various ways to be directly ancestral to the Pleistocene and living musk oxen [125].

The Musk Ox, *Ovibos moschatus* Linné. On the basis of two finds from Süssenborn and Obergünzburg in Germany, Kahlke has recently

[125] shown that this species is of great antiquity, dating back in the scheme used here to the 1-Günz II. An English specimen, unfortunately without locality data but with the fossilization typical of the Forest Bed may perhaps date from the C-Cromerian. In the later Middle Pleistocene and Late Pleistocene the musk ox was quite common in Europe and it did not become extinct there until the end of the Ice Age. Its geographic range was circumpolar in the late Ice Age; it may have reached North America in the 4-Würm. Since the Ice Age, however, it has become extinct in most of its former range and it now lives only in northeastern Canada and Greenland. The musk ox appears to be one of the species that now live in an entirely different part of the world than its place of origin.

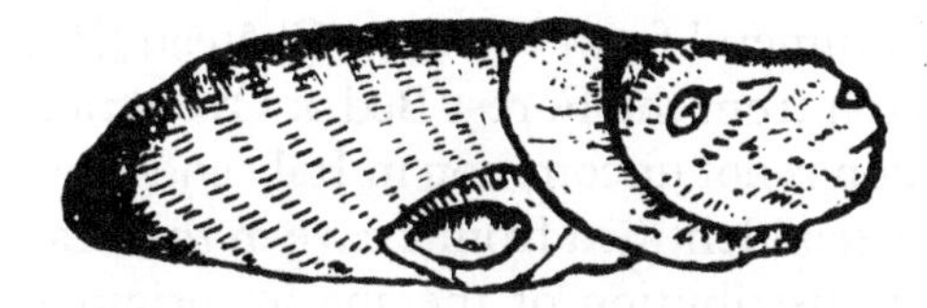

Figure 79. Figurine of Musk Ox, *Ovibos moschatus*. Kesslerloch at Thayngen, Switzerland. After Koby.

Everybody knows the peculiar appearance of the musk ox with its hanging fur, its thick-set build and its doubled-down horns, enormously thick at the base and with points turned upward. The adaptation to tundra life is completed by the strong gregarious instinct of these animals. The adults shelter their young in the form of a living wall, not only against enemies but against wind and cold as well. In the summer, the musk ox puts on a thick layer of fat, which helps it to withstand the rigours of the Arctic winter.

The ecological success of the musk ox may be the explanation of its great longevity as a species; it is interesting that its first appearance coincided with that of another well-adapted Arctic herbivore, the reindeer. But the 'living wall' defence, though effective against most carnivores, is harmful rather than the opposite for a species confronting man; a small group of hunters with long-range weapons, spear-throwers or bows and arrows, are able to slaughter whole herds making such a stubborn stand. Perhaps this contributed to the extinction of the species in the Old World [125; 230].

The Giant Sheep, *Megalovis latifrons* Schaub. This rare species dates from the Villafranchian and has, somewhat doubtfully, been identified as early as Villaroya; the undoubted records are from Senèze,

Erpfingen and a Roumanian locality. Allied forms, perhaps the same species in actual fact, are known from China.

In size and also in the general shape of the head, the giant sheep resembled the musk ox; the horns, however, grew out in sheep fashion, behind the eyes at the side of the head. The dentition, too, resembles that of the sheep. Limb bones indicate a powerfully built animal. Perhaps this species was an Alpine form like so many other Villafranchian bovids in Europe, like the living mountain sheep [233].

The Mouflon, *Ovis musimon* Schreber and other true sheep. Fossils of the genus *Ovis* are among the most rare in the Pleistocene of Europe. A late Villafranchian specimen from Senèze, probably from the uppermost deposits, indicates the presence of a true sheep at this date. Other fossils from the Middle Pleistocene have been described under different names: *O. antiqua* Pommerol from Pont-du-Château (Puy-de-Dôme) and *O. savini* Newton from the Forest Bed at Overstrand. In the Late Pleistocene, mouflon is not uncommon in Italy and North Africa and one specimen has been identified from a cave in the Franconian Jura. The present-day distribution of the species originally comprised only Corsica and Sardinia but it has been successfully introduced on the mainland.

The domestic sheep is known in Europe since the Neolithic but the earliest domestic sheep known so far come from Zawi Chemi Shanidar in northern Iraq, where Perkins has found evidence of domestication in deposits from about 9000 BC. On this basis it would appear that the sheep is the oldest domestic animal in the world [216].

Soergel's Goat, *Soergelia elisabethae* Schaub. This species from the earlier Middle Pleistocene (1-Günz II, Süssenborn; Erpfingen; 2-Mindel, Koneprusy) is not very well known. It is a goat of the size of a cow, with short, thick, lateral horns which bend downward and forward but do not show true spiralling. The cannon bones indicate a thick-set, heavy build. Perhaps this extinct giant goat was an ecological successor of the giant sheep, *Megalovis* [244].

Merla's Goat, *Hesperoceras merlae* Villalta & Crusafont. This form is only known from Villaroya, where an incomplete skull has been found. It was a goat- or sheep-like animal the size of a modern goat but its horns, which grew obliquely upward and backward, were curved forward. According to the Spanish authors, this species probably represents an African element in the Villafranchian fauna of Spain [299].

The Ibex, *Capra ibex* Linné. Although various Middle Pleistocene (Solilhac, Hundsheim, etc.) and even earlier finds (a large *Capra* sp. at Saint-Estève) have been regarded as related to the ibex, the documented history of this species begins with the early part of 3-Riss (glacial terraces in Thuringia; Achenheim). The oldest finds are often put in a distinct species, *C. camburgensis* Toepfer. Whether this species is upheld or not depends on how the systematics of the living ibex are interpreted. The Camburg ibex combines characters that nowadays separate different local 'species' or subspecies such as the Pyrenean ibex (*Capra pyrenaica* Schinz), the Alpine ibex and others. If these are regarded as distinct species, the Camburg ibex must obviously be a distinct species too; but if they are only given the status of local subspecies in keeping with modern trends in taxonomy, it is best included as a subspecies of *C. ibex*. However this may be, it seems evident that the Camburg ibex is ancestral to the whole group of modern forms; so that they must have diverged and differentiated during the lapse of time since the 3-Riss glaciation.

In the F-Eemian, the ibex is found in the Alps, for instance in the Wildkirchli and Drachenloch caves. Like the chamois it was widely distributed in the mountainous areas of Europe during the F-Eemian and 4-Würm. In Spain it ranged to Gibraltar and it was also present in southern Italy, the south of France, Belgium, Germany to the Harz Mountains and in England as far north as Yorkshire (Victoria Cave, etc.). Abundant remains have been found in and around the Alps. The species extended eastward to Austria, Moravia, Transsylvania and so on to the Crimea and Palestine.

In modern times this species is an even more extreme Alpine form than the chamois. It had been exterminated in most of its European habitat except the Italian Alps and the Pyrenees, but was then reintroduced in Switzerland, Bavaria, Austria and Yugoslavia. Outside Europe the ibex is common in the mountains of Asia and eastern North Africa, where it ranges to Ethiopia. The male is characterized by its immense, curved horns which may reach a length of 1 metre or more and are slightly recurved. As a mountaineer the ibex is quite unparalleled; it seems, for instance, to be the only mammal able to ascend between two vertical cliff faces by successive leaps from side to side.

The menu of the ibex is about the same as that of the chamois, except that more grass is eaten by the ibex.

The domestic goat is descended from the wild goat *C. aegagrus* Erxleben, which is distributed from the Greek Islands and Asia

Minor eastward to India. Various finds of this species have been reported from caves in Europe, e.g. Ireland, Belgium, France, Austria, Italy, Bulgaria and Hungary, but in many and perhaps all instances

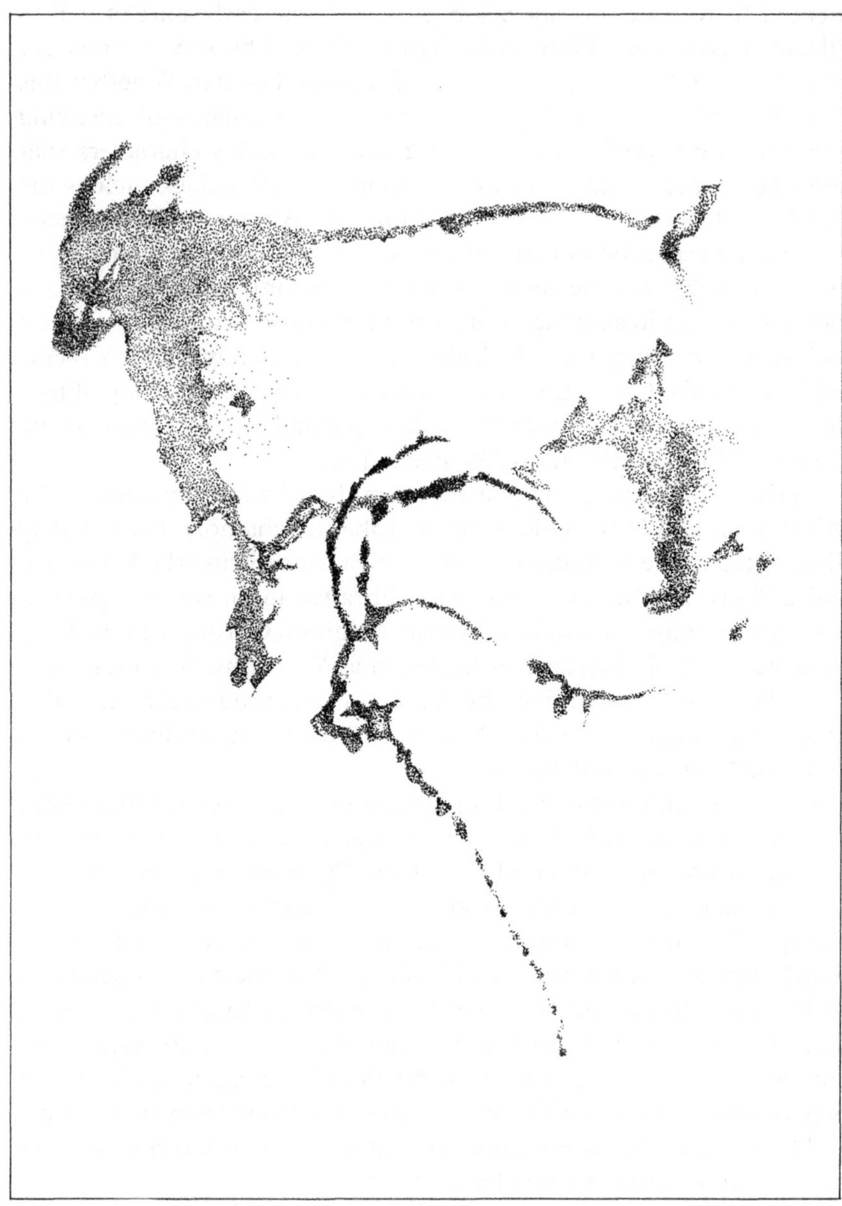

Figure 80. European Tahr, *Hemitragus bonali* (upper figure) and Ibex, *Capra ibex* (below). Cave painting, Cougnac. After Koby.

either the specific determination or the Pleistocene date is doubtful [284]. The earliest domestic goats have been recovered at Jarmo and Jericho and date from about 6500 BC [222].

The European Tahr, *Hemitragus bonali* Harlé & Stehlin. Fossil goats of tahr type make sporadic appearances in Europe at two different points in time. There seems to have been an early invasion from the end of the Astian to the early Middle Pleistocene. Most of the finds are from Hungary (Csarnóta, Villány, Nagyharsányhegy) but the species ranged into Austria in the 1-Günz II (Hundsheim). This early form of the tahr may be a distinct species, *H. stehlini* Freudenberg, but its status is uncertain. A second tahr invasion occurred in the Late Pleistocene at the end of the F-Eemian or the beginning of the 4-Würm. This time the species penetrated into France (a cave in the Céou Valley near Sarlat; also a couple of uncertain finds). A cave painting in Cougnac seems also to represent a typical tahr [131]; if this identification is correct, the species persisted well into 4-Würm II.

The European tahr was considerably larger than the living Himalaya tahr, *H. jemlahicus* H. Smith, which attains a shoulder height of about 100 cm.; the horns are of the same general type as in the ibex but much shorter (about 30 cm. in length) and somewhat more curved. The hanging hair gives the animal an entirely different appearance, very realistically shown in the Cougnac painting it would seem. The modern species lives on the southern slopes of the Himalaya from Kashmir to Sikkim and moves skilfully over the rocky ground [237].

The Perrier Ox, *Leptobos elatus* Pomel. The true cattle and allied forms, constituting the subfamily Bovinae, were represented in Europe during the Villafranchian by the genus *Leptobos*. The true Perrier ox is only known from Etouaires but its successor in the middle Villafranchian, *L. stenometopon* Rütimeyer, is rather closely similar and probably no more than a subspecies of *L. elatus*. It persisted to the late Villafranchian; it has been found at Villaroya, Saint-Vallier, Senèze and Val d'Arno.

The *Leptobos* cattle were probably rather reminiscent of large antelopes in the flesh, for they were smaller and lighter in build than most living bovines. The females were hornless. The cranial anatomy indicates a close relationship with the bison group.

The bulls carried rather long horns, which in this species were strongly curved forward and upward so that their points actually faced inward. At the base the horns pointed almost straight out to the side, but in the Etouaires form they were slightly inclined upward and in *stenometopon* slightly depressed. The teeth had rather high crowns but were less robust than in later bovines and generally speaking the species makes an 'archaic' impression, even compared to other *Leptobos*.

The genus *Leptobos* is also found in the Villafranchian of Asia – for instance in the Siwalik mountains – and is generally thought to be Asiatic in origin [302].

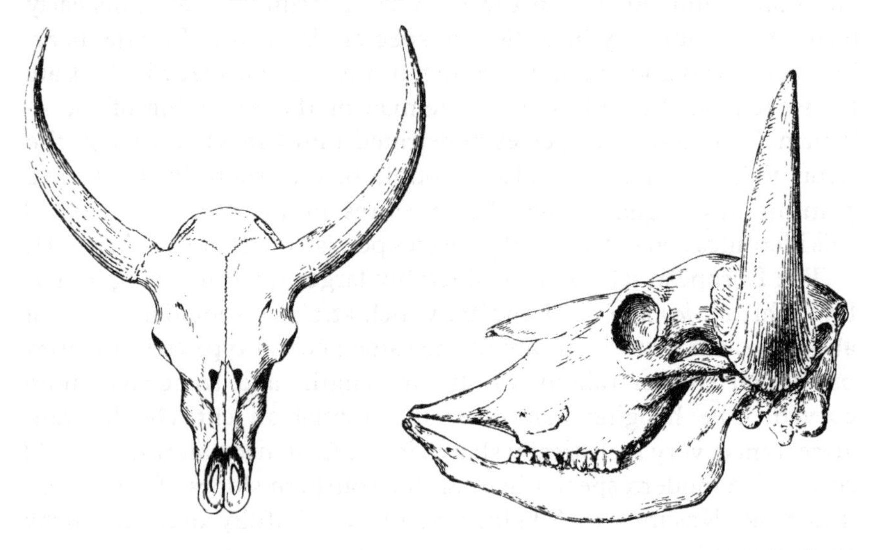

Figure 81. Skulls of Etruscan Ox, *Leptobos etruscus*, Villafranchian, Val d'Arno (left) and Steppe Wisent, *Bison priscus*, Pleistocene, USSR (right). Both one-twelfth natural size. After Gromova.

The Etruscan Ox, *Leptobos etruscus* Falconer. In this species the horns of the male sweep backward in a long curve; they lie in the same plane as the frontal profile, the total effect being somewhat reminiscent of water buffalo. This form is somewhat heavier of build than the Perrier ox and on the whole more progressive and bison-like. An allied Chinese form seems already to represent the genus *Bison*.

The Etruscan ox has only been found in the late Villafranchian (Senèze, Val d'Arno, Tegelen) and possibly the C-Cromerian (Abbeville). Recently another species with lower tooth crowns has been described from Val d'Arno under the name *L. vallisarni* Merla. This may have been a browsing woodland form [242].

The Steppe Wisent, *Bison priscus* Bojanus. The earliest members of the genus *Bison* are known from the Villafranchian of India (*B. sivalensis* Lydekker) and China (*B. palaeosinensis* Teilhard & Piveteau). At least the latter form is late Villafranchian in age. They were comparatively small, slender forms not far removed from the *Leptobos* type. In Europe the first bisons appear in the Middle Pleistocene, evidently as invaders from Asia.

It is generally reckoned that there were two species of this genus in the Pleistocene of Europe, the steppe wisent and the woodland wisent, of which the former became extinct at the end of the Ice Age, while the subsequent history of the latter has been debated, as will be noted later on.

Figure 82. Restoration of Steppe Wisent, *Bison priscus*. After a painting by Franz Roubal.

The steppe wisent, *Bison priscus*, is very common in Pleistocene deposits, starting with 1-Günz II (Hundsheim, Süssenborn, etc.). It ranges through the 2-Mindel (Mosbach, etc.), D-Holsteinian (Heppenloch, Lunel-Viel, etc.), 3-Riss (Achenheim, etc.), F-Eemian and 4-Würm (numerous cave, loess and travertine deposits ranging from Yorkshire to Spain, Italy, Palestine, and Siberia). It vanished in Europe at the end of the 4-Würm, but in Palestine it survived into Mesolithic times.

It was a most powerful animal, larger than the living wisent which may stand 195 cm. at the withers and reach a head-and-body length of some 270 cm. The horn-cores of the steppe wisent have a span of up to 120 cm., which is almost twice as much as in the living wisent. These broad horns (the outer, horny sheath of course increased the spread greatly) would have been cumbersome in the forest, so it seems likely that *B. priscus* kept to open ground.

It was apparently the steppe wisent that was portrayed by the Palaeolithic artists during the later part of 4-Würm, for instance in the famous paintings of the Altamira Cave in northern Spain or the ivory sculpture from the Vogelherd in Wurttemberg. The long horns, the gigantic shoulders and the slender hind quarters are very characteristically rendered.

The mode of life of the steppe wisent probably resembled that of the modern American bison, *B. bison* Linné. Only a century ago enormous herds of these animals roamed the prairies, where they made long seasonal treks, foraging for food. The abundance of fossil steppe wisent bones in European Pleistocene deposits may indicate that this animal too congregated into large herds [4; 188; 253].

The Woodland Wisent, *Bison schoetensacki* Freudenberg. This species, which is much rarer than the steppe wisent, appeared in the early Middle Pleistocene where it has been recorded from the B-Waalian (Episcopia), 1-Günz II (Gombasek), C-Cromerian (Mauer) and 2-Mindel (Mosbach 2; Koneprusy). Sporadic later records date from the D-Holsteinian (Steinheim), 3-Riss (Achenheim) and even 4-Würm I (Wallertheim). The species is often but not always associated with a fauna of woodland type. It was much smaller than the steppe wisent, with shorter horns. It has generally been regarded as the immediate ancestor of the modern wisent and has even been called a subspecies of *B. bonasus*; but the most recent studies indicate that the Pleistocene woodland wisent became extinct without issue and that the modern wisent is a late immigrant from North America [188; 253].

The Wisent, *Bison bonasus* Linné. Several finds of Pleistocene date have from time to time been referred to this species but it seems that only the Postglacial material in Europe may be definitely so determined. A specimen from the 4-Würm has been regarded as transitional between *B. schoetensacki* and the living species and has been described under the name *B. bonasus mediator* Hilzheimer. However, it now seems more probable that the wisent is a late immigrant from North America.

A great number of different species of *Bison* are known from the North American Pleistocene; they apparently stem from an ancestral form that got there from Asia in the 3-Riss or perhaps even earlier.

They developed a variety of forms: large and small, long-horned and short-horned, with straight or curved horns.

One of the sequences in North America shows in some detail the evolution of the modern *B. bison* out of a somewhat larger ancestral form (*B. crassicornis* Richardson). It is known that the living wisent of Europe and the American bison are really very closely related, although their external appearances are rather different; many authors contend that the two forms are but subspecies of a single species. The subspecies *B. bison occidentalis* Lucas, which was in existence some 10,000 years ago, is regarded as ancestral to both the American and the European form. Thus it would seem that the living wisent in Europe is the descendant of a line that invaded the New World in the 3-Riss and returned to the Old at the end of 4-Würm.

The modern European wisent is a typical forest animal, unlike its American ally, but this mode of life was evidently adopted at a fairly recent date. As has been shown by Degerbøl and Iversen [55], the earliest Postglacial wisent in Denmark lived during the younger Dryas age in a so-called park tundra – that is to say a tundra with occasional groves in sheltered places. When this environment was succeeded by forest in Pre-Boreal times the wisent, far from multiplying in this seemingly more congenial milieu, became extinct; so this form was certainly not a woodland animal. Yet it is anatomically identical with *B. bonasus*.

Very near extinction in the recent wars, a population of the wisent has survived in the Bielowice Forest shared by Poland and the USSR and in various zoological gardens and parks [55; 253].

The Murr Buffalo, *Bubalus murrensis* Berckhemer. This European water buffalo takes its name after the river Murr, a tributary to the Rhine, in the waters of which it swam during the D-Holsteinian interglacial. It has been found in deposits of this age at Steinheim and Schönebeck. It is closely allied to the modern Asiatic water buffalo *B. bubalus* Linné and evidently was an invader from Asia. The living water buffalo prefers muddy shores and never strays far from rivers, lakes or the seacoast [29].

Stehlin's Buffalo, *Syncerus iselini* Stehlin. A buffalo of African type has been described from Val d'Arno. Unfortunately it is an isolated find without associated fauna, so that we cannot be entirely sure that it dates from the Villafranchian [267].

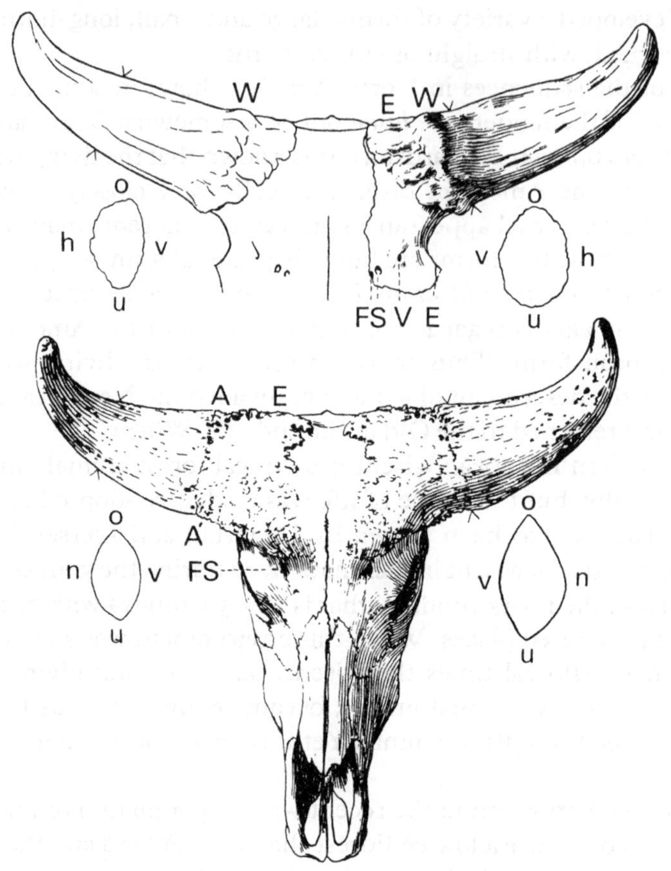

Figure 83. Horn-cores of Val d'Arno Buffalo, *Syncerus iselini*, compared with skull of Recent African Buffalo, *S. caffer*. The age of *S. iselini* may be Early or Middle Pleistocene. One-ninth natural size. After Stehlin.

The Aurochs, *Bos primigenius* Bojanus. In contrast with the wisent, the aurochs is relatively rare in Pleistocene deposits but very common in the Postglacial. The apparently low numbers of aurochs in the Pleistocene may have some connection with its more stationary habits compared with the bison. It would not have made the long seasonal migrations that may have made it possible for the latter to inhabit the steppes and tundras in spite of the low food production of these barren areas. The aurochs would have been confined to more densely overgrown grasslands and open woodlands. In historical times in Europe the aurochs lived in forests but this cannot be taken as the primary biotope of the species; more probably it was a last refuge, just as the

Scandinavian brown bear was driven out of the woods into the fields in the present century.

The earliest aurochs in Europe date from the D-Holsteinian interglacial (Steinheim, Swanscombe, etc.). These were gigantic animals, much larger than the specimens populating Scandinavia in the Postglacial and the latter were certainly of no mean size. The enormous horns of the Steinheim aurochs were less curved than in more recent forms and also tended to be somewhat flatter in cross section.

During the interglacials that followed (E-Ilford and F-Eem) the size of the aurochs gradually decreased, but as late as the 4-Würm immense forms of this animal were still in existence south of the Alps. In the Postglacial this species was particularly common. It survived in Scandinavia to the times of the Vikings, while the last aurochs, a cow in the Jaktorow Forest near Warsaw, died in 1627.

The aurochs has been portrayed both by Palaeolithic man and in historical times. The cave paintings, for instance at Lascaux, tend to emphasize the difference in size and colour between the bulls and the cows; the former are often depicted as black and the latter as red. Apparently the cows were in fact a reddish brown, while the bulls were much darker.

Figure 84. Aurochs, *Bos primigenius*, cow and bull. Teyjat, Dordogne. After Breuil.

The aurochs may have descended from the Asiatic *B. namadicus* Falconer of the early Middle Pleistocene, which in turn originated from the Villafranchian species *B. planifrons* Lydekker of the Siwaliks. The last-mentioned species was a relatively small animal for a *Bos*, but there was a gradual size increase through the *namadicus* stage. The bodily magnificence of the line culminated in the great Middle Pleistocene form that invaded Europe; then, as we have seen, it gradually decreased in size again.

The species is also found in the Pleistocene of North Africa and to the east of the Mediterranean. Domestic cattle, easily identified by their smaller dimensions, are met with for the first time at Banahilk in northern Iraq in deposits that seem to date from at least 5000 BC [32; 185; 186; 222].

Chapter 14

Order Rodentia

THE Rodentia are richer in species and individuals than any other order of mammals and so in a way may be termed the most successful mammalian order. The earliest known Rodentia date from the Paleocene, but it was only in the Middle Tertiary that the rodents multiplied to overshadow other orders in numbers and taxonomic variety.

The Rodentia are characterized by a single pair of chisel-formed, gnawing incisors in both the upper and lower jaws and by various other characters correlated with this basic adaptive trait. Though uniform in this respect, the order is greatly varied in many others and contains mammals of aquatic, subterranean and arboreal habits as well as forms adapted to every conceivable terrestrial environment. The range of size is also great, from some of the smallest of mammals to extinct rodents as large as bears.

The infraordinal classification is difficult; at present there is no agreement on how to group the various superfamilies and families. As regards the fossil Rodentia, taxonomy on the species level is also in its infancy. We know for the time being very little about the infraspecific variation in osteology, dental morphology and size that may exist in rodent populations. The great number of species in many genera increases the difficulties and makes this order singularly difficult to survey. Much of the current taxonomy on fossil rodents is evidently typological classification that cuts through natural populations in an artificial way. This is perhaps an unavoidable early stage in the study of any group of organisms; it will in time be superseded by a more realistic classification.

When this has been achieved, the Rodentia will probably surpass all other mammalian groups in stratigraphic importance because of their ubiquity, large numbers, rapid evolution and great ecological significance. They will probably furnish better evidence on the details of past environments than any other mammals because of the narrow adaptive ranges of most species of the order. The study of the Rodentia is thus expected to gain greatly in breadth and impact in the future. It is the more regrettable that only the barest outline can be given here.

Family Sciuridae, Squirrels and Marmots

One of the smaller rodent families, the Sciuridae are incompletely known in the Pleistocene. The tree squirrels are seldom found in the fossil state. The marmots have a fairly good record but only for the Late Pleistocene. The only group with an extended representation are the sousliks.

The sciurid family dates back to the Miocene.

White's Squirrel, *Sciurus whitei* Hinton. This squirrel, originally described on the basis of a single fourth upper premolar from the top of the Forest Bed (C-Cromer), would seem to be ancestral to the living red squirrel [117]. The difference between the two species lies mainly in the anterior part of the mentioned premolar. Material from the late 2-Mindel or early D-Holstein (Tarkö) is transitional between *S. whitei* and *S. vulgaris*.

The genus *Sciurus* has apparently existed in Europe since the Miocene. An Astian form, somewhat larger than *S. vulgaris*, has been described from Poland as *S. warthae* Sulimski; this form may have survived in the Villafranchian (Rebielice). Various other records are only determinable as *Sciurus* sp. (Senèze, Sackdilling, Episcopia, Schernfeld, etc.).

The Red Squirrel, *Sciurus vulgaris* Linné. The modern species appears in the F-Eemian (Lambrecht Cave, Cotencher, etc.) and has been found sporadically in the 4-Würm and Postglacial, but is not common in the fossil state presumably because its arboreal habits are not conducive to fossilization.

The red squirrel is now found throughout the temperate forest belt of Eurasia and is the most commonly seen wild animal in this area. Numerous different subspecies may be separated on the basis of size and colour; probably a true appreciation of the position of *S. whitei* must await a study of the odontological variation throughout the modern population.

The European form has a body length of about 25 cm. with an additional 20 cm. for the long, bushy tail. The squirrel is predominantly a woodland animal, although vagrant individuals have been noticed on the moors. In coniferous forests the squirrel subsists on cones, spruce buds and the like, while a more varied diet is available in mixed forests; although vegetable food predominates, grubs, insects, eggs, etc. are also eaten. The size of the population is variable, mass occurrences being produced at odd intervals. They may lead to massive migrations of the same type as in the lemming, sometimes on a broad front tens of kilometres across.

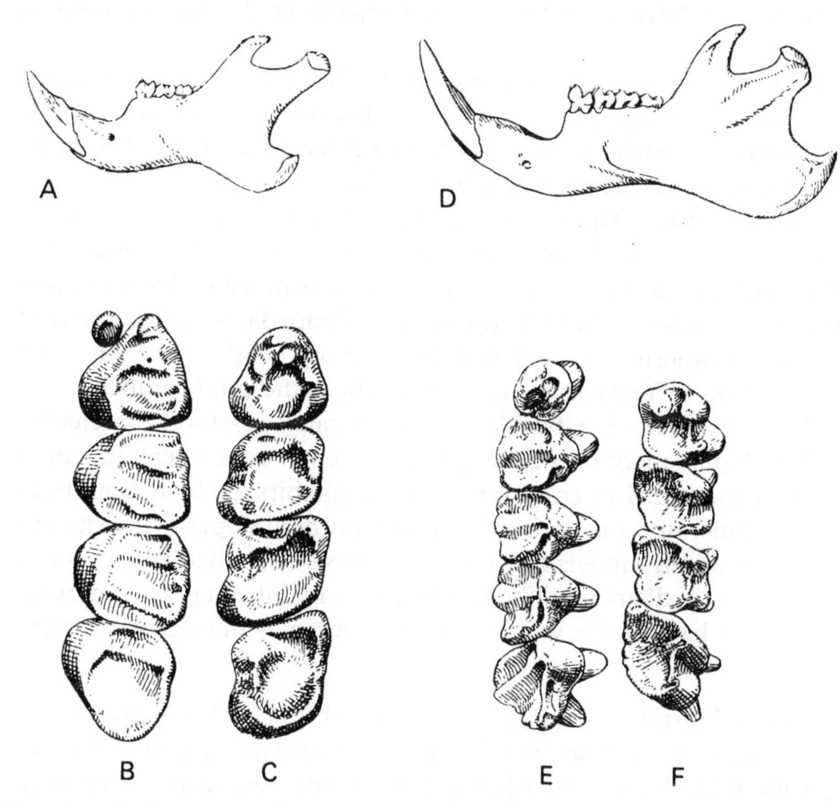

Figure 85. A–C, Red Squirrel, *Sciurus vulgaris*; D–F, Alpine Marmot, *Marmota marmota*. A, D, mandibles; B, E, upper tooth rows; C, F, lower tooth rows. A natural size; B–C five times enlarged; D three-quarters natural size; E–F $1\frac{1}{4}$ times natural size. After Gromova.

In the British Isles the red squirrel has been ousted from part of its range by the introduced grey squirrel (*S. carolinensis* Gmelin), externally separable by its colour and tuftless ears [119].

The Alpine Marmot, *Marmota marmota* Linné. This species is found as early as 3-Riss (Châtillon-Saint-Jean, Achenheim, etc.) but becomes common only in the Late Pleistocene [185]. A useful osteological character is the course of the temporal lines, which form extended curves in *M. marmota* but are more nearly transverse in the related *M. bobak*. The Pleistocene marmot used to be regarded as a species of its own (*M. primigenia* Kaup) ancestral to both *M. marmota* and *M. bobac*, but it is now clear that the separation between the two modern species goes back to their first appearance in the fossil record in Europe.

Evidently the marmots are immigrants from Asia, where representatives of the group appear in the late Villafranchian and early Middle Pleistocene (Choukoutien). This form is referred to *M. bobak*. Originally the marmots arose in North America.

At present the Alpine marmot inhabits disjointed and restricted areas in the Carpathians, the Alps and (introduced by man) the Pyrenees, but in the Pleistocene its range was much wider, including large parts of Spain down to the coast of Granada, the greater part of France, southern Germany and the territory south of the Alps and Carpathians to Yugoslavia and Italy at about the 42nd parallel.

The species also inhabits Asia, where it ranges east to Kamtchatka.

The Alpine marmot is a fairly large rodent, about 55 cm. in length with an additional 15 cm. for the tail. It inhabits the high mountains above timber line, preferring the sunny southern slopes, where it dens in subterranean tunnels forming the centres of individual territories up to 50 m. in diameter. The marmot lives mainly on grass and herbs and stores hay for the winter; but it hibernates during most of the winter.

The Grey Marmot (*M. caligata* Eschscholz) which inhabits the mountainous areas of northwestern North America is now regarded as conspecific with *M. marmota* [221]. No information on the date of its migration across the Bering Strait is available. The related woodchuck (*M. monax* Linné) is known in North America since the 3-Riss [200].

The Bobak Marmot, *Marmota bobak* Müller. This species appears well back in the early Middle Pleistocene in Asia (Choukoutien) and seems then to have gradually progressed to the west, attaining its maximal range in the 4-Würm when it entered Central Europe. At that time the area inhabited by the bobak marmot extended in a corridor from southern Russia through Poland and Germany to the

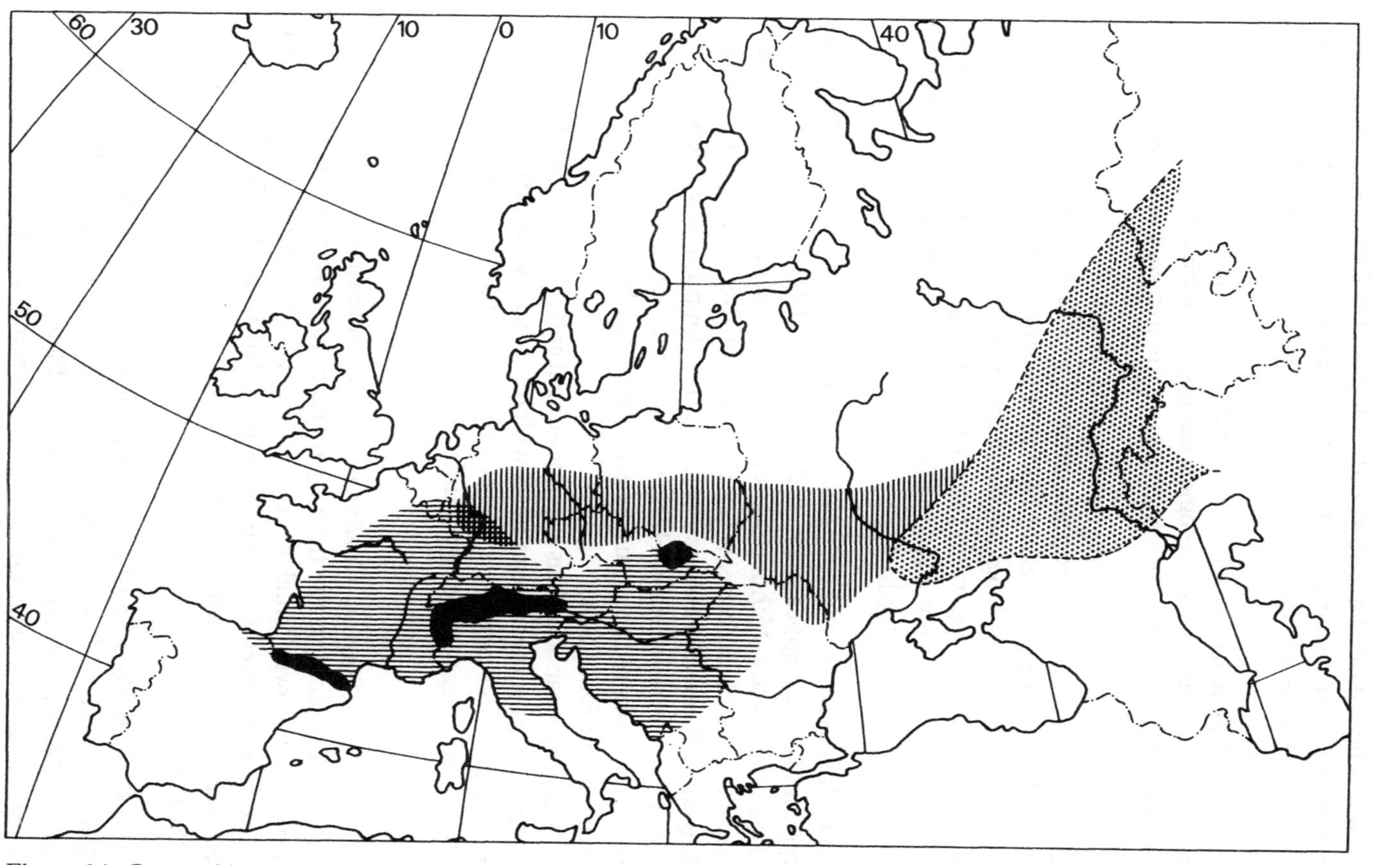

Figure 86. Geographic ranges of European Marmots. Alpine Marmot, *Marmota marmota*: black, present range; horizontal hatching, Pleistocene range. Bobak Marmot, *M. bobak*: stippled, present range; vertical hatching, Pleistocene range. Data from Mottl.

Rhine, where it overlapped with the range of the Alpine marmot. The bobak has later become extinct in this area and it is now found only north of the Black Sea and further east to Himalaya, China and Mongolia.

The bobak marmot, which is slightly larger than *M. marmota* and differs from it in some osteological details, resembles the latter species in its mode of life. Its main enemies are the fox, wolf and eagle [200].

Kormos's Souslik, *Citellus primigenius* Kormos. This early form differs from modern European souslik by its larger size. It occurs in the Villafranchian and early Middle Pleistocene of eastern Europe (Villány; Episcopia; Nagyharsányhegy; Chlum). Like other sousliks it probably was a gregarious form with a biology not unlike that of the marmots. These they also resemble externally, though the sousliks are much smaller [145].

The European Souslik, *Citellus citellus* Linné. Sousliks of modern type make their appearance in the 1-Günz II (Hundsheim) and are also found in the 2-Mindel (Tarkö), 3-Riss (Achenheim) and throughout the Late Pleistocene, when the range of this species extended to England (Kent; Somersetshire); it is also found in cave deposits in Germany, Austria, Italy, Hungary, Bulgaria, etc. Its modern distribution is more reduced and extends from southeast Germany to the Aegean and Black Sea.

The European souslik is a steppe animal and also inhabits cultivated areas. It digs subterranean nests which it stocks with provisions carried in the cheek pouches; it lives on grass, herbs, roots and grains but also insects, eggs and other animal foods and it hibernates in winter [312].

The Spotted Souslik, *Citellus suslicus* Gueldenstaedt. This species, easily recognized in the flesh by its numerous white spots on a grey (males) or brown (females) ground-colour, is much rarer in the fossil state than *C. citellus*. It has been recorded from the Kesslerloch in Switzerland (4-Würm). It is now found in southeastern Europe (Poland, Roumania, southern Russia) [312].

The Red-cheeked Souslik, *Citellus major* Pallas (*C. rufescens* Keyserling & Blasius). This species, now living in Transvolgan Russia and parts of Siberia, occurs sporadically in Pleistocene deposits in Hungary, Czechoslovakia, Germany, Switzerland and France, evidently mainly or only in the 4-Würm. Its most dangerous predator is the steppe polecat [312].

The Flying Squirrel, *Pteromys volans* Linné. This species is now distributed eastward from the Baltic Sea through the wooded parts of Siberia to Sakhalin, Korea and China. Its fossil history is almost unknown; it is found in cave deposits in Siberia and a related form has been described from Choukoutien (2-Mindel). The small size and the habits of this animal combine to render it very rare in the fossil state, so that negative evidence is most unreliable in this case.

Like other flying squirrels this species is nocturnal in habits. It lives on nuts, fruits, green bark, insects and grubs.

Another genus and species of flying squirrel, *Petauria helleri* Dehm, has been described on scanty material from the early Middle Pleistocene of Schernfeld, while Astian and Villafranchian deposits in Poland have yielded still another genus and species (*Pliosciuropterus schaubi* Sulimski) [56; 270].

Family Castoridae, Beavers

Unlike most other rodent families, the beavers have an excellent fossil record. This is presumably due to the mode of life of these animals as well as their comparatively large size, factors that combine to make them common in river and lake deposits. Beavers occur as early as the Oligocene. The living beavers all belong to the genus *Castor* and form the culmination of a long line of dam-building forms, while another (extinct) group seems to have been somewhat intermediate between castorine beavers and sea-cows in mode of life; these latter include some of the largest known rodents, for instance the American *Castoroides* which reached the size of a black bear.

The Beaver, *Castor fiber* Linné. All the Pleistocene members of the genus *Castor* in Europe seem referable to the living species, though Villafranchian forms are occasionally separated as *C. plicidens* Major. The species may be present at Etouaires and is certainly so at Hajnacka, Villaroya, Saint-Vallier, Val d'Arno and various other late Villafranchian deposits (Tegelen; Erpfingen). There is then a continuous record all through the Middle and Late Pleistocene (e.g. Gombasek, Süssenborn, Forest Bed, Mauer, Mosbach 2, Koneprusy, Tarkö, Grays, Châtillon-Saint-Jean, Achenheim, Montmaurin, the Eemian travertines, various 4-Würm caves and numerous Postglacial deposits). The beaver has long been hunted for its fur but also for its meat and the medicinal value of the castoreum. Intense hunting has seriously reduced its modern range. The original population has survived in some parts of Scandinavia, especially Norway, in the rivers Rhône and Elbe and in eastern Europe; the species is being reintroduced in various areas. In Asia its range extends to Siberia and Mongolia. The closely related North American *C. canadensis* Kuhl is also found in the Late Pleistocene.

The beaver is the largest living European rodent; it is about 80 cm. long, with an additional 32 cm. for the tail, and weighs up to 30 kg. The famous dams serve to regulate the water level, keeping the exits of the beaver lodges safely flooded. In winter, when the river is icebound, the beavers will make a notch in the dam so that an air reservoir is created between the ice and the water. The construction of the dam is, of course, wholly instinctive; it has been found that the beaver reacts by dam-building on the sound of gushing water and goes on building until the noise stops.

Remains of fossil beaver dams are not uncommon, for instance in the Forest Bed, and it can even be shown that the dams have trapped floating carcasses of various animals and thus helped to create accumulations of fossils. The beaver ponds will gradually silt up and the site may finally wind up as a meadow.

The beaver collects branches of willow, aspen and other trees, the bark of which forms its main food; in the summer it will also feed on shore vegetation. It always gets out of the water to feed [219; 247; 302].

The European Giant Beaver, *Trogontherium cuvieri* Fischer (*Conodontes boisvilletti* Laugel). Several species of *Trogontherium* have been described from time to time, but a comprehensive modern study of population variation must be made before any definite taxonomic conclusions can be drawn; until then I prefer to follow the writers who regard *boisvilletti* as a western subspecies of *T. cuvieri*. Remains of trogontheres have been found both in the Villafranchian (Chagny; Erpfingen; Tegelen; the Crags) and especially in the early Middle Pleistocene (Episcopia; Saint-Prest; Gombasek; Süssenborn; Forest Bed; Abbeville; Mauer; Mosbach 2). The last records come from the D-Holsteinian (Swanscombe lower gravels; Bruges; Ingress Vale).

Although no complete skeleton of *Trogontherium* has been found, the rich material from Tegelen represents all the skeletal parts. It has not been possible to determine the length of this animal but it was apparently slightly larger than a modern beaver.

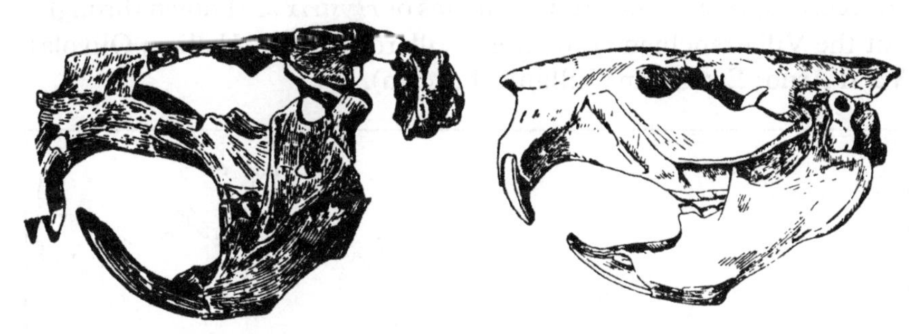

Figure 87. Skull and mandible of Giant Beaver, *Trogontherium cuvieri*, Villafranchian, Tegelen (left) and Recent Beaver, *Castor fiber* (right). Both one-quarter natural size. After Van der Vlerk & Florschütz.

It has been suggested that the bipartite form and rough surface of the premaxillary beneath the nasal opening indicate the presence of a muscular, prehensile, bipartite upper lip. The giant beaver may have used this lip, manatee fashion, to snatch the water plants while swimming in the pools and abandoned channels of the great meandering river delta [224]. It is highly suggestive that the Tegelen flora includes various succulent aquatic forms (*Euryale*, *Potamogeton*, *Najas*, *Trapa natans*, *Sagittaria*, *Alisma*, *Stratiotes*, water-ranunculus, water-mint, etc.). It is thus likely that the trogontheres, unlike true beavers, fed while swimming. However, *Trogontherium* may also have fed on rhizomes, tubers, and the bark of trees like *Castor*, especially in winter.

Trogontherium also occurs in the Villafranchian and Middle Pleistocene of Asia (e.g. Choukoutien). It is not found in the New World [109; 189; 247].

Family Hystricidae, Old World Porcupines

The history of the porcupines in the Old World goes back to the Oligocene. In these animals the hairs on the upper side of the body have been partially or entirely developed as spines, while the limbs carry strong digging claws. Porcupines often gnaw bones, leaving characteristic markings, so that their presence may sometimes be documented even if their own bones have not been found.

The Crested Porcupine, *Hystrix cristata* Linné. Various species of porcupine have been described from the Villafranchian (*H. refossa* Gervais, *H. etrusca* Bosco) and their relationship to each other and to the recent species is not clear. Remains of *Hystrix* are known throughout the Villafranchian (Etouaires; Villaroya; Saint-Vallier; Olivola; Val d'Arno; Erpfingen; Villány, Tegelen).

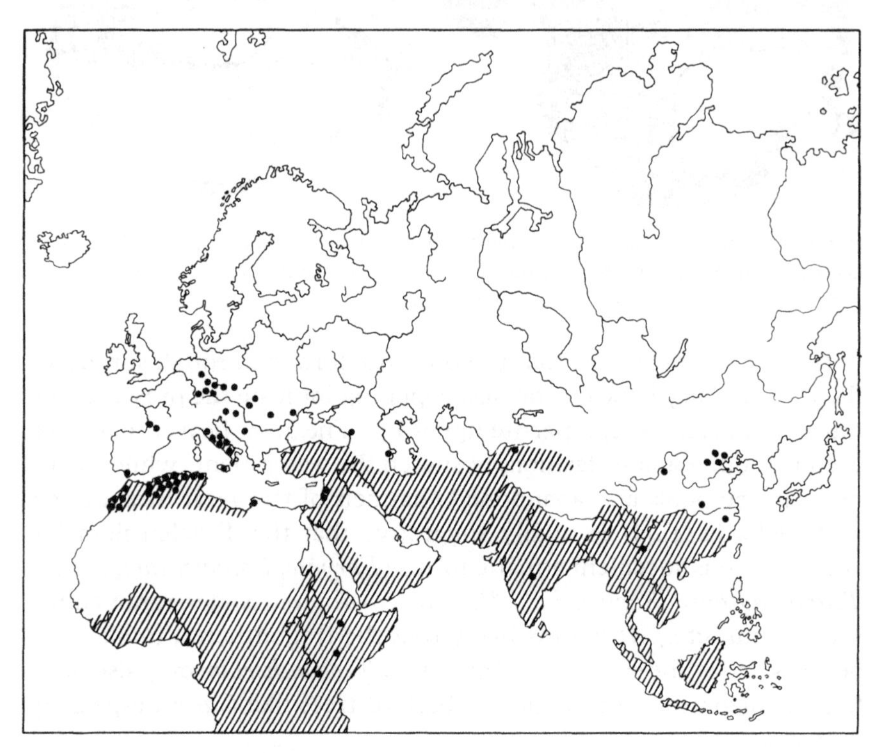

Figure 88. Geographic distribution of Porcupines of the genus *Hystrix*. Recent range hatched; dots indicate Pleistocene finds. After Jánossy.

In the Late Pleistocene a relatively small form is predominant; it may be a distinct species (*H. vinogradovi* Argyropulos). The earliest find of the small form comes from Brassó and thus dates from the early Middle Pleistocene, but other finds are F-Eemian and 4-Würmian. European records of porcupine come from Gibraltar, the Pyrenees, Italy, Germany, Austria, Yugoslavia, Czechoslovakia, Hungary and southern USSR. Most of these are of the small species, but occasional larger specimens are also found, e.g. at Montréjeau in the Hautes-Pyrénées. Outside Europe, *Hystrix* fossils occur in North and

East Africa and at various localities in Asia, including Choukoutien [119].

The living crested porcupine, which may or may not have descended from the Pleistocene form in Europe, is now only distributed in part of Italy (with Sicily) and in northern Greece; its main range lies in northern and eastern Africa. The crest or mane is formed by the ordinary hairs covering the shoulders and back of the neck; the hindmost two-thirds of the back are covered with powerful spines up to 40 cm. in length, forming a highly effective defensive armour. The length without tail is 57–68 cm. The crested porcupine frequents dry, open ground and hill slopes. It is nocturnal in habits and grunts and rattles its spines to frighten off predators; it returns to its hole to spend the day in sleep [119; 302].

Family Zapodidae, Birch Mice and Jumping Mice

This family comprises the Eurasian birch mice, Sicistinae, and the jumping mice or Zapodinae which are mainly distributed in the New World. The birch mice appear as early as the Oligocene in Europe.

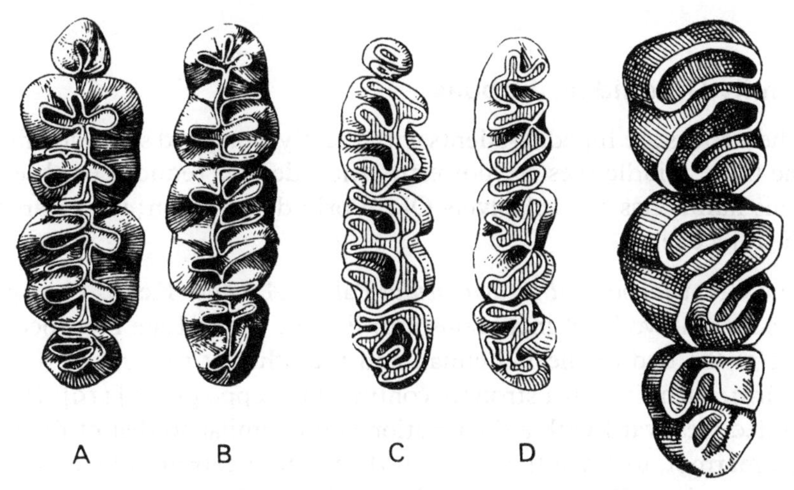

Figure 89. Teeth of Birch Mice, Jerboas, and Mole Rats. A, B, upper and lower cheek teeth of Southern Birch Mouse, *Sicista subtilis*, Recent, thirteen times enlarged. C, D, upper and lower cheek teeth of Great Jerboa, *Allactaga jaculus*, Recent, five times enlarged. E, upper cheek teeth of Primitive Mole Rat, *Prospalax priscus*, Pleistocene, ten times enlarged. After Gromova.

The Northern Birch Mouse, *Sicista betulina* Pallas, and the Southern Birch Mouse, *Sicista subtilis* Pallas. It is almost never possible to separate these two species in a fossil assemblage. The earliest Pleistocene finds in Europe are referred to a third species, *S. praeloriger* Kormos, and come from the early Middle Pleistocene (Episcopia; Sackdilling; Nagyharsányhegy; Kövesvarad; Koneprusy, etc.). There are also a few finds in the later Middle Pleistocene (Tarkö, etc.); it then occurs in the F-Eemian (Lambrecht Cave; Subalyuk) and 4-Würm in Germany, Switzerland, Austria, Hungary and Czechoslovakia. The very small teeth of this animal will usually escape detection unless the matrix is washed and screened. Fossil *Sicista* are also found in the Pleistocene of Asia.

The length of the Northern birch mouse is 52–70 mm., tail 79–106 mm.; the Southern birch mouse is about the same size but has a slightly shorter tail. The Northern species occurs in more or less isolated populations in Scandinavia, Finland, Germany, Poland, Russia and Siberia; the Southern form in steppes of southeastern Europe and Asia. The Northern form also prefers open country, especially hilly and mountainous land. It lives on buds, shoots, grains, berries and insects, and hibernates in nests underground [117; 118; 119].

Family Dipodidae, Jerboas

The jerboas are bipedal rodents with greatly elongated shins and feet; the three middle toes are powerful, the side toes reduced or absent. This family has an exclusively Old World distribution; it appears in the Pliocene.

The Great Jerboa, *Allactaga jaculus* Pallas (*A. major* Kerr). This species makes two brief incursions into Europe in the Late Pleistocene, one at the end of the F-Eemian and the other at the end of the 4-Würm, each time in a strongly continental steppe phase [119]. It extended westward with a distribution about similar to that of the bobak marmot, to reach the Rhine as the western extreme of a corridor from southern Russia north of the Black Sea. Its present-day distribution lies east of the Volga in a belt to the Altai steppe.

This species is the largest of the jerboas, with a body length of 180 mm. and a tail of 260 mm. The powerful hind feet still carry the vestigial first and fifth digits; the claws are hoof-like. The great jerboa is a nocturnal form which prefers clayey ground; its food consists mainly of bulbs, especially *Gagea*, which it digs out with its teeth, but it will also eat grains and bark. It moves in rapid bipedal jumps with the short fore-feet tucked in under the head. It uses its teeth to dig a subterranean nest with an additional emergency exit. Often several individuals hibernate in the same nest.

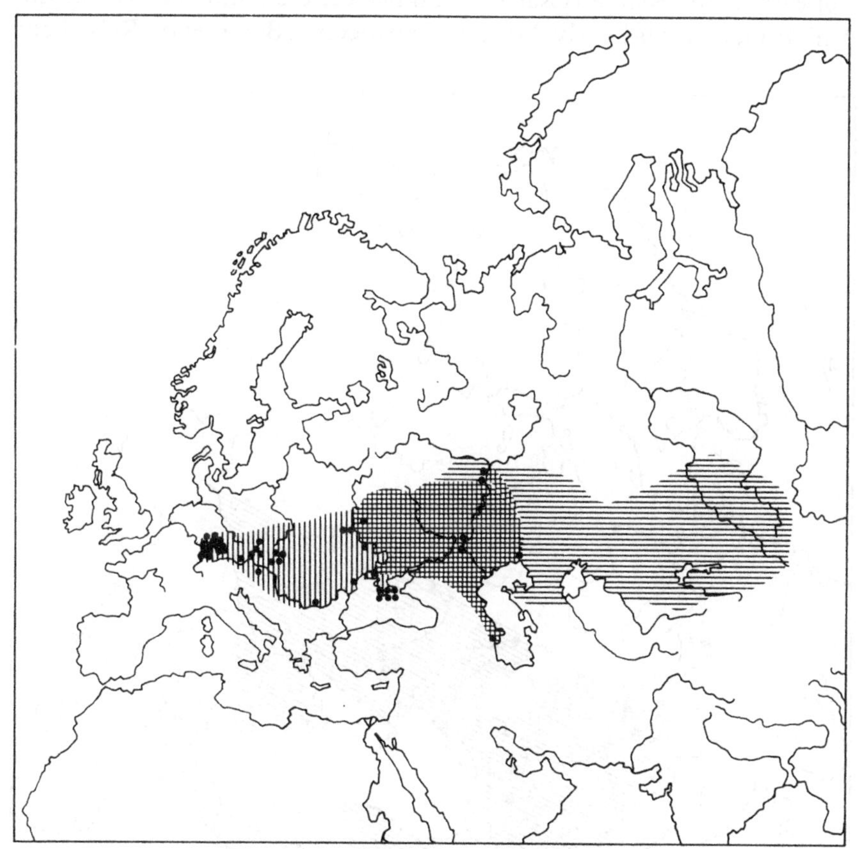

Figure 90. Geographic distribution of Great Jerboa, *Allactaga jaculus*. Recent range horizontally hatched; dots and vertical hatching indicate Late Pleistocene finds and range. After Jánossy.

Various other species of *Allactaga* live in Asia, where the history of this genus goes back to the Villafranchian.

Family Spalacidae, Mole Rats

This family has a short history, involving only the Pliocene and Quaternary. The mole rats are highly adapted for a subterranean life.

The arms and legs are short; the digging is performed with the teeth and hands. The eyes are covered with skin as in some moles and the external ears are reduced.

The Primitive Mole Rat, *Prospalax priscus* Nehring. This species appears in the Astian (Csarnóta, Podlesice, etc.) and survives in the Villafranchian and early Middle Pleistocene (Beremend; Rebielice;

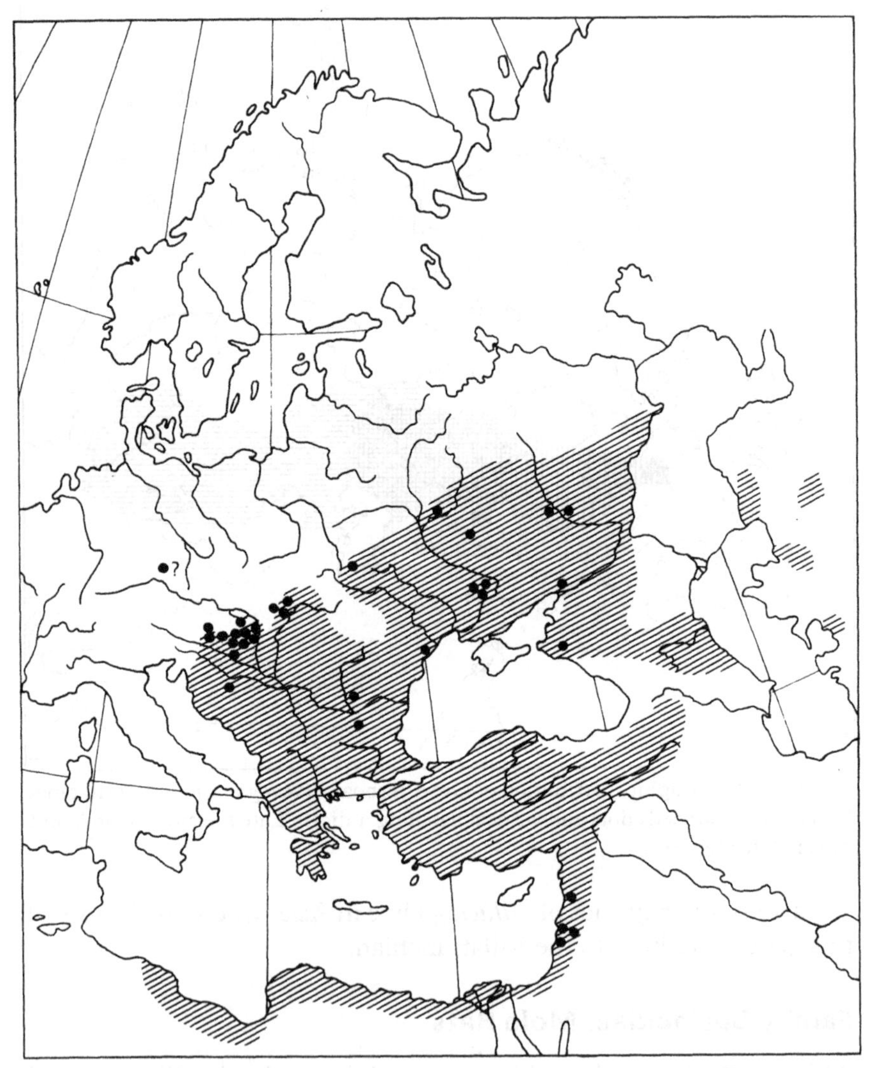

Figure 91. Geographic distribution of Mole Rats of the genus *Spalax*. Recent range hatched; dots indicate Pleistocene finds. After Jánossy.

Villány; Nagyharsányhegy). It has some primitive characters, mainly in the shape of its mandibular processes [153].

The Lesser Mole Rat, *Spalax leucodon* Nordmann. Specific determination of fossil mole rat specimens is difficult and the Greater mole rat, *S. microphthalmus* Gueldenstaedt, may also be represented in the fossil material. The mole rat is not uncommon in the Late Pleistocene within its present-day area of distribution; finds have been made in Germany, Austria, Hungary, Yugoslavia, Roumania, Bulgaria and USSR. The earliest finds of modern *Spalax* come from the later part of the early Middle Pleistocene (Villány 8, Tarkö, etc.). The similarity between the fossil and recent ranges of the species is striking and presumably is correlated with its subterranean habitat, making the nature of the soil the most important environmental factor in its life. The mole rat feeds mainly on roots and bulbs, and requires fertile, dry steppe or garden soil. The length of *S. leucodon* is 185–270 mm.; *S. microphthalmus* is a little larger, 242–310 mm. [119].

Family Gliridae, Dormice

The dormice form one of the most ancient rodent families; they arose in the Eocene. They are small to medium sized rodents, mostly arboreal, and are restricted to the Old World. The name alludes to the fact that all the glirids of the temperate area hibernate. The fossil record of this group is relatively good, probably because these animals are frequently caught by owls so that their remains find their way into the deposits formed by owl pellets.

In the Eocene and Oligocene, Europe was inhabited by primitive forms belonging to the subfamily Gliravinae. The modern subfamily Glirinae arose in the Oligocene and is particularly well represented in the Miocene, when its first great radiation took place. More modernized glirines evolved in the Pliocene and Quaternary, but relicts of the Miocene radiation survived locally in the Pleistocene, for instance on some of the Mediterranean islands.

Heller's Dormouse, *Glirulus pusillus* Heller (*Amphidyromys pusillus*). The genus *Glirulus* is at present restricted to Japan but was present in Europe from the early Miocene to the Villafranchian. This species is known from the Astian (Podlesice; Csarnóta) and Villafranchian (Gundersheim; Schernfeld). Members of *Glirulus* are very small and the scarcity of finds may be due in part to inadequacy of collecting technique [157].

The Pliocene Dormouse, *Muscardinus pliocaenicus* Kowalski. This species is so far known only in the Astian (Weze; Podlesice) and early Villafranchian (Rebielice) of Poland; it is a member of *Muscardinus* slightly smaller than the living *M. avellanarius*, which it otherwise resembles; it seems clearly ancestral to the modern form, since the material from Rebielice is transitional between the Astian and the living Dormouse [149; 157].

The Dormouse, *Muscardinus avellanarius* Linné. This species, which is descended from the Pliocene dormouse, makes its first appearance in the early Middle Pleistocene (Kadzielnia; Sackdilling; Koneprusy; Moggaster Cave); it is then recorded in the later Middle Pleistocene (Tarkö, Breitenberg) and Late Pleistocene (various fissures and caves in France, Germany, Italy, Switzerland, Austria).

The living dormouse is 60–90 mm. long, tail 55–75 mm.; it weighs 15–40 g. and is the smallest of the living European glirids. Its modern distribution extends from England, southern Sweden, the Baltic countries and the Moscow region in the north to the Pyrenees, southern Italy and Asia Minor in the south; it ranges eastward to the Volga. It prefers shrub vegetation and is a skilful climber; its food consists of nuts, acorns, fruits, berries, shoots and tender bark. It builds elaborate nests in the shrubs or trees [157].

The Dacian Dormouse, *Muscardinus dacicus* Kormos. An extinct species slightly larger than *M. avellanarius* and dating from the early Middle Pleistocene (Episcopia, Kövesvarad); a specimen from the Astian of Podlesice has also been tentatively referred to this species [137].

The Fat Dormouse, *Glis glis* Linné. This or a closely related species (*G. sackdillingensis* Heller) is fairly common in the early Middle Pleistocene (Sackdilling; Schernfeld; Gombasek; Süssenborn; Hundsheim; Podumci; Verona; Koneprusy; Kövesvarad; Kamyk; Ukraine; Dalmatia, etc.); the early form is slightly smaller than the recent. Numerous finds of the modern form come from the later interglacials (D-Holstein, F-Eem) in France, Belgium, Germany, Poland, Czechoslovakia, Hungary, Switzerland, Italy, Yugoslavia and Roumania. Records of *G. glis* in a tundra faunal association probably result from stratigraphic contamination and should not be taken seriously.

The species is now found in southern and central Europe (only in the north of Spain; introduced in England) and ranges eastward to the Volga and south to Asia Minor, Palestine and northern Persia. It is the largest of the living glirids with a body length of 130–190 mm. and a tail of 110–150 mm. It resembles a squirrel to some extent because of its bushy tail, but the ears are rounded and tuftless. The fat dormouse is originally an inhabitant of mixed and leafy woods but is also found in parks and gardens. Its diet is varied and includes chestnuts, acorns, young shoots, berries, insects, eggs, etc.; it is mainly nocturnal. Drowsiness is rapidly induced by sinking temperature and the fat dormouse hibernates longer than any other mammal in its geographic range.

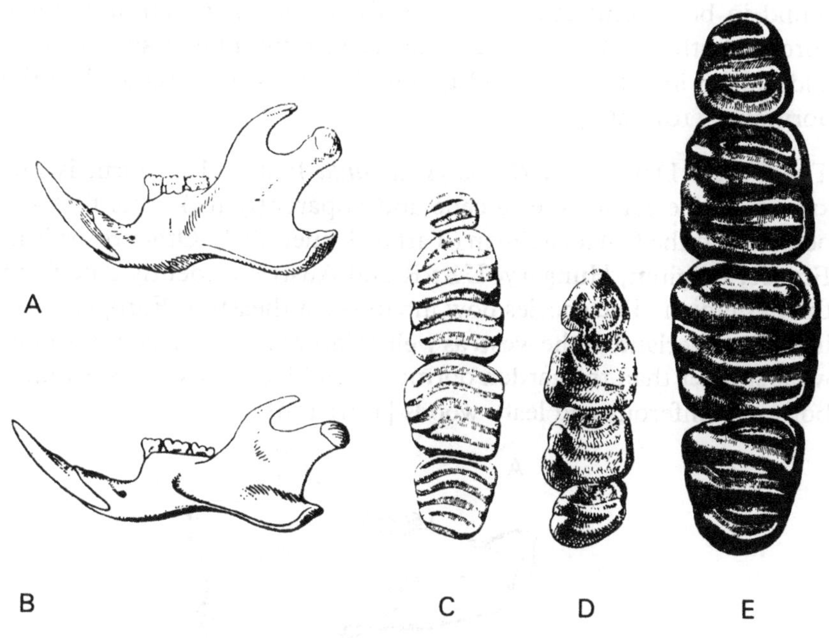

Figure 92. Mandibles of A, Fat Dormouse, *Glis glis*, $1\frac{1}{2}$ times natural size; B, Forest Dormouse, *Dryomys nitedula*, $2\frac{1}{2}$ times natural size. Lower cheek teeth of C, Dormouse, *Muscardinus avellanarius*, ten times enlarged; D, Forest Dormouse, *Dryomys nitedula*, $7\frac{1}{2}$ times enlarged; E, Fat Dormouse, *Glis glis*, ten times enlarged. A, B, D after Gromova; C, E after Miller.

The origin of the fat dormouse is not certain. A smaller species, *G. minor* Kowalski, is present in the Astian (Podlesice; Weze; Csarnóta) and early Villafranchian (Rebielice) but seems to be more progressive than *G. glis* in that its P_4 and M_3 are relatively more reduced, so that it is not likely to be ancestral [157].

The Garden Dormouse, *Eliomys quercinus* Linné. Early (Astian to early Middle Pleistocene) finds of *Eliomys* have not been identified as to species (Sète; Villány; Erpfingen 2). Later records, probably of the living species, come from the D-Holstein (Tarkö, Lunel-Viel), F-Eemian and 4-Würm (various caves in France, Germany, Spain, Italy, Switzerland, Czechoslovakia). The present-day distribution covers almost all of Europe except the British Isles, Fennoscandia and the coasts of the North Sea and the Baltic; it ranges east to the Ural Mountains.

The garden dormouse is 100–170 mm. long, tail 90–125 mm. It is found in both coniferous and leafy woods and is practically omnivorous; both small birds and mice are eagerly hunted and eaten in addition to insects, fruits and grains. It builds nests rather like the dormouse [102; 265].

The Forest Dormouse, *Dryomys nitedula* Pallas. This form is unknown in the earlier Pleistocene and apparently makes its first appearance in the D-Holsteinian (Tarkö; Lunel-Viel). Other records in France, Belgium, Hungary, USSR and Austria evidently date from the F-Eemian. The species now inhabits southeastern Europe, ranging to Switzerland in the west, as well as large areas in Asia. It is somewhat smaller than the garden dormouse and like that species is found both in coniferous and leafy woods [116; 119].

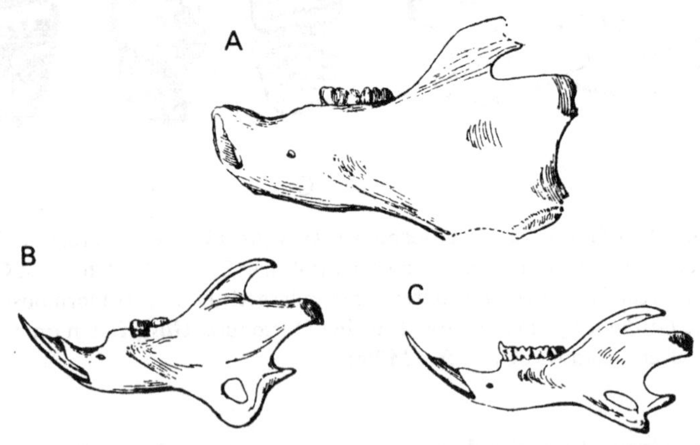

Figure 93. Mandibles of A, Maltese Dormouse, *Leithia melitensis*, Pleistocene, Malta; B, Balearic Dormouse, *Hypnomys morphaeus*, Pleistocene, Mallorca; C, Garden Dormouse, *Eliomys quercinus*, Recent. All 1½ times natural size. After Bate.

The Maltese Dormouse, *Leithia melitensis* Leith Adams. A giant form, probably with a body size like that of the hamster, this is an aberrant species, probably a relict from the Miocene radiation of the dormice. The species is present at several sites in Malta, probably both from the Middle and Late Pleistocene, and is also found in Sicily [21].

The Balearic Dormouse, *Hypnomys mahonensis* Bate and *H. morphaeus* Bate. Two closely related species differing only in size: *H. morphaeus*, on Mallorca, was about as large as *Glis glis*, while *H. mahonensis*, on Menorca, was still larger, intermediate in size between *Glis* and *Leithia*. These species are probably relict forms like *Leithia*. It is an interesting fact that, while large mammals tend to be dwarfed when isolated on oceanic islands, small mammals tend rather to increase in size [21; 192].

Family Cricetidae, Hamsters

We now come to the large group of muroid rodents *sensu stricto*. Many authors combine the mice, voles and hamsters in a single family. However, the number of species and genera in that single family would then become so enormous that it would seem more useful to divide it up.

True hamsters are known since the Early Pliocene; they probably descended from the Oligocene to Pliocene cricetodonts. The hamsters are small to medium-sized rodents with a short tail and three cheek teeth in each jaw half. They have a fairly good Pleistocene record.

The Common Hamster, *Cricetus cricetus* Linné. The earliest members of the genus *Cricetus* appear in the Pliocene of Europe. The modern species is first found in the late Villafranchian or early Middle Pleistocene. At this time no less than three distinct species of the genus *Cricetus* were present in Europe. Originally described as so many subspecies of *C. cricetus* [236], they form three well-separated size distributions; the evidence shows that the three forms coexisted without crossing and hence by definition belong to separate species [173].

The largest of the three Middle Pleistocene forms has been tentatively referred to the modern species as an early, large subspecies (*C. cricetus runtonensis* Hinton). It occurs at Villány, Erpfingen, Episcopia, Gombasek, Hundsheim, Sackdilling, Forest Bed, Koneprusy,

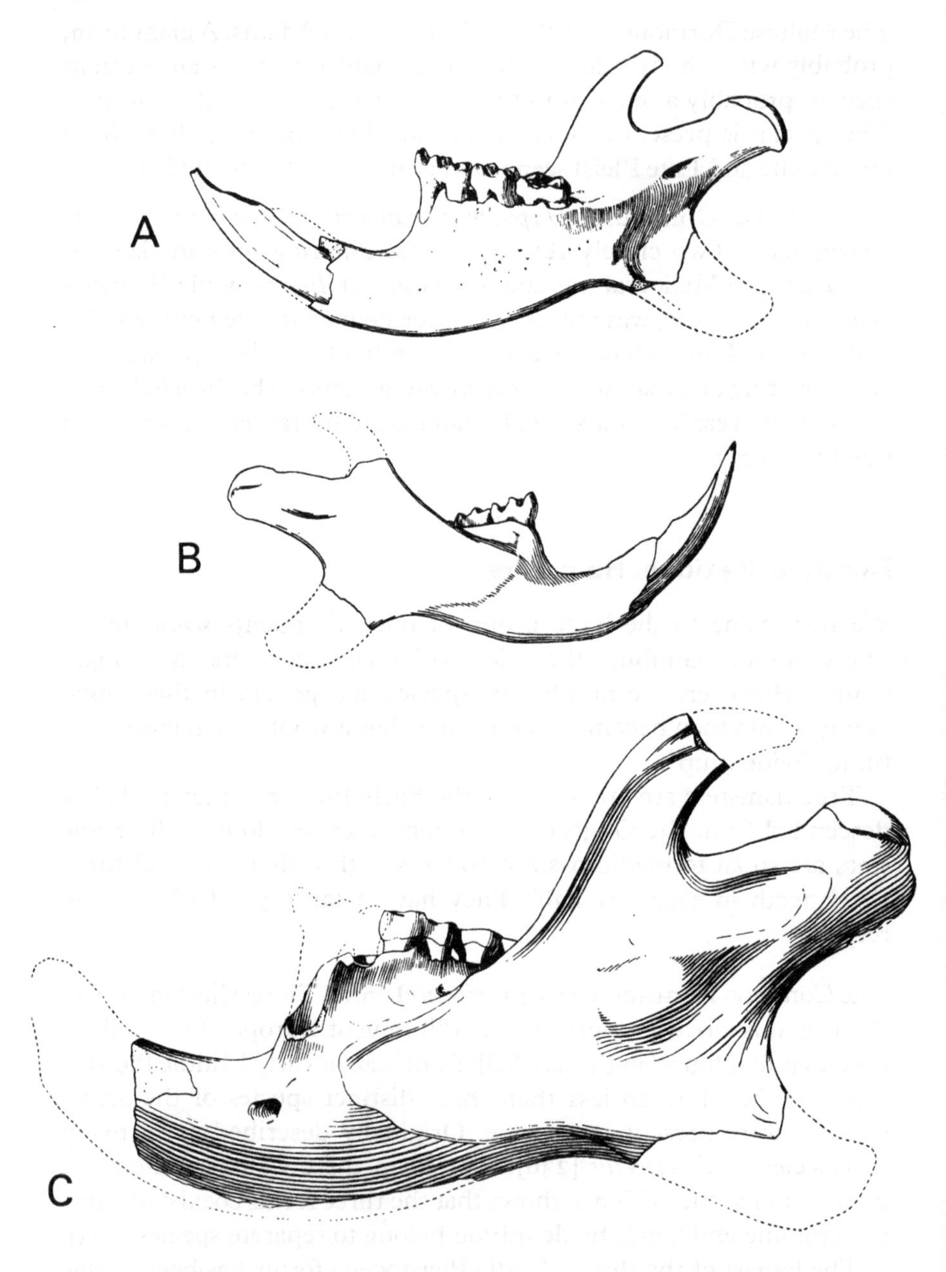

Figure 94. Mandibles of Cricetidae. A, Dwarf Hamster, *Cricetus nanus*, B-Waalian, Episcopia; B, Schaub's Dwarf Hamster, *Cricetulus bursae*, Late Villafranchian, Villány; C, Giant Hamster, *Cricetus cricetus runtonensis*, l-Günz II, Brassó. All 1·5 times natural size. After Schaub.

Tarkö, Heppenloch, etc., and thus ranges well into the D-Holsteinian. Another large subspecies, the giant hamster (*C. cricetus major* Woldrich) is found in the F-Eemian and 4-Würm at numerous localities from Belgium and France to Siberia; it grades into the modern type of *C. cricetus*. The fossil members of the species tend to vary according to Bergmann's rule, the largest forms occurring in the coldest phases.

The living common hamster is found in steppe areas in central and eastern Europe and eastward to Yenisei. Its body length is 215–280 mm., tail 28–60 mm. It is a heavily built, short-legged animal which stocks its underground nest with large winter stores. It is practically omnivorous and eagerly hunts mice, voles, young birds and other small animals to supplement the vegetable diet; stores are carried in the big cheek pouches [97; 99; 159].

The Preglacial Hamster, *Cricetus praeglacialis* Schaub. This species is about as large as the living common hamster from which it differs, however, in some dental characters. It was distinctly smaller than the large race *C. cricetus runtonensis* with which it was contemporary. It occurs from the late Villafranchian or early Middle Pleistocene to the D-Holsteinian (Villány; Episcopia; Sackdilling; Nagyharsányhegy; Kövesvarad; Koneprusy; Chlum; Gaisloch; Heppenloch) [173; 236].

The Dwarf Hamster, *Cricetus nanus* Schaub. This form, the third of the 'preglacial' species of *Cricetus*, has so far only been identified in the early Middle Pleistocene (Episcopia; Nagyharsányhegy; Chlum) and possibly Villafranchian (Villány) [173; 236].

Schaub's Dwarf Hamster, *Cricetulus bursae* Schaub. The species was originally made the type of a genus *Allocricetus*, which however appears to be identical with *Tscherskia*, a subgenus of *Cricetulus*. This is a small form, about the size of *C. migratorius* (see below), and quite common in the fossil record. It appears in the late Villafranchian and early Middle Pleistocene (Villány; Beremend and numerous other localities in Roumania, Hungary, Austria, Czechoslovakia, Poland and Germany). Later records come from the D-Holstein (Tarkö) and F-Eemian (Compaña del Pinar in the Pyrenees; Cotencher in Switzerland; and Lambrecht Cave, Hungary). The only living member of the subgenus *Tscherskia* is *C. triton* de Winton from northeastern Asia [119; 236].

The Migratory Hamster, *Cricetulus migratorius* Pallas. A small form (body 87–117 mm., tail 22–28 mm.), little known in the fossil state; material from the Late Pleistocene in England, Germany, Switzerland, Czechoslovakia, Hungary and USSR has been referred to this species. According to Jánossy [116], this or a related species also made brief incursions in eastern Europe in the early Middle Pleistocene and D-Holsteinian. Its modern distribution extends from Greece through southern Russia to the east. It inhabits park steppes, gardens and cultivated fields; it does not hibernate.

Ehik's Dwarf Hamster, *Rhinocricetus ehiki* Schaub. This species may belong in one of the living genera, but until this is settled it may remain in the genus *Rhinocricetus* created for it by Kretzoi. It appears in the late Villafranchian (Villány; Beremend) and persists to I-Günz II (Episcopia; Nagyharsányhegy; Brassó). It is intermediate in size between *Cricetus nanus* and the small *Cricetulus bursae* and was originally referred to the genus *Allocricetus* [236].

Family Microtidae, Voles

The Microtidae arose in the later Tertiary, perhaps in North America, but they reached their apogee in the Quaternary and dominate the Pleistocene and Recent rodent faunas of Europe in numbers and variety.

A central evolutionary trend in the Microtidae is the gradual development of rootless, very high-crowned cheek teeth that grow throughout life and are well adapted to a highly abrasive diet. The rootlessness was achieved by increasing the age at which roots began to form and this change can be followed in the fossil record. In the Astian and Villafranchian mostly forms with rooted teeth are found, while in later faunas the dominance passed to progressive voles with rootless cheek teeth.

Nehring's Snow Vole, *Dolomys milleri* Nehring. An Astian and Villafranchian species, at present little known (Csarnóta; Beremend). It has been thought that this genus and species survived in a living form, but the identification of the species is evidently incorrect and the identification of the genus doubtful [116].

Martino's Snow Vole, *Dolomys bogdanovi* Martino. This species, now found only in the mountains of Dalmatia and obviously a primitive relict, has been tentatively referred to the genus *Dolomys* (and even to the species *D. milleri*). Others contend that the modern species belongs in a separate genus, *Dinaromys*, provided by Kretzoi. It is a medium sized vole (body 99–148 mm., tail 74–119 mm.) [296].

The Episcopal Vole, *Pliomys episcopalis* Méhely. This is one of the most common voles in the earlier part of the Pleistocene (Villány; Episcopia; Sackdilling; Brassó; Hundsheim; Podumci; Kadzielnia; Koneprusy, etc.). The genus *Pliomys* is sometimes regarded as a subgenus of *Dolomys*; its members differ from that genus by their somewhat smaller size and details in M_1. A second species, *P. coronensis* Méheley (*P. lenki* Heller) appeared in the early Middle Pleistocene and probably survived to the D-Holstein (Episcopia; Sackdilling; Brassó; Kövesvarad; Tarkö; Breitenberg Cave) [150].

The Bank Vole, *Clethrionomys glareolus* Schreber. While the genus *Pliomys* became extinct in the Middle Pleistocene, the allied genus *Clethrionomys*, also with closed roots on the cheek teeth, survives to the present day. The bank vole is known in the early Middle Pleistocene (Forest Bed; Koneprusy; Kövesvarad, etc.) and both Villafranchian and Astian fossils have also been tentatively referred to this species or to the '*glareolus* group' (Csarnóta; Beremend; Villány). Several other species of this genus have been described from the earlier Pleistocene; of these, *C. acrorhiza* Kormos (Erpfingen 2, Brassó, Kövesvarad) is a close relative of the bank vole.

In the Late Pleistocene, the bank vole is found at many sites from England to Poland and USSR. The species now occurs throughout Europe except Ireland, northernmost Scandinavia and the Mediterranean peninsulas; it ranges eastward in Asia to Lake Baikal.

The bank vole is a medium sized vole (body 81–123 mm., tail 36–72 mm.) which prefers open woodlands. It is partially tree-living and eats nuts, fruits, berries, bark, etc., and also various kinds of animal foods.

The other Recent European species of *Clethrionomys*, the Large-toothed Redbacked Vole (*C. rufocanus* Sundevall) and the Northern Redbacked Vole (*C. rutilus* Pallas) are absent in the Pleistocene of Europe; their present distribution is Arctic and Subarctic. An early Middle Pleistocene form, *C. esperi* Heller (Sackdilling) is thought to be related to the redbacked voles. A redbacked vole has been identified at Choukoutien in China (2-Mindel). The Northern redbacked vole is also a member of the modern fauna of North America [97; 98; 118].

The *Mimomys* Voles. The predominant vole genus of the Astian and Villafranchian was *Mimomys*, which then played the same role as *Microtus* today. A very great number of species of this genus have been described but it is most unlikely that all of them, or even a majority, will withstand critical revision. A primitive form, *M. stehlini* Kormos, is found only in the Astian and earliest Villafranchian. Among later species, some forms of different size and comparatively great range may be mentioned. *M. newtoni* Major is a small form from the later Villafranchian and early Middle Pleistocene to the C-Cromer

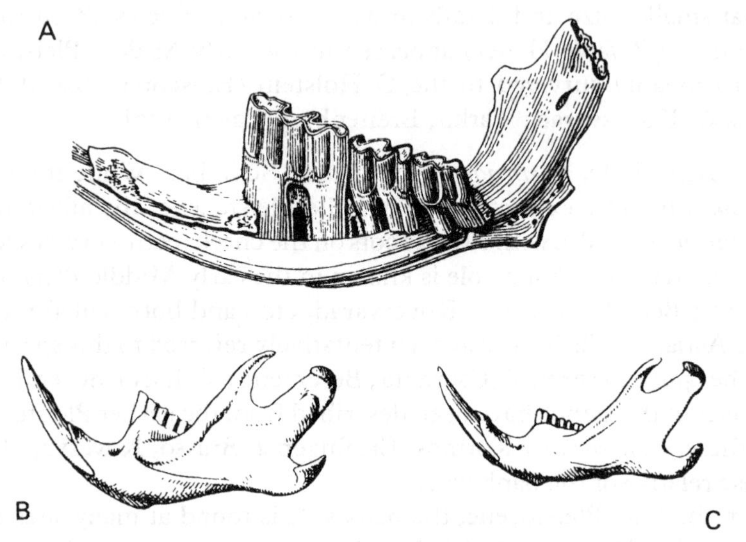

Figure 95. Mandibles of Microtidae. A, *Mimomys pliocaenicus*, nine times enlarged, to show rooted cheek teeth; Villafranchian. B, Water Vole, *Arvicola terrestris*, Recent, 1½ times natural size. C, Field Vole, *Microtus agrestis*, Recent, twice natural size. After Gromova.

(Senèze; Villány; the Crags; Tegelen; Kadzielnia; Kamyk; Kisláng, etc.). *M. pusillus* Méhely, also a small form, ranges from the early Villafranchian to the 1-Günz II (Etouaires; Pardines; Senèze; Villány; Episcopia; Sackdilling, etc.). *M. reidi* Hinton, another small form, is early Villafranchian to C-Cromerian (Rebielice; Villány; the Crags; Verona; Kisláng; Kadzielnia; Kamyk, etc.). *M. intermedius* Newton is a medium sized form ranging from the late Villafranchian to the 2-Mindel (Val d'Arno; Villány; Beremend; Tegelen; Episcopia; Forest Bed; Kövesvarad; Koneprusy; Kadzielnia; Kamyk, etc.). *M. pliocaenicus* Major appears in the Astian (Sète) and persists through the Villafranchian and to the C-Cromerian (Villaroya; Saint-Vallier; Villány; Senèze; Val d'Arno; Beremend; Tegelen; the Crags; Episcopia; Kadzielnia; Kamyk, etc.). *M. rex* Kormos was a large but not very common form (Villány; Nagyharsányhegy). The North American *Cosomys* closely resembles *Mimomys* but may represent a separate Nearctic line of evolution [224].

The genus *Mimomys* was becoming rare in the C-Cromerian and at the same time the first members of *Arvicola* were appearing, apparently descendants of *Mimomys*. A small form of *Mimomys*, *M. cantianus* Hinton, survived into the D-Holsteinian (High Terrace at Ingress Vale, Greenhithe, Kent) [101; 105].

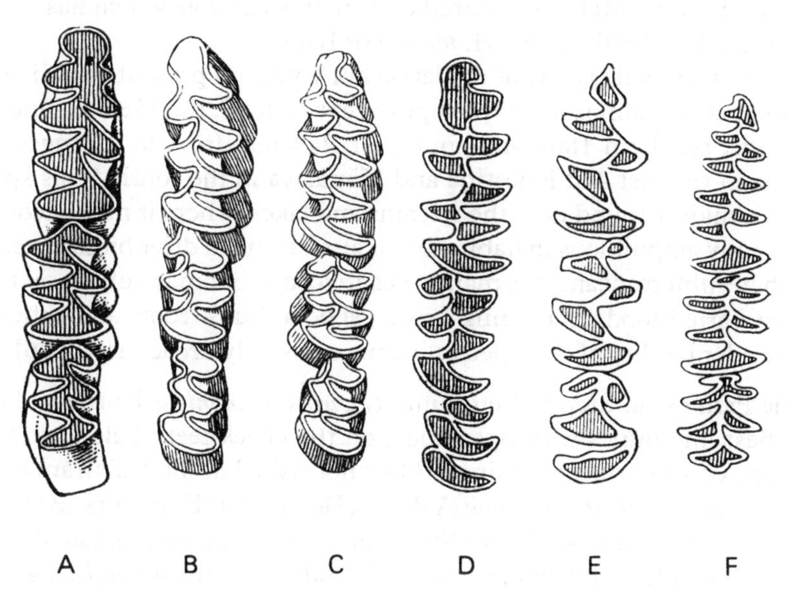

Figure 96. Lower cheek teeth of Microtidae. A, *Mimomys pliocaenicus*, Villafranchian; B, Water Vole, *Arvicola terrestris*, Recent; C, Field Vole, *Microtus agrestis*, Recent; D, Gregarious Vole, *Microtus gregalis*, Recent; E, Norway Lemming, *Lemmus lemmus*, Recent; F, Arctic Lemming, *Dicrostonyx torquatus*, Recent. A, C, nine times enlarged; B, E, F, six times enlarged; D, eleven times enlarged. After Gromova.

The Water Voles, *Arvicola terrestris* Linné and *A. amphibius* Linné (*A. scherman* Shaw). Authorities still debate whether the western form (*A. amphibius*) is specifically distinct from *A. terrestris*. They differ a little in mean size, *A. amphibius* being slightly larger (body 162–220 mm. *versus* 120–200 mm.; tail 98–144 mm. *versus* 56–104 mm.), but are similar in morphology as well as habits and are usually not separable in fossil material.

A number of fossil species have been described; the earliest are *A. greeni* Hinton, which ranges from 1-Günz II to D-Holstein (Hundsheim; Bacton; Mauer; Mosbach 2; Koneprusy; Heppenloch) and *A. mosbachensis* Schmidtgen with approximately the same range (Hundsheim; Bacton; Brassó; Sackdilling; Mauer; Mosbach 2; Koneprusy; Swanscombe). Later Pleistocene material is mostly referred to the living species or to *A. antiquus* Pomel (*A. abbotti* Hinton) which may be identical with the living subspecies *A. terrestris scherman*. Water vole is found in Late Pleistocene deposits within most of its modern range. Malta harboured a form of water vole which has been described under the name *A. melitensis* Bate.

The water vole is one of the largest of voles. Its present-day distribution covers almost all of Europe except Ireland, the Mediterranean islands, southern Italy and southeast Balkans. In Asia it ranges to Lena in the east and Palestine and Himalaya in the south. This species is mainly found near the streams and lakes, where it may become perfectly amphibious in habits, but it also occurs in drier biotopes and is then subterranean. It is mainly vegetivorous but may subsist on the most varied foods, including leaves, shoots, buds, roots and tubers. The adaptability of this species is thus unusually broad [105; 116].

The Pannonian Steppe Lemming, *Lagurus pannonicus* Kormos. The earliest microtines with open roots on the cheek teeth belong to the steppe or sagebrush lemmings. They flourished in the Villafranchian and early Middle Pleistocene (Villány; Beremend; Episcopia; Brassó; Kisláng; Nagyharsányhegy; Stránská Skála). The Pannonian steppe lemming differs only in minor dental details from the living *L. lagurus* and is almost certainly ancestral to the latter; a transitional form has been found at Tarkö (D-Holstein) [117; 148].

The Steppe Lemming, *Lagurus lagurus* Pallas. This species, of which an ancestral form is present in the D-Holsteinian (Tarkö), is common in the Carpathian Basin up to the end of the F-Eemian (Subalyuk, etc.) but temporarily ranged as far west as Devon (Tornewton Cave). Steppe lemming also occurs in the Pleistocene of Asia. The present distribution of this species is in the steppes of southern Russia and western Siberia; another species of *Lagurus* takes over further east and the genus is also represented in North America. The steppe lemmings resemble the true lemmings externally, with their thickset bodies, short tails and small ears; but they are generally regarded as more closely related to voles such as *Mimomys* and *Microtus* than to lemmings. The lack of cement in the folds of the evergrowing cheek teeth distinguishes *Lagurus* from *Microtus* [117; 159].

Kormos's Steppe Lemming, *Allophaiomys pliocaenicus* Kormos. Various species of the extinct genus *Allophaiomys* have been described; it belongs to the *Lagurus* group. All the known specimens (from Italy, Roumania, Yugoslavia, Hungary, Czechoslovakia, Ukraine) date from the Villafranchian and early Middle Pleistocene [116; 139].

The Common Vole, *Microtus arvalis* Pallas. The genus *Microtus* superseded *Mimomys* during the Middle Pleistocene as the leading genus of voles. A great number of extinct species have been described; many of them may prove to be only subspecifically distinct from living populations.

Forms closely related to *M. arvalis* (*M. arvalinus* Hinton, *M. subarvalis* Heller, *M. coronensis* Kormos) are common in the Middle Pleistocene, beginning with 1-Günz II (Brassó; Nagyharsányhegy; Gombasek; Hundsheim; Forest Bed; Kövesvarad; Koneprusy; Mosbach 2; Tarkö; Lunel-Viel; Swanscombe; Heppenloch). The modern species has been found at numerous Late Pleistocene sites from England, France and Spain in the west to eastern Europe; it has since become extinct in the British Isles except for a relict population in the Orkneys and also in Spain and Italy except the north. It now inhabits the remaining part of Europe except the north and ranges east to Mongolia.

A comparatively small species (body 95–120 mm., tail 30–45 mm.), the common vole inhabits open woodland, glades and cultivated land; it feeds on grass, sedge and roots and may store large amounts of food in its subterranean burrows [116].

The Field Vole, *Microtus agrestis* Linné. This species, which is closely related to *M. arvalis*, is difficult to separate from its relative in fossil assemblages. More or less definite identifications of *M. agrestis* have been made at F-Eemian and 4-Würm sites within the same range as *M. arvalis*. Earlier records are mostly given as '*arvalis-agrestis*' group, but a form from the D-Holsteinian of Grays has been described as *M. agrestoides* Hinton.

At present the field vole inhabits Europe north of the Pyrenees,

Alps and Carpathians (but is absent in Ireland and present in north-western Spain). It is of the same size as *M. arvalis*, from which it differs externally by its slightly darker colour and longer fur. It prefers moist, open ground and open woodland and feeds on grass, leaves and seeds, of which it may store large amounts in its burrows [119].

The Mediterranean Vole, *Microtus guentheri* Danford & Alston. In this species are collected various comparatively large voles found in the mountains of Spain, the Balkans, Asia Minor, Syria, Palestine and North Africa [296]. The Mediterranean vole is somewhat larger than *M. arvalis*, which it otherwise resembles. It has been recorded from the Late Pleistocene of Palestine.

The Root Vole or Tundra Vole, *Microtus ratticeps* Keyserling & Blasius (*M. oeconomus* Pallas). A closely related form, *M. ratticepoides* Hinton, occurs in the Middle Pleistocene from 1-Günz II to D-Holstein (Sackdilling; Nagyharsányhegy; Kövesvarad; Forest Bed; Swanscombe). It may be noted that the records from Hungary and southern Germany fall in the cold phases only, while in England the species is found in interglacial deposits. Late Pleistocene finds of *M. ratticeps* (from England to eastern Europe north of the Alps) seem to date from cold phases only (4-Würm).

The modern distribution of the root vole in Europe is of relict type with various isolated populations (the Frisian coast, the south coast of the Baltic, patches on the Danube and in Scandinavia). It ranges eastward through Asia and into North America; a form closely related to *M. ratticepoides* occurs in the Middle Pleistocene of North America (*M. paroperarius* Hibbard). The distribution pattern in Europe is probably typical of earlier interglacials as well.

The root vole is a fairly large form (body 118–148 mm., tail 40–64 mm.). It resembles *M. agrestis* externally but has a characteristically shaped M_1. It keeps to moist, reedy or boggy ground and swims or digs with equal facility. It feeds mainly on herbs and roots and stores food for the winter [115].

The Snow Vole, *Microtus nivalis* Martins. Early members of the snow vole group have been referred to a series of extinct species (*M. nivalinus* Hinton, *M. nivaloides* Major, *M. subnivalis* Pasa). Sites include Nagyharsányhegy, Gombasek, Sackdilling, Kövesvarad, Forest Bed, Koneprusy, etc. Late Pleistocene records are widespread from

England and France in the west to Italy in the south and Czechoslovakia and Hungary in the east. The species is found at present in the mountains of central Europe from the Pyrenees to the Balkans and in the Caucasus, Asia Minor and Palestine. Its habitat is the high slopes above timber line and it has been observed at an altitude of 4700 m. near the summit of Mont Blanc (4810 m., the highest peak in Europe). The snow vole is relatively large (body 117–140 mm., tail 50–75 mm.). It lives on roots, seeds and the like [105; 118].

The Gregarious Vole, *Microtus gregalis* Pallas (*M. anglicus* Hinton). This species is present during cold phases in Europe, beginning with the 2-Mindel (Kövesvarad; Koneprusy; Tarkö). Late Pleistocene records date from the 4-Würm of Germany, England, Switzerland, Czechoslovakia, Hungary and Poland. At present this species, with an Arctic distribution, ranges from the shores of the White Sea in the west to the Bering Straits. A related form inhabits North America. Osteologically the gregarious vole is characterized by its narrow skull [117; 118].

The Gregarious Pine Vole, *Pitymys gregaloides* Hinton. This and various other species of pine voles (*P. arvaloides* Hinton, *P. dehmi* Heller, *P. hintoni* Kretzoi, etc.) have been described from the Middle Pleistocene. They first appear in the B-Waalian (Episcopia) and the genus is then very common from the 1-Günz II on (Hundsheim; Sackdilling; Nagyharsányhegy; Brassó; Gombasek; Kövesvarad; Koneprusy; Marjan Peninsula; Heppenloch; Tarkö, etc.). A record from Kent's Cavern probably comes from the Middle Pleistocene basal horizon in the cavern [105].

The Pine Vole, *Pitymys subterraneus* De Sélys Longchamps. Fossil remains of this species (or of the related *P. multiplex* Fatio, which is morphologically similar but differs in chromosome number) occur in the F-Eemian of central and eastern Europe; in the 4-Würm *Pitymys* is absent in the east, but present in England and Germany. Its modern distribution forms a comparatively narrow zone to the east from France and Belgium through southern Russia to the Caucasus, Asia Minor and Persia. The pine vole is a small form (length 75–106 mm., tail 25–39 mm.) which inhabits high ground and open woodlands; it is both a good swimmer and a burrowing form.

The related modern species, *P. duodecimcostatus* De Sélys Longchamps and *P. savii* De Sélys Longchamps, have a Mediterranean distribution. Late Pleistocene finds from Spain and Italy may represent these species; they are difficult to separate from *P. subterraneus*, which they resemble also as regards biotope and habits.

The genus *Pitymys* has been represented in North America since the Middle Pleistocene [116].

The Norway Lemming, *Lemmus lemmus* Linné. The earliest finds of this species date from 1-Günz II (Sackdilling); it has also been recorded from the 2-Mindel (Koneprusy) and the Middle Pleistocene of Ukraine. In the Late Pleistocene the Norway lemming occurs in cold associations in Ireland, England, Belgium, Germany, Austria, Czechoslovakia and Poland but it did not reach Hungary [106]. Its modern distribution is restricted to the mountains in Fennoscandia and the Kola Peninsula; it may have had a similar relict distribution in earlier interglacials. Other species of *Lemmus* occur in Siberia (also in the Pleistocene) and North America.

The Norway lemming is a fairly large form (body 130–150 mm., tail 15–19 mm.). It is characterized, like other lemmings, by a thickset body, very short tail, high-crowned cheek teeth and powerful zygomatic arches.

This species is found mostly in the mountain birch woods and the zone immediately above timber line, where it feeds on grass, mosses, lichens, leaves of various trees and shrubs, bark, fungi and insects. The lemming is famous for its mass occurrences, which occur with an average periodicity of about 4 years and may lead to a spectacular mass exodus that often results in the death of almost all the migrating animals. In other cases the migration may lead to colonization of a new area, while the original area may be left almost devoid of lemmings, to be repopulated in another lemming year [97; 98].

The Wood Lemming, *Myopus schisticolor* Lilljeborg. This species is almost unknown in the fossil state. In a cave in the Irkutsk region a specimen was discovered in frozen earth with soft parts preserved. The wood lemming has a patchy distribution in Fennoscandia, northern Russia and Siberia. It is a very small form (body 85–95 mm., tail 15–19 mm.) which inhabits forests of taiga type.

The Arctic or Varying Lemming, *Dicrostonyx torquatus* Pallas. The genus *Dicrostonyx* occurs in Europe in the 4-Würm, when it is a common element of the periglacial tundra faunas from Ireland and France in the west (there is also a Scottish record from the Uamh

Cave, Sutherland) to Hungary, Czechoslovakia and Poland in the east. It has been assumed that the Pleistocene Arctic lemming of Europe differed specifically from the living form and in fact two different species used to be separated (*D. guilelmi* Sanford and *D. henseli* Hinton) on some details in the dentition. It has since been shown that these are only individual variants of a single population and that the European form is inseparable from the living species.

The Arctic lemming now inhabits the Arctic coasts of the USSR from the White Sea eastward; the North American form, which used to be regarded as a separate species (*D. groenlandicus* Traill) is now referred to the same species [221]. The habitat of the Arctic lemming is the treeless tundra; it feeds on *Salix*, *Polygonum*, etc. The nest of the American form is frequently lined with musk ox wool, otherwise with grass. Mass occurrences like those of *Lemmus* have also been observed in the Arctic lemming [152; 265].

The Tyrrhenian Vole, *Tyrrhenicola henseli* Forsyth Major. This is also a member of the endemic Tyrrhenian fauna, whose remains have been recovered in great numbers from cave breccias in Corsica and Sardinia, ranging in date from the Middle Pleistocene to the Postglacial. Originally described as a member of *Arvicola*, it is now regarded as more closely related to *Microtus* and *Pitymys*, of which latter genus *Tyrrhenicola* may be a subgenus. There seems to have been a tendency to size increase in this species during the later half of the Pleistocene; Postglacial forms average about ten per cent larger than those of the Middle Pleistocene [193; 289].

Family Muridae, Mice and Rats

The murids are less common in the Pleistocene than the microtids, but seem to have increased greatly in numbers as late as in the Postglacial. It is now an intensely vigorous family with some 90 extant genera but only a few extinct ones. The murids are characterized externally by their pointed nose, large ears and long, scale-covered tail. The cheek teeth are three in each jaw half but differ completely from those of the voles in being comparatively low crowned with closed roots and rounded cusps. The family arose in the Old World, probably in the Early Pliocene.

Schaub's Field Mouse, *Parapodemus coronensis* Schaub. The genus *Parapodemus* arose in the Early or Middle Pliocene; early members of the genus are close to the ancestry of *Apodemus*. The genus *Parapodemus* flourished with several species in the Astian, among them the very small *P. coronensis* (Podlesice); later records come from Schernfeld and Brassó, indicating that the species survived to 1-Günz II [238].

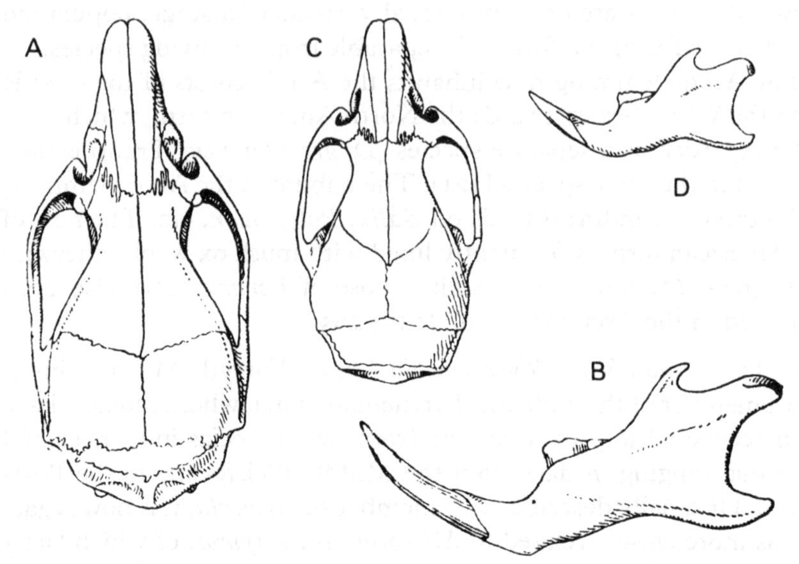

Figure 97. Skull and mandible of A, B, Striped Field Mouse, *Apodemus agrarius*, Recent; C, D, House Mouse, *Mus musculus*, Recent. $2\frac{1}{4}$ times enlarged. After Gromova.

The Common Field Mouse, *Apodemus sylvaticus* Linné. The genus *Apodemus*, which probably arose from Pliocene members of *Parapodemus*, appears in the Astian. The earliest finds of *A. sylvaticus* seem to be late Astian (Sète; Csarnóta); later records come from the Villafranchian and early Middle Pleistocene (Hajnacka 2; Villány; Schernfeld; Episcopia; Sackdilling; Verona; Forest Bed; Koneprusy, etc.), D-Holsteinian (Tarkö), 3-Riss (Châtillon-Saint-Jean) and Late Pleistocene (many localities from Ireland and England to Gibraltar, Italy, Hungary, Czechoslovakia and Poland). The field mouse is absent in the cold-continental phases of 4-Würm in the Carpathian Basin.

The species is now distributed in the British Isles, Iceland and continental Europe, but is absent in the north. It also has a wide Asiatic range to Korea, Japan and Formosa. It differs from the closely related *A. flavicollis* in being slightly smaller (body 77–110 mm., tail

69–115 mm.). The common field mouse inhabits open ground, dunes and shrubs along the edges of woods; it eats mainly seeds and the like, but also insects [97].

The Yellow-necked Field Mouse, *Apodemus flavicollis* Melchior. This species is larger than the common field mouse (body 88–135 mm., tail 92–138 mm.). Large field mice of this type have been described from the Villafranchian and early Middle Pleistocene (Villány; Schernfeld; Beremend; Kadzielnia) under the name *A. alsomyoides* Schaub. The modern species occurs in the Late Pleistocene of England and Germany. It is now found in central, northern and eastern Europe except the Arctic, but is mostly lacking in the west (Spain to Holland, Britain) except for relict pockets that have been variously interpreted. In Asia it has a wide distribution to China in the east and Kashmir in the south.

The habits of this species resemble those of *A. sylvaticus* but it is also found in forests, where it leads a semi-arboreal life. It lives on acorns, nuts, cones, etc., as well as insects, grubs and spiders [56; 238].

The Broad-toothed Field Mouse, *Apodemus mystacinus* Danford & Alston. This species, which now inhabits a limited range in Yugoslavia, Greece, Asia Minor and Palestine, is very rare in the fossil state but has been recorded from the early Middle Pleistocene of Podumci. It is a comparatively large form (body 128–150 mm., tail 115–146 mm.) which lives in woods and shrubs on rocky ground [151].

The Striped Field Mouse, *Apodemus agrarius* Pallas. Though this species is very rare in the fossil state, there are a few records from the Late Pleistocene in Germany and Czechoslovakia and recently the species was identified in the early D-Holstein at Tarkö [117]. This species is about as large as *A. flavicollis* but has a shorter tail (body 97–122 mm., tail 66–88 mm.) and short ears. It inhabits birch woods, forest skirts, steppes and tilled land; it is now widely distributed in Asia and eastern Europe and also inhabits northern Germany as well as adjoining parts of Jutland and Holland.

The Harvest Mouse, *Micromys minutus*. This is the smallest of the murids and extremely rare in the fossil state. It has been identified in the Breitenberg Cave (D-Holsteinian?). Its modern distribution extends from France and Great Britain in the west to Korea, China and Japan in the east. It inhabits grassland and shrub country, where it uses its prehensile tail in climbing. It feeds on seeds, grains, insects, etc. [45].

Hensel's Field Mouse, *Rhagamys orthodon* Hensel. This is an extinct form related to the field mice rather than the rats, but intermediate between the two groups in size; it differs from other murids in the development of high-crowned cheek teeth. Remains of this species have been found in caves on Corsica and Sardinia, where it seems to have been a common faunal element from the Middle Pleistocene well into the Postglacial. This species probably arose from *Apodemus*-like ancestors but evolved in a quite different direction during its isolation in the ancient Tyrrhenis [238].

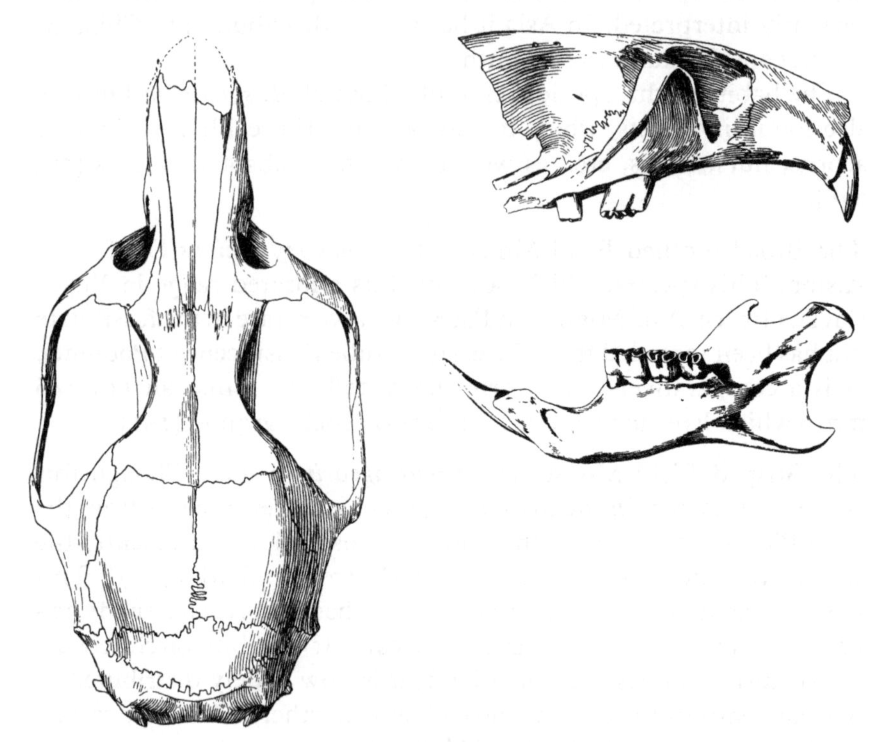

Figure 98. Skull and mandible of Hensel's Field Mouse, *Rhagamys orthodon*, Pleistocene, Sardinia; twice natural size. After Schaub.

The House Mouse, *Mus musculus* Linné. Though extremely rare, remains of this species have now been found in Middle Pleistocene deposits probably of D-Holsteinian age (Tarkö; Breitenberg; travertines at Budapest). Other finds may date from the F-Eemian (Kirkdale Cave; the Binagady asphalt deposits), but most of the remains are evidently subfossil and date from the Postglacial. The wild form occurs in eastern Europe and Asia from the Volga to the Yellow Sea; the commensal form is world wide through introduction by man. In the wild state the house mouse lives mainly on seeds and grains; the commensal form is omnivorous.

Early members of the genus *Mus* have been found in the Pleistocene of Asia (Choukoutien). Several species of the genus now live in Asia [117].

The Black Rat, *Rattus rattus* Linné, and the Brown Rat, *R. norvegicus* Berkenhout. Both species have probably entered Europe in Postglacial times as human commensals. The genus *Rattus* is known in the Middle Pleistocene of Choukoutien and at Late Pleistocene sites in Uzbekistan. Wild forms of both species exist in Asia; *R. rattus* comes from southeastern Asia, *R. norvegicus* from China and eastern Siberia.

Chapter 15

Order Lagomorpha

THE Lagomorpha, or 'duplicidentate' rodents, are characterized by the presence of two pairs of gnawing incisors in the upper jaw. They first appeared in the Paleocene in Asia and thus seem to be of ultimate Old World origin, whereas the Rodentia may have arisen in the New World. Successful on a more modest scale, the lagomorphs never attained anything like the variety and numbers of the true rodents.

All the Lagomorpha except the earliest primitive forms may be referred to one or the other of the two extant families, the Ochotonidae and Leporidae.

Family Ochotonidae, Pikas

The pika family has been in existence since the Oligocene. The main part of its history is confined to the Old World, mainly Eurasia, though stray forms have invaded Africa. In the Pleistocene, pikas entered North America. The pikas of the present day are burrowing mountain and steppe forms, short-legged and short-eared in contrast with the Leporidae. The only living genus is *Ochotona*, which dates back to the Pliocene, but the genus *Prolagus* survived well into the Postglacial and probably in historical times.

The Sardinian Pika, *Prolagus sardus* Wagner. The genus *Prolagus* arose in the Miocene. A species in the early Pliocene, *P. elsanus* Forsyth Major (from Tuscany), is probably ancestral to *P. sardus*. The Sardinian pika has been found at numerous localities in Sardinia and Corsica, ranging stratigraphically from the earlier Middle Pleistocene to the Postglacial. The most recent remains are from historical times and it is possible that the species survived as late as the eighteenth century, since F. Cetti in 1774 mentions the presence 'des rats géants dont les terriers sont si abondants, qu'on croirait la surface du sol récemment remuée par des porcs' on the small island of Tavolara off the Sardinian coast.

A gradual size increase seems to have taken place in the history of *Prolagus*, for the Miocene forms were much smaller than *P. sardus*, which probably attained a length of between 20 and 25 cm. This change may also be traced within the time span of the Sardinian pika, for the Postglacial forms average some 15–20 per cent larger than the Middle Pleistocene ones. In the same interval the microtid *Tyrrhenicola henseli* also increased in size, although at a more moderate rate, while in contrast the murid *Rhagamys orthodon* tended to decrease in size [289].

The Steppe Pika or Mouse-Hare, *Ochotona pusilla* Pallas. The genus *Ochotona* is not uncommon in the time range from the Astian to the Middle Pleistocene in Europe (Podlesice; Weze; Schernfeld; Kadzielnia; Kamyk; Kövesvarad; Mosbach 2; Koneprusy; Tarkö, etc.) but identification with *O. pusilla* is not certain and it is possible that an additional species, perhaps related to the Northern pika (*O. hyperborea* Pallas), may also be represented. Later Pleistocene finds (F-Eemian and 4-Würm) seem to belong only to steppe pika. The Late Pleistocene expansion of this species carried it westward to England and southward to Switzerland, Austria and Roumania. Crimean Pleistocene material also belongs to this species.

The steppe pika now ranges from the Volga to western Siberia. It is a very small form, less than 15 cm. long (the northern pika is only slightly larger) which inhabits brushy valleys. It is nocturnal and lives on grass and herbs.

Several other species of *Ochotona* are now found in Asia and North America; the migration across the Bering Bridge cannot yet be dated [115].

Family Leporidae, Rabbits and Hares

The Leporidae differ from the Ochotonidae in the development of long hind legs and long ears; they have three molars in the upper jaw, the pikas only two. The primitive Palaeolaginae, of which a few relicts still survive, appeared in the Eocene; *Hypolagus*, a member of this subfamily, survived in the earlier part of the Pleistocene of Europe. (Some authors recognize an additional subfamily, the Archaeolaginae, which would include *Hypolagus*.) The modernized forms, the Leporinae, arose in the Pliocene.

The Beremend 'rabbit', *Hypolagus brachygnathus* Kormos (*Pliolagus beremendensis* Kormos; *P. tothi* Kretzoi). The genus *Hypolagus* arose in North America in the Miocene and invaded the Old World in Pliocene times; it survived in North America well into the Middle Pleistocene.

The Beremend rabbit ranges from the Astian through the Villafranchian and Middle Pleistocene up to the C-Cromer (Ivanovce 1; Podlesice; Weze; Casrnóta; Rebielice; Schernfeld; Beremend; Kadzielnia; Villány; Tegelen; Episcopia; Kövesvarad; Kamyk; Marjan Peninsula; Chlum, etc.); it is absent in the southwest, where *Oryctolagus lacosti* acted as its vicar. Several different species and even genera have been described, but it would now seem that they are nothing but artificially separated variants of a single population.

The Beremend rabbit was intermediate in size between the modern European rabbit and the brown hare, being not unlike the latter in general proportions, but its fore foot was adapted for digging and scraping as in the rabbit. There was a tendency to gradual size increase in this species from the Astian to the Middle Pleistocene [273].

The Arno Rabbit, *Oryctolagus lacosti* Pomel (*Lepus etruscus* Bosco, *L. valdarnensis* Weithofer). This species ranges from the beginning of the Villafranchian almost to the end of this stage (Etouaires; Pardines; Saint-Vallier; Senèze; Val d'Arno). It was about as large as the brown hare but belongs to the rabbit genus. The lower incisors are of rabbit type, i.e. they are less curved and relatively broader in cross section than in *Lepus*. The length relationship between fore- and hind limb also agrees with *Oryctolagus* (the rabbits actually have longer arms in relation to the leg, when the fore- and hind foot are not considered; the great length of the hind leg in the hare is due to the elongation of the foot, especially its phalanges) [133].

The Rabbit, *Oryctolagus cuniculus* Linné. Some records of the modern rabbit evidently date back as far as the D-Holstein (e.g. Lunel-Viel in southern France) but more definite identifications of this species refer to the F-Eemian deposits (e.g. the Pinar Cave, Granada). In general the identification of fossil Leporidae is tricky even for a specialist on the group and although the list of localities containing Late Pleistocene rabbit is very long (with a geographic range from Ireland to Siberia, Hungary, Italy, Spain and Portugal) the possibility of confusion with other forms cannot be excluded, especially as regards the older literature. Several records from North Africa would seem to give reliable evidence of the presence of this species in the Pleistocene.

The rabbit probably had its primary modern range in southwestern Europe and North Africa but has since advanced to the north and east, partly with the help of human agency. It now ranges to the British Isles, southern Sweden, Poland, the Carpathian Basin and Italy.

The body length of the rabbit is 60–71 cm. It prefers dry, sandy ground, in which it digs a complicated system of burrows. It feeds on a variety of vegetable foods, including grass, roots, leaves, bark, nuts, acorns, etc. The various races of domestic rabbit are all descended from *O. cuniculus* [312].

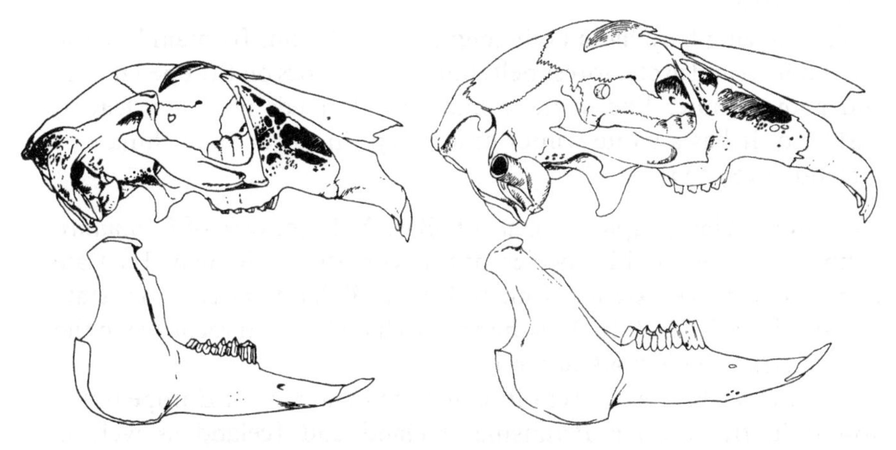

Figure 99. Skull and mandible of Varying Hare, *Lepus timidus* (left) and Brown Hare, *L. europaeus* (right). Two-thirds natural size. After Miller.

The Varying Hare, *Lepus timidus* Linné. Members of the genus *Lepus* appear in the early Middle Pleistocene (Hundsheim; Sackdilling; Gombasek; Koneprusy; Mosbach 2, etc.). A form at Beremend, Villány and Nagyharsányhegy has been named *L. terraerubrae* Kretzoi. The relationships of these early members of *Lepus* are not yet well understood; they are in some respects intermediate between modern *L. timidus* and *L. europaeus*, in others closer to one or the other of these species.

In the F-Eemian, material definitely referable to the varying hare is known (Lambrecht Cave, Cotencher, etc.). The species is quite common in the 4-Würm. However, the difficulties in separating the two species of hare on the basis of fragmentary fossil material are often insuperable and some published records, especially in the older literature, are doubtless incorrect. Reliable records of *L. timidus* include those already mentioned and, for instance, numerous Magdalenian stations from the Pyrenees in the west to Hungary in the east.

The varying hare retreated northward in the Postglacial but a local population maintained itself in the Alpine area, so that there is now a relict subspecies, *L. timidus varronis* Miller in the mountains of central Europe. The main population extends in a northern belt including Iceland, Ireland, northern Great Britain, Fennoscandia and Europe east of the Baltic Sea to Kamtchatka, Korea and Japan. The related North American snowshoe hare (*L. americanus* Erxleben) dates back to the 3-Riss.

The varying hare has a body length of 52–60 cm. Its main biotope is the boreal and arctic forest belt, but it also occurs on the open moors and above timber line in the mountains. It feeds on grass, leaves, bark, etc. It has a white winter coat except in the oceanic climate of Ireland [132; 133].

The Brown Hare, *Lepus europaeus* Pallas. As in the case of *L. timidus*, definite records of this species begin with the F-Eemian. Its Late Pleistocene range extends from Ireland to Poland, Yugoslavia, Italy and southern USSR. In Asia the species has also been recorded in the Late Pleistocene (Choukoutien Upper Cave).

The brown hare at present inhabits the main part of Europe but is absent in the Iberian Peninsula, Ireland and Iceland as well as northern Fennoscandia; it is now advancing east of the Ural Mountains and ranges southward to Palestine, Syria and Persia. In Fennoscandia the species is gradually ousting the varying hare except in the deep forest, where the latter is superior. Introduction by human agency has contributed to the success of the brown hare.

The body length of *L. europaeus* is 48–68 cm. It prefers open fields and its expansion has been greatly accelerated by its ecological success in areas under cultivation [132; 133].

The Cape or Tolai hare, *Lepus capensis* Linné. (*L. tolai* Pallas). If Ellerman [73] is right in uniting the Asiatic Tolai hare with the African Cape hare in a single species, the modern geographic range of that species is tremendous; it includes great areas of Africa down to the Cape; Spain, Portugal and Sardinia in Europe; and southwestern, central and eastern Asia to China and Mongolia.

This is a relatively small species with a body length of 40–54 cm. It inhabits fields and mountain slopes but is also found in woodlands. Fossil finds in Europe have occasionally been referred to this species, e.g. Fuchsloch in Germany, the Lambrecht Cave in Hungary and Grotta Reale in Italy, all of these Late Pleistocene in age. It is also recorded from Asia, e.g. the Mousterian fauna of Teschik-Tasch (Uzbekistan).

Even more uncertain are records from the Middle Pleistocene of small forms that may belong in the ancestry of this species (Verona; Podumci; Monrupino) [119].

Part Three
The Changing Fauna

LOOKING back at the survey of the Ice Age mammals of Europe, it must of course be remembered that it is still far from complete. Many forms are still unknown to us and others have left but few traces of their presence. Species like *Hesperoceras merlae* or *Syncerus iselini* must have had a long evolutionary history and may have lived in Europe for 100,000 years or more although we have only one or a few skeletal fragments of each, perhaps representing only a point in time.

Still, we do know much of the life histories of a multitude of species. We have seen them in perpetual change, migrating back and forth as the climatic belts moved over the continent; we have seen them evolve and become extinct, immigrate from distant areas or give rise to descendants that populated other continents. A survey of this kind becomes an impressive demonstration of evolution, not as a theory but as a fact of record [194]. Zeuner wrote as follows: 'I am convinced that a thorough investigation of the Pleistocene fauna will, in the long run, provide most valuable information concerning the evolution of new species' [318].

Individual instances of evolution and migration have been described in the preceding part; in this part some general aspects of evolution and migration will be touched upon.

Chapter 16

The Species Problem in the Quaternary

IT has long been a tradition to use the genus as the basic taxonomic unit in palaeomammalogy, especially in the Tertiary. Stratigraphic zones in the Tertiary generally represent lapses of time long enough to make the genus usable in correlation, although of course the use of species will give greater precision and an increasing emphasis on that taxonomic level is highly desirable. Actually, many Tertiary genera are monotypic (that is to say they contain only one valid species each), so that through over-splitting on the genus level the genus is doing the work of the species.

The Quaternary divisions of time are much shorter than the average in the Tertiary and far shorter than the length of life of most mammalian genera. This makes the species, and even in some cases the subspecies, the basic unit for correlation in the Quaternary and we have to concentrate on this category to a much greater extent than tradition has prescribed for the study of fossil mammals in general.

There is a great literature on the topic of how to define a species and the role of the species in paleontology [114; 251; 274]. In the past biologists used what is now called a typological species concept, taking little heed of the variation in morphological characters found in any natural population. This primitive concept has now been long defunct and the species is now regarded as a population or interbreeding group of populations, prohibited from gene exchange with other populations by the existence of isolating mechanisms.

Although this definition is genetical in principle, practical work on most living and all fossil mammals is based on morphological and ecological criteria rather than on direct observation of breeding behaviour, so that this species concept may also be used in palaeontology. The main difference between neontology and palaeontology is the introduction of the time dimension and the fact that species change in time. The species of neozoology are 'transient species', practically frozen into immobility by their restriction to a single plane in time. In palaeontology, too, we may find transient species: our entire material of a species then comes from a single horizon. Sometimes we may find a sequence of transient species from successive levels, each ancestral to the next but sufficiently separated from it in time to have evolved into a new species. This type of record is very common in the Tertiary and older deposits.

In the Pleistocene on the other hand our information on evolving lineages is much more extensive and may approach the ideal condition when we have an unbroken sequence of temporal populations showing the change from species to species; these are termed 'successional species'. This raises the acute practical problem of how to draw the boundary between the species.

The problem might be overcome deftly by uniting the whole lineage into a single species and this is sometimes done with some justification when the ancestral and descendant forms are not too unlike each other. Yet there are many cases when the ancestral and descendant forms differ from each other fully as much as two related present-day species and to unite them in a single species would falsify the record of evolution and reduce their usefulness in correlation.

A lineage evolving steadily in one direction as indicated in figure 100A may be divided on morphological criteria (line a–a) or temporally (b–b). The former is a typological method and it results in the recognition of two distinct species during the entire period of transition, although in fact there is only a single population at any given time. The latter method must be preferred, although it makes the record show the new species to arise through a sudden 'jump' or saltation rather than a gradual transition. We must still use a morphological criterion but we now classify the population as a whole, depending on whether its mode (c–c) has crossed the morphological boundary (a–a) or not.

Practical considerations may help in fixing the morphological boundary for the population mode. If the record should show an episodic increase of the rate of change, this might well be selected (figure 100B). On the other hand episodes with reduction or reversal of change should be avoided (figure 100C). Even with the best selection of the boundary for each transition there will be a period when single specimens from an unknown horizon cannot be certainly classified as to species. A statistically respectable sample, however, can always be classified and used in correlation.

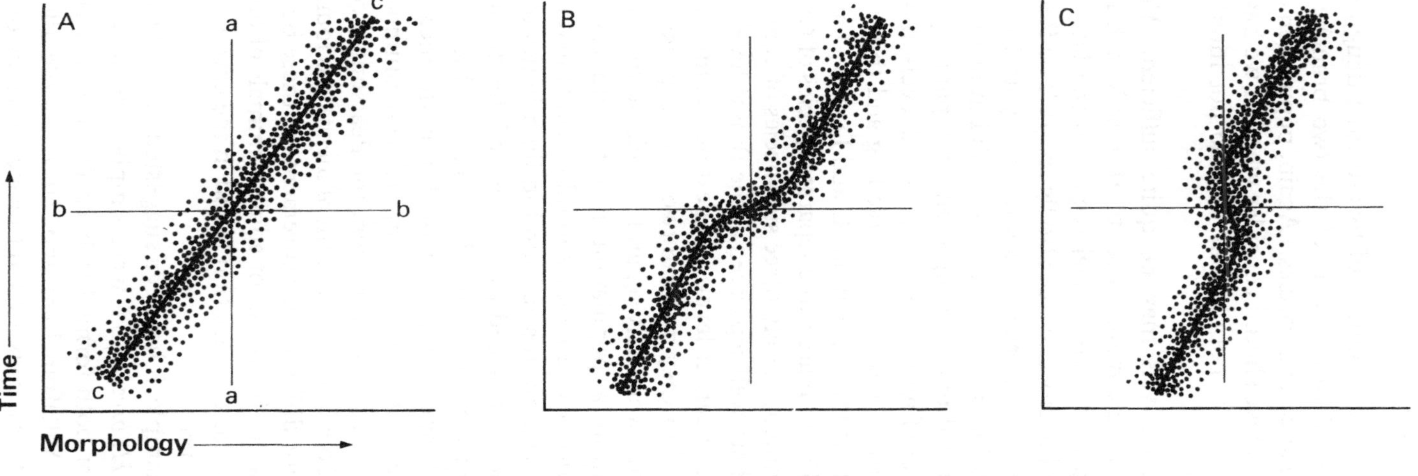

Figure 100. A, diagrammatic representation of a continuously evolving population, changing evenly in mean character as it moves through time. The line c–c is generated by the population mean; dots represent individuals. The axis a–a represents the boundary between successive morphotypes; b–b the boundary between successive species. B, analogous instance with an episode of rapid evolution, suitable for species boundary. C, instance with temporary reversal of evolution, unsuitable for species boundary.

Another problem is the evolution of two distinct daughter species out of a single ancestral form. (One of the two branches may, of course, remain in the ancestral species.) Again, population analysis is necessary. At each level we should consider whether we have a single variable population or two populations that do not interbreed any more.

Simple univariate analysis may be quite sufficient. Figure 101 shows distributions for the mandibular tooth row length in various cricetid (hamster) species with special regard to the late Villafranchian – early Middle Pleistocene (see also figure 9). The Recent sample of common hamster (*C. cricetus*) shows the typical pattern of a homogeneous population in a cricetid species and indicates that each of the peaks in the fossil material represents a separate population. Populations A and B are found together at several levels, indisputable proof that they coexisted without interbreeding and thus were good species; the same is true for populations C and D, and D and E. It is thus evident that C, D and E are three separate species of *Cricetus* and not, as originally assumed, merely three subspecies of common hamster. Species A and B are well separated at the Brassó level but merge more closely together at earlier levels, suggesting a common ancestry at a still earlier stage; but they do form two distinct peaks at Villány and Beremend and thus were fully separate species at that stage.

The variation within a population is often under-estimated; modern texts in quantitative zoology should be consulted [252]. For instance, the great variation in size found in limb bones of the cave lion (*Felis leo*) has been regarded as evidence of the division into a larger and a smaller race of this form; the extreme values in a sample of 14 third metatarsals and 10 radii diverged from the mean by some 8 or 9 per cent. However, the data form in each case unimodal distributions and the best measure of variation, the Pearsonian coefficient of variation (standard deviation in per cent of mean) was found to vary between 5·2 and 8·2 for different measurements of these bones (table 2). Generally such values for bones of adult mammals of one species range from 4 to 9, so that the variation found in the cave lion is quite normal for a wild species.

To take another example, it has been suggested that the two species *Hyaena perrieri* and *H. brevirostris* were simply variants of one species [302]. Material of both forms has been found in a series of deposits from the late Villafranchian and early Middle Pleistocene, showing that the two forms coexisted in Europe for a long time. If, for

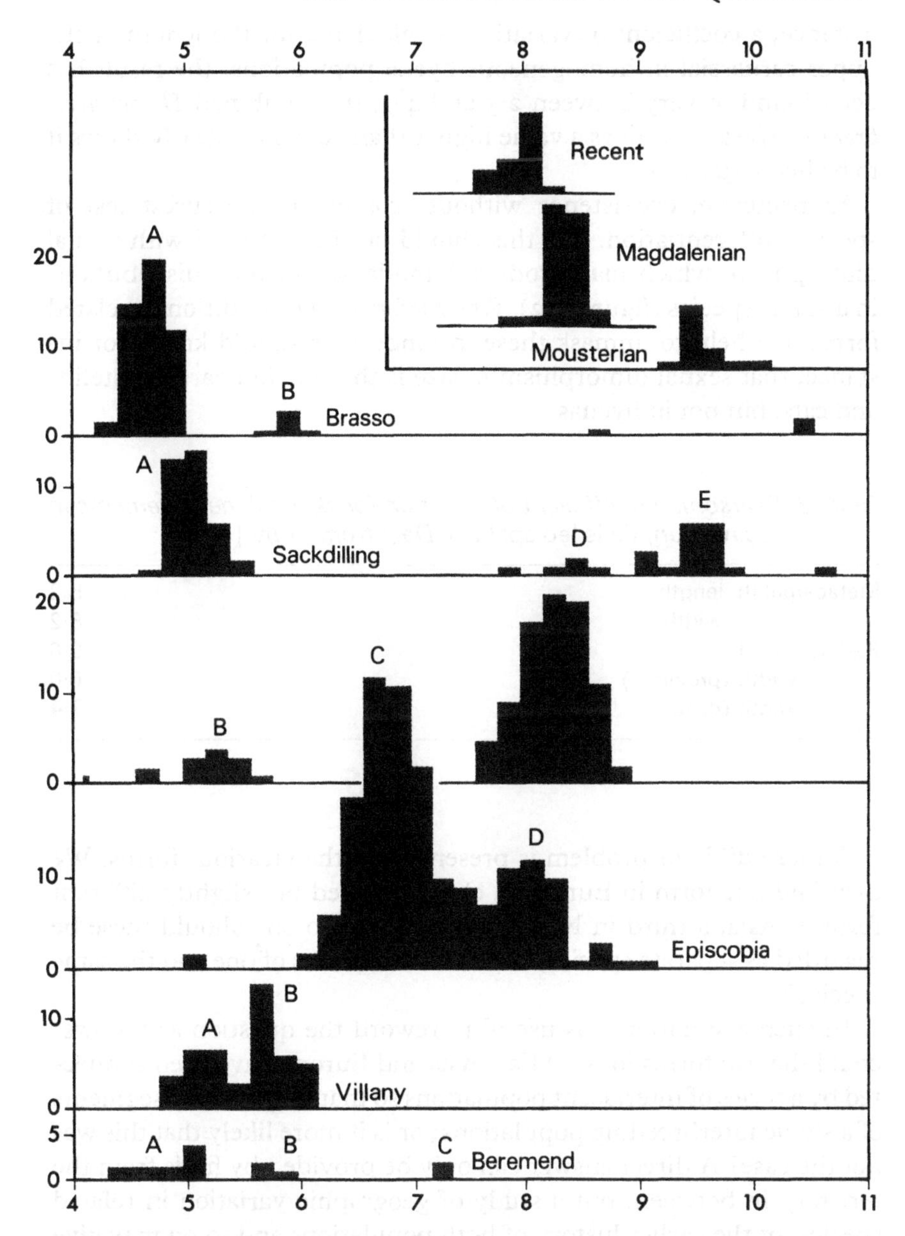

Figure 101. Distributions of the length of the mandibular tooth row in temporal samples of five cricetid species, arranged in stratigraphic sequence from late Villafranchian (Beremend) to 1-Günz II (Brassó). Inserted, Late Pleistocene and Recent samples of *C. cricetus*. Identification of distribution modes: A, *Cricetulus bursae*; B, *Rhinocricetus ehiki*; C, *Cricetus nanus*; D, *C. praeglacialis*; E, *C. cricetus*.

instance, a coefficient of variation is calculated for the length of the upper carnassial in homogeneous hyena populations, the result has been found to vary between 2·5 and 4·9; the combined *H. perrieri-brevirostris* sample gives a value higher than 10 which clearly shows it to be heterogeneous.

In principle, coexistence without crossing is the surest test of specific differentiation. But this should not be confused with sexual dimorphism, which may produce bimodal or separate distributions in a single species (figure 102). Knowledge of the condition in related forms will help to unmask these instances; we should know, for instance, that sexual dimorphism in size is the rule in bears, mustelids and cats, but not in hyenas.

Table 2. Pearsonian coefficient of variation for skeletal measurements in cave lion, Felis leo spelaea. *Data from Koby [128].*

Metacarpal III, length	5·2
width	8·2
Radius, length	5·6
width (proximal)	6·0
width (distal)	7·4

A more difficult problem is presented by the vicarious forms. We may find one form in Europe, a closely related but slightly different form in Asia, a third in North America and so on. Should these be regarded as separate species, or just as subspecies of one and the same species?

In such a situation it is useful to reword the question as follows: could the two forms in (say) East Asia and Europe have been connected by a series of interjacent populations so that they formed segments of a single interbreeding populations; or is it more likely that this was not the case? A direct answer can only be provided by finds from the territory in between, but a study of geographic variation in related species, of the earlier history of both populations and so on may give useful hints. It is also possible to supplement the orthodox morphological investigation by the use of various indices of similarity or differentiation [258] (two types of indices specially devised for fossil mammals are described in Kurtén [170; 181]).

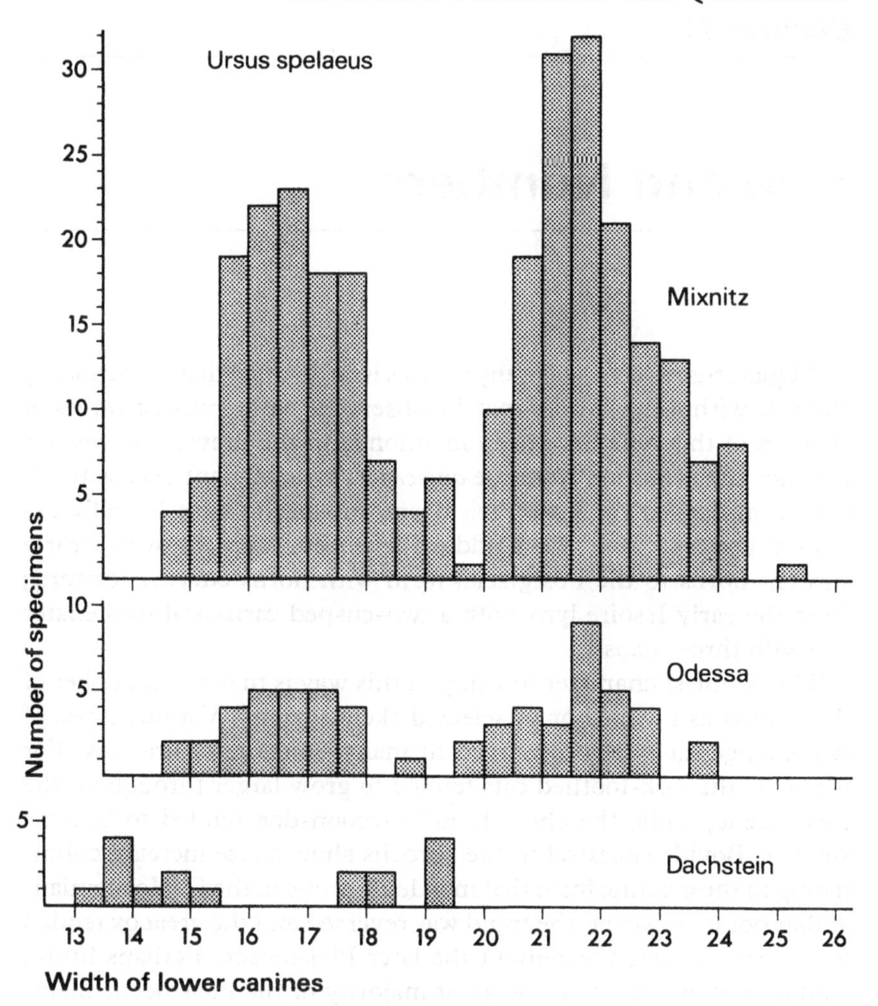

Figure 102. Distributions for the width of the lower canine teeth in local samples of Cave Bear, *Ursus spelaeus*, showing division into two groups representing males and females. In the small form of cave bear from Dachstein a similar sex dimorphism is found, but these specimens average smaller than the normal males and females from Mixnitz and Odessa. After Kurtén.

Chapter 17

Size and Numbers

THE Quaternary record of many species is so detailed that evolutionary changes within the species may be observed. Numerous examples of changes of this type have been mentioned in the previous pages, for instance the evolution from the *antecedens* type of giant deer with recurved antlers to the Late Pleistocene subspecies with laterally extended antlers; from the Middle Pleistocene aurochs with nearly straight horns to the Postglacial form with horns curving forward; from the early Issoire lynx with a two-cusped carnassial to the later one with three cusps.

The simplest character to study in this way is that of size, either of the animal as a whole or of selected skeletal parts. Various types of size change have been observed in many Quaternary lineages. For instance, the dirk-toothed cats tended to grow larger throughout the Pleistocene, while the cheetah and raccoon-dog tended to become smaller. Bovids ancestral to the aurochs show a size increase culminating in the gigantic form that invaded Europe in the D-Holsteinian; at that point, however, the trend was reversed and the great ox tended to decrease in size throughout the later Pleistocene. Perhaps future studies will indicate that the great majority of the Pleistocene mammals tended to change in size in some manner.

An excellent example of size change in the animal as a whole against an absolute time scale is given by Heintz & Garutt [94]. These authors compared the skeletons of radiocarbon dated mammoth carcasses preserved in the permafrost of Siberia (figure 103). From the size of the front leg as figured the total shoulder height (and with fair accuracy also the weight) of these animals may be estimated. There is a distinct oscillation in size, the mammoth of the interstadial (4-Würm I-II) being up to 20 or 25 per cent taller than the ancestral and descendant forms that lived during 4-Würm I and 4-Würm II respectively; it probably weighed about twice as much.

Oscillation in relation to the temperature is probably very common. In this case the 'cold' form was smaller than the 'warm' form. The converse relationship is somewhat more common (Bergmann's rule). It is related to heat-loss, since the increase in size will reduce the ratio between the surface and the bulk of the animal so that the relative area from which heat is given off becomes smaller. The cricetids (figures 9, 101) are good examples. In the Late Pleistocene and Post-glacial history of the common hamster a gradual reduction may be observed through the sequence Mousterian – Magdalenian – Recent (figure 101). The oscillation in the Middle Pleistocene was of the same relative magnitude.

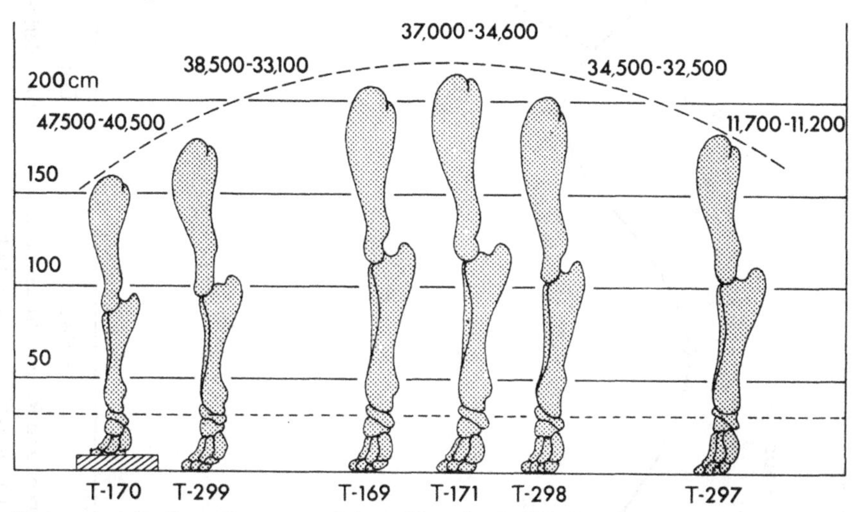

Figure 103. Radiocarbon age and size of fore-leg in Woolly Mammoth, *Mammuthus primigenius*, from Siberia, showing oscillation in size. The small forms T-170 and T-299 date from 4-Würm I; the large forms T-169, T-171 and T-298 from the interstadial; the final specimen T-297 dates from the final 4-Würm stadial. After Heintz & Garutt.

The history of the brown bear (*Ursus arctos*) in Europe shows size oscillation in relation to the alternation between glacials and inter-glacials (figure 104). But this history is valid only for Europe. The brown bear population of Asia must have had an entirely different history; the form now inhabiting northeastern Asia is almost as large as the largest Pleistocene brown bears in Europe. On the other hand the size of the brown bear in Palestine culminated in the F-Eemian at a time when the brown bear in Europe was much smaller; these inter-glacial bears in Palestine were giants comparable to the great 4-Würm

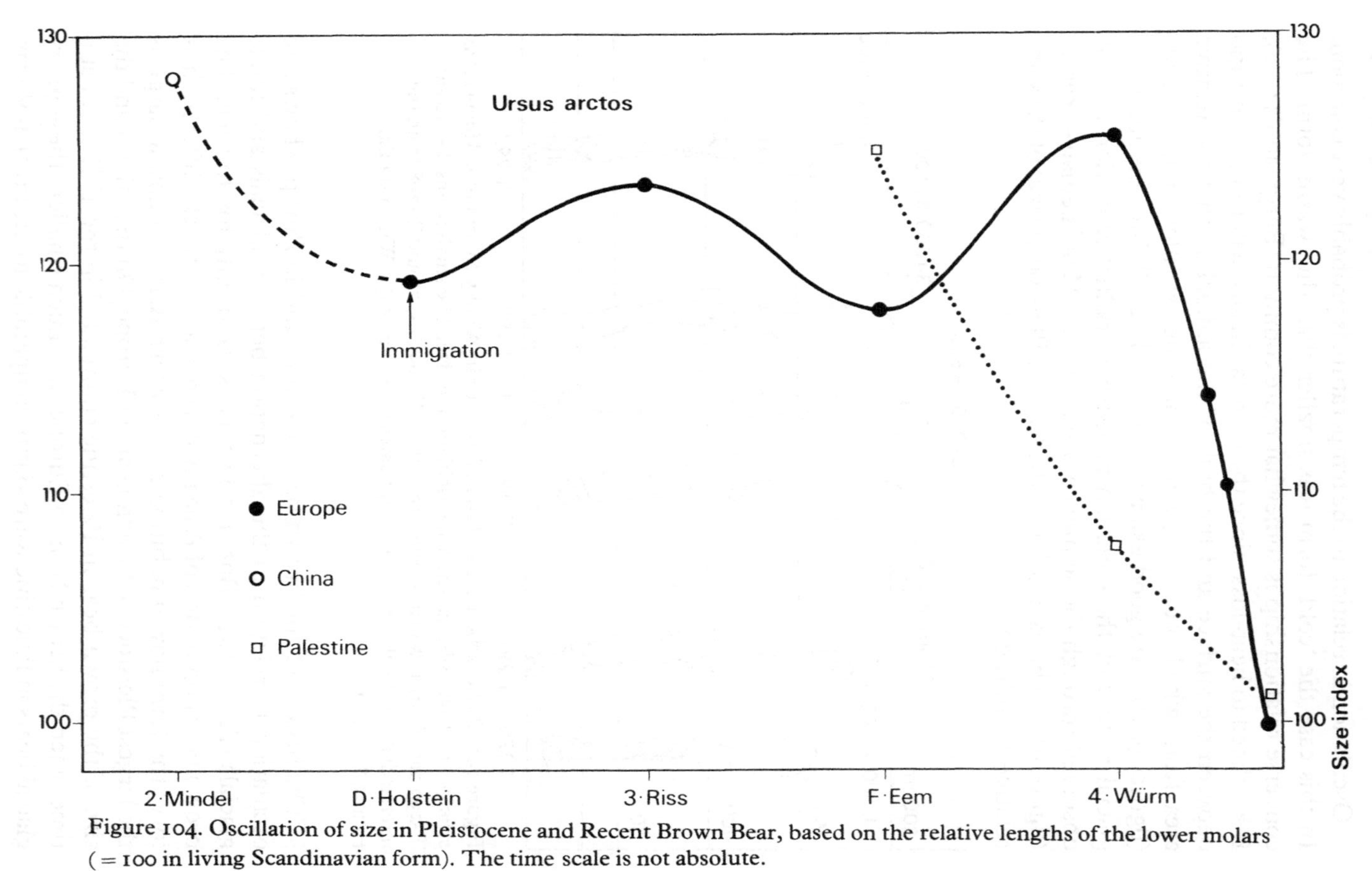

Figure 104. Oscillation of size in Pleistocene and Recent Brown Bear, based on the relative lengths of the lower molars (= 100 in living Scandinavian form). The time scale is not absolute.

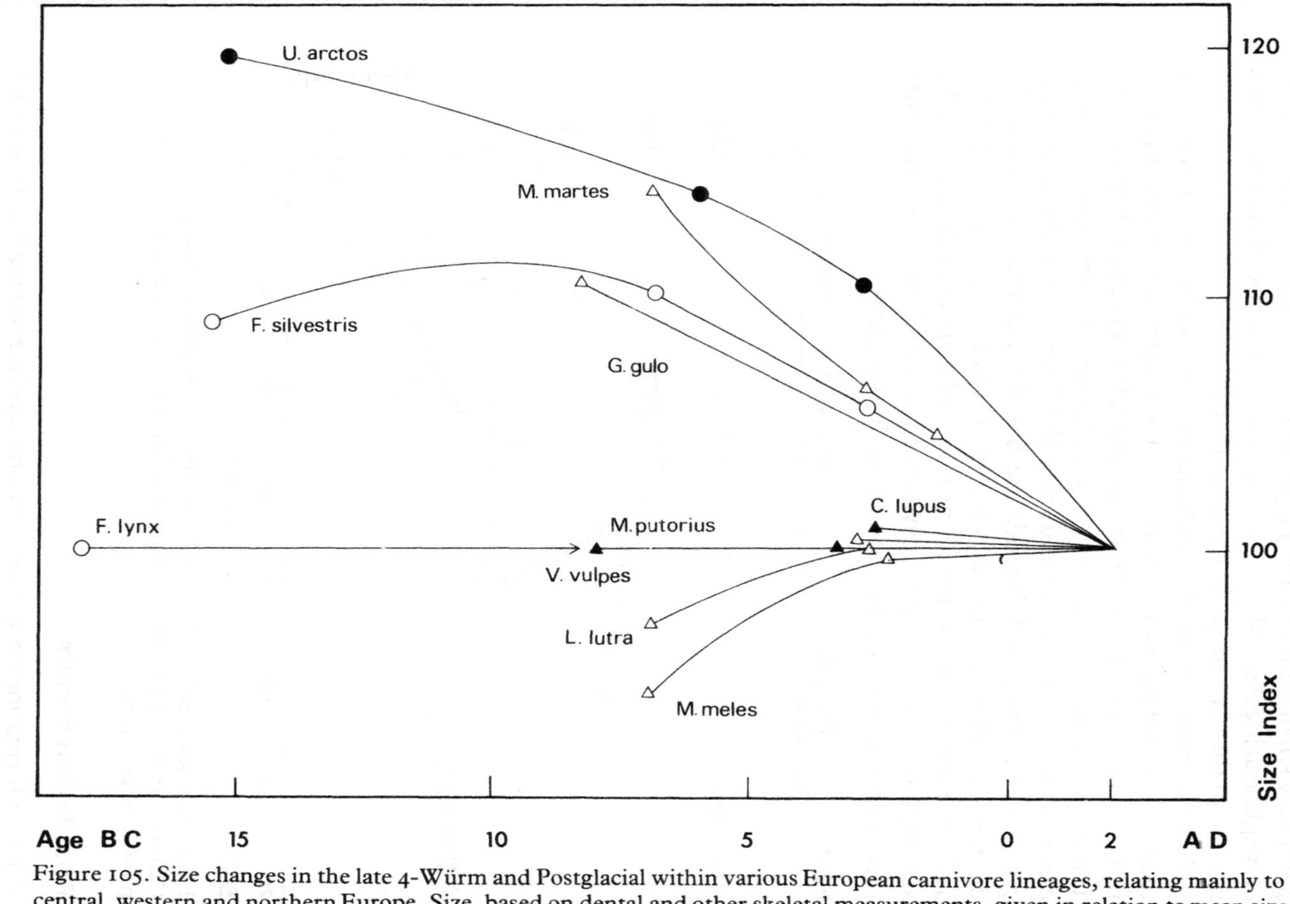

Figure 105. Size changes in the late 4-Würm and Postglacial within various European carnivore lineages, relating mainly to central, western and northern Europe. Size, based on dental and other skeletal measurements, given in relation to mean size in living form.

bear of Europe. In contrast, the 4-Würm bear in Palestine was comparatively small, only slightly larger than the present-day brown bear of neighbouring Asia Minor. So it seems that the size change here was inverse to that in Europe in the Late Pleistocene.

A comparative study of size changes in various lineages at the same time in different parts of the world is an interesting and promising project, but data for such a study are not available at present. Figures 105–106 show a comparison between size changes in some carnivore species in Europe and Palestine during the Late Pleistocene and Postglacial [174; 179; 181]. Trends are based on changes in size in the dentition or other skeletal parts and are brought to a common scale by assigning a size index of 100 to the terminal form of each lineage. The significance of such trends may sometimes be illuminated by other types of population studies, as will be shown below.

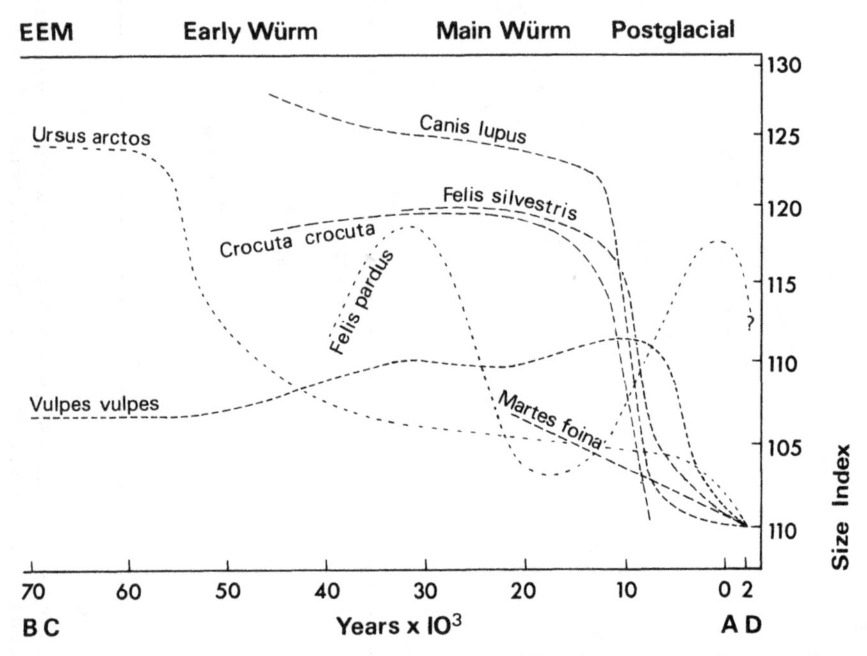

Figure 106. Size trends in the Late Pleistocene and Postglacial among lineages of Carnivora in Palestine; representation as in figure 105 but time scale more compressed. After Kurtén.

Population trends

While information on size and size changes in evolving lineages is easy enough to obtain – in principle at least – measurement of the standing population at any time is difficult. The relative number of individuals of different species represented in the Tischofer Cave, a typical bear cave in the northern Alps, is shown in table 3; the values are based on a total of about 420 individuals. But it is clear that these numbers say nothing whatever of the actual size of the populations of these animals. Europe was not flooded by cave bears in the Pleistocene; the numbers reflect the preferences and habits of the mammals and the ecology of the cave.

Table 3. Relative number of individuals of different species represented by fossil remains in Cave Earth Layer C, Tischofer Cave, probably dating from the interstadial 4-Würm I-II. Data from Gross [88].

Ursus spelaeus (Cave Bear)	91·1%
Vulpes vulpes (Red Fox)	2·9
Capra ibex (Ibex)	2·9
Canis lupus (Wolf)	1·4
Rangifer tarandus (Reindeer)	0·7
Crocuta crocuta (Cave Hyena)	0·5
Felis leo (Cave Lion)	0·2
Rupicapra rupicapra (Chamois)	0·2

On the other hand, other factors being equal, trends in the relative representation of a single population (or of ecologically similar populations) may give valid evidence of actual changes in the size of the standing population. Alternation between 'cold' and 'warm' or forest and steppe forms indicate oscillations of the local climate, while the petering out of old forms and the increase of new will help to date the sequence. Very good results have been obtained by such studies of microtine rodents, the so-called 'vole spectra' [162] of which an example is shown in figure 107; it is based on the succession in the early Middle Pleistocene deposits of the C718 cave near Koneprusy [82].

Layers B–F on top of the sequence are loessic in origin and contain a cool fauna with a predominance of pine vole of the *Pitymys gregaloides* group; here are also found snow vole (*Microtus nivalinus* group), tundra vole (*M. ratticepoides*) and gregarious vole (*M. gregalis*). On the other hand the representation of *M. arvalinus*, a relative of the common vole, declines in these deposits.

In contrast the layers G–H7, which consist of a fossil soil, contain a temperate fauna with voles of *M. arvalis* type (*M. arvalinus*, *coronensis*), bank vole (*Clethrionomys*) and the genera *Mimomys* and *Pliomys*.

Norway lemming however also occurs at these levels. With layer H8 (bottom) we enter an early cold episode with tundra vole and Arctic lemming (*Dicrostonyx*) in a loess matrix. Although dominated by the oscillation from cold to warm and back to cold the sequence also shows the gradual petering out of the archaic *Mimomys* and *Pliomys* forms.

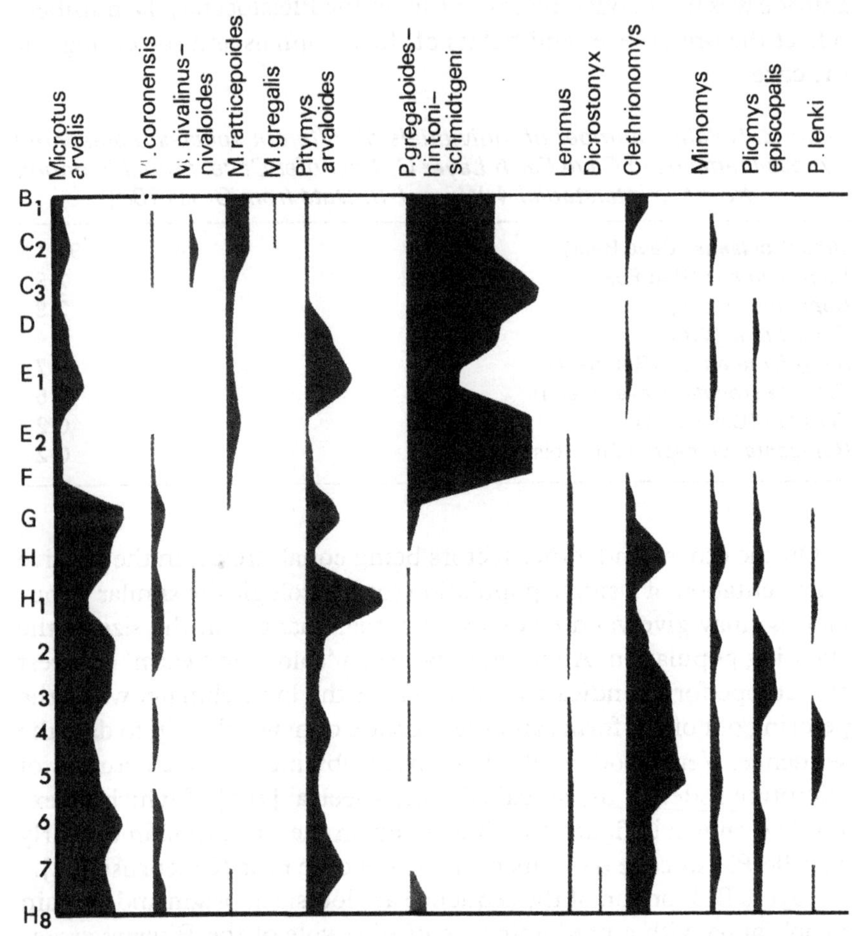

Figure 107. Vole spectrum from the stratified sediments of the early Middle Pleistocene cave C718 at Koneprusy. After Fejfar.

Another local sequence, this time based on the Carnivora in a number of caves in Palestine covering a time span from the F-Eemian to the Postglacial, is shown in figure 108 [179]. Various types of population trends are evident. One of the most interesting is the alternation between ecologically related, probably competing species, of which now one and now the other gains the upper hand. The spotted hyena (*C. crocuta*) versus the striped hyena (*H. hyaena*) is a good example. The former entered the area in the 4-Würm I and rapidly crowded out the latter, but became extinct in the Postglacial, whereupon *H. hyaena* returned and is still present. A second case is that of the wolf (*Canis lupus*) and wolf jackal (*C. lupaster*). These two species are ecological vicars[1] at the present day; the boundary between their ranges is now not far from Palestine. Oscillation of this boundary back and forth across the area would produce an alternation like that seen in the record.

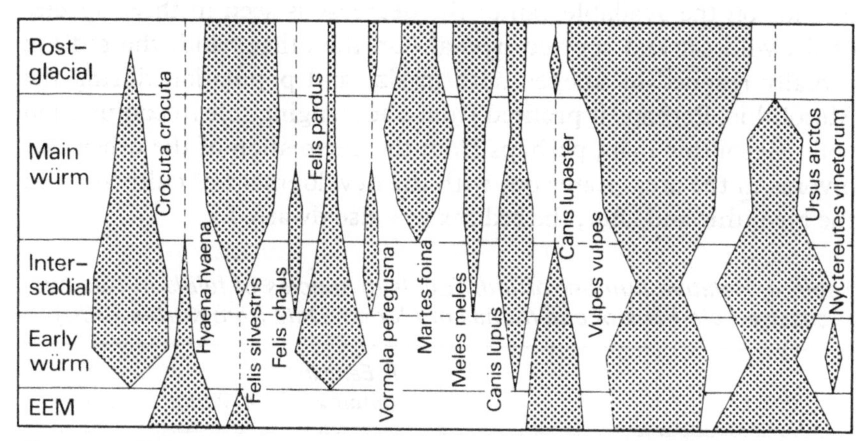

Figure 108. Relative representation of species of Carnivora at successive levels in Palestine caves from the Late Pleistocene and Postglacial. After Kurtén.

Several different species show a long-range increase, like badger (*M. meles*) and red fox (*V. vulpes*), or decrease, like the spotted hyena in the interval from the Interstadial to the Postglacial, the leopard (*F. pardus*) and the brown bear (*U. arctos*). All of the Carnivora that are now extinct in Palestine have a record of decline dating back in some cases to the beginning of the 4-Würm and certainly to the interstadial. Analogous results were obtained in a study of the Late Pleistocene Felidae of Florida [180]. Two successive faunas could be distinguished (see table 4); three felid species were numerous enough to be analysed in this manner. Of these, the bobcat (*Felis rufus*) shows population increase and survives in the modern fauna, while the jaguar (*F. onca*) and sabre-tooth (*Smilodon fatalis*) both show population decline and are now extinct in Florida.

[1] See note on p. 45.

A combined study of changes in size and population may be very illuminating. In some instances a population decline is clearly associated with dwarfing, for instance in the Post-glacial spotted hyena of Palestine (figures 106, 108), the brown bear in the same area and the jaguar in Florida. It may be suspected that the main factor at work here is the same that resulted in the evolution of dwarf forms on islands: the necessity to keep up an adequate population density in spite of a severe limitation of habitat and/or food supply. In the case of Palestine a remarkable decrease in size is evident during the Mesolithic, a time when human hunter-gatherers probably made serious inroads on the available game: the decrease is seen in three species (wolf, wild cat and spotted hyena). On the other hand, the red fox actually reached its apogee both in size and population during the Mesolithic. Perhaps it profited from a scavenging role in conjunction with Mesolithic man, perhaps from the decimation of the other carnivores in the area. Later on, with the devélopment of cultivation in the Neolithic and later, the red fox was also dwarfed.

Table 4. Relative number of individuals in species of felids in two successive Late Pleistocene faunas of Florida. Data from Kurtén [180].

	F-Eemian (approx.)	*4-Würm*	*Recent*
Felis rufus (Bobcat)	17%	41%	Survives
F. onca (Jaguar)	50%	31%	Extinct
Smilodon fatalis (Sabre-Tooth)	23	12	Extinct
Others	10	16	—

Chapter 18

Origination of Species

Arctic forms

VARIOUS other aspects of evolutionary trends might be mentioned, but one peculiar to the Pleistocene deserves special study: the rise of various species adapted to a cold environment. Palaeomammalogists have stated repeatedly that only the 4-Würm fauna has a decidedly High Arctic stamp, while there is little evidence of cold climates in earlier mammalian faunas [146; 265].

In this reasoning it has been overlooked that there probably were few if any Arctic land mammals before the Ice Age, so that this ecological type had to evolve in various separate lineages before we can expect to meet it in the fauna. Environments for Arctic land mammals must have been very restricted in the time before the Pleistocene. The climate was apparently warmer over the entire earth than now and the seas stood higher, flooding the northern margins of the continents. It was not until the Pleistocene that a sudden enormous expansion of Arctic biotopes took place, thereby stimulating the evolution of suitable adaptations among both plants and animals. The adaptation could not be produced at a moment's notice though and so the lack of some Arctic mammals in the faunas of the 2-Mindel and 3-Riss may simply be due to evolutionary lag.

In actual fact the great majority of Arctic or Boreal species make their appearance well before the 4-Würm. As early as 1-Günz II we meet the first Arctic mammals: the reindeer, musk ox, a form related to tundra vole, the Norway lemming and possibly also a member of the Arctic lemming genus *Dicrostonyx*. The steppe mammoth also appears at this time, as well as the snow vole, a cold-adapted Alpine form. This is quite a large collection to have before the 2-Mindel, usually regarded as the first great continental glaciation.

The roster of northern forms was increased during the 2-Mindel by the appearance of glutton, woolly rhinoceros (the latter in Asia) and Alpine shrew. Additions in the 3-Riss are woolly mammoth, Arctic fox and elk (or moose); this is also the time when the woolly rhinoceros entered Europe. Varying hare may also be present in the 3-Riss, but this requires further study.

This leaves only one species on our list to appear in the 4-Würm: the polar bear (*U. maritimus*) of which the earliest known specimens date from the late F-Eemian or early 4-Würm. The evidence on the cold forms is summarized in table 5, which clearly indicates that the Arctic faunal type is of long standing in Europe.

Table 5. First appearances of Northern forms in the Pleistocene (mainly in Europe).

Phase	*Appearances*
4-Würm or F-Eem	1 or 2
3-Riss	4 or 3
2-Mindel	3
1-Günz II	7

Age of living species

This brings us to the question of when the living species of mammals in general originated, not only those adapted to a cold climate. The first appearance of species that are still in existence seems to have occurred in the Astian, when we find pygmy shrew, pond bat, long-winged bat, common field mouse, and possibly bank vole and steppe pika – a total of six species out of the 119 living forms with a fossil record; these species are probably over 3 million years old. Some seven species are added during the Villafranchian, most of them in the later part of the stage (beaver, yellow-necked field mouse, hedgehog – though uncertain if recent species –, Mediterranean horseshoe bat, hippopotamus, badger and striped hyena, the two last-mentioned outside Europe). These, too, are oldtimers in comparison with the remainder of the fauna; they are probably between one and three million years old.

About 51 modern species make their appearance in the early Middle Pleistocene (1-Günz to 2-Mindel inclusive); they represent some 43 per cent of the recent species, while about 11 per cent arose in pre-Günz times. Twenty-seven modern species, or 23 per cent, were added in the late Middle Pleistocene (D-Holstein and 3-Riss); the remaining twenty-eight species (24 per cent) with a Pleistocene record appear in the Late Pleistocene. A more detailed listing, with a cumulative percentage showing the growth of the modern fauna, is given in table 6. In addition to the species listed here there are some 24 living species of mammals in Europe without a fossil record; if included, they would bring the total up to 120·2 per cent.

Table 6. First appearance of modern species in the Pleistocene faunas (including appearances outside Europe).

	First Appearances	Percentage, Cumulative
4-Würm	6	100·0
F-Eem	22	95·0
3-Riss	5	76·5
D-Holstein	22	72·3
2-Mindel	12	53·8
C-Cromer	9	43·7
1-Günz II	17	36·1
B-Waal	7	21·8
1-Günz I	6	16·0
Late Villafranchian	5	10·9
Early Villafranchian	2	6·7
Astian	6	5·0

If it is assumed that the present-day species originated at a constant rate, the cumulative numbers would increase in accordance with the curve shown in figure 109 in which the rate of increase is measured by the time of duplication, i.e. the time in which a given number of species is doubled. The successive stratigraphic phases would then be spaced temporally as shown in the figure, according to the percentage values in table 6.

If two points on the curve can be given an absolute date, a chronology for the entire succession can be constructed (still assuming that the rate of origination is constant). The final phase or 4-Würm may be given a mean age of about 30,000 years; for earlier phases a few radiometric dates are available (table 7). In alternative A the absolute chronology has been constructed on the basis of the D-Holstein date of 230,000 years; alternative B shows a chronology based on the early Villafranchian date of 2·9 million years (average of two dates).

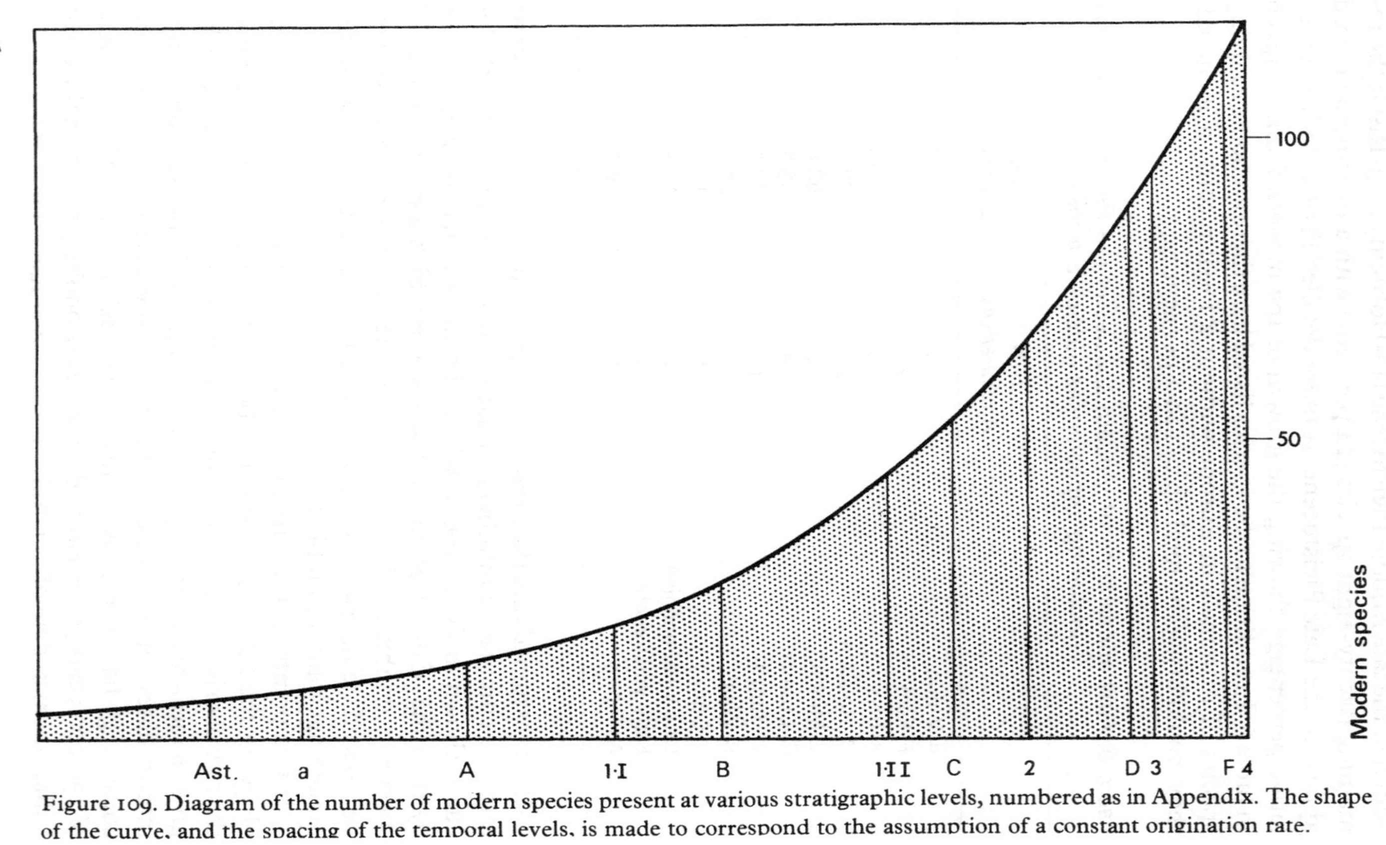

Figure 109. Diagram of the number of modern species present at various stratigraphic levels, numbered as in Appendix. The shape of the curve, and the spacing of the temporal levels, is made to correspond to the assumption of a constant origination rate.

The two chronologies are obviously not equal. That based on the D-Holsteinian date gives good values for the Late and Middle Pleistocene back to the 2-Mindel but much too low values for the Villafranchian, while that based on the Villafranchian date tends to make the Late Pleistocene too long. If the radiometric ages are correct, the assumption of a constant rate of origination must be wrong. Evidently the evolution of present-day species was relatively slower in the Villafranchian than in the Middle and Late Pleistocene.

The mean age of the living species is approximately equal to 1-Günz II: about 650,000 years in the short chronology and 1·1 million in the long.

Table 7. Alternative absolute chronologies for the Pleistocene based on the appearance of modern mammalian species, as explained in the text.

	Absolute Age		
	Radiometric	*Alternative A*	*Alternative B*
4-Würm	30,000	30,000	30,000
F-Eem	70,000	60,000	84,000
3-Riss	—	190,000	310,000
D-Holstein	230,000	230,000	370,000
2-Mindel	>400,000	410,000	690,000
C-Cromer	—	540,000	900,000
1–Günz II	—	660,000	1,100,000
B-Waal	—	970,000	1,650,000
1-Günz I	—	1,160,000	2,000,000
Late Villafranchian	—	1,400,000	2,400,000
Early Villafranchian	2,900,000	1,700,000	2,900,000
Astian	—	1,900,000	3,200,000

The evolution of new species

Do species evolve gradually or by sudden steps? The old idea that species evolved by sudden great 'systemic' mutations has been totally discredited, but there seems to be no theoretical difficulty in the idea that new species may arise as a result of a rapid, episodic 'spurt' bringing the population to a new adaptive plane [250].

That rates of evolution may vary has long been established. If all the present-day species evolved at the same rate they would have originated at exactly the same time, which is obviously not the case. In actual fact some of them have persisted unchanged for several million years, while others have arisen from ancestral forms within the last 200,000 years. It follows that the rate of evolution may be markedly different in different lineages. The rate within a single lineage may also change from time to time, as has been shown in many cases and notably in the history of the horse family [250]. Population geneticists have devised various models showing the possibility of greatly increased rates of evolution [195].

The Pleistocene fossil record may in future solve the problem of how species originate. In the following instances the transition between two species is actually recorded:

Macaca florentina → *M. sylvana*
Felis issiodorensis → *F.* cf. *pardina*
Gulo schlosseri → *G. gulo*
Cuon majori → *C. alpinus*
Ursus minimus → *U. etruscus*
Ursus deningeri → *U. spelaeus*
Ursus etruscus → *U. thibetanus*
Archidiskodon planifrons → *A. meridionalis*
Archidiskodon meridionalis → *Palaeoloxodon antiquus*
Archidiskodon meridionalis → *Mammuthus trogontherii*
Mammuthus trogontherii → *M. primigenius*
Dicerorhinus megarhinus → *D. etruscus*
Dama clactoniana → *D. dama*
Lagurus pannonicus → *L. lagurus*

In many other cases the phyletic connection between ancestral and descendant species is clear enough, although the actual transition has not yet been recorded. Instances are as follows:

Sorex praealpinus → *S. alpinus*
Talpa fossilis → *T. europaea*
Talpa minor → *T. caeca*
Rhinolophus delphinensis → *R. ferrum-equinum*
Homo heidelbergensis → *H. neanderthalensis*
Hyaena perrieri → *H. brevirostris* (as side branch)
Felis lunensis → *F. silvestris*
Martes vetus → *M. martes*
Martes vetus → *M. foina*
Vormela beremendensis → *V. peregusna*
Mustela stromeri → *M. putorius*
Mustela palerminea → *M. erminea*
Mustela praenivalis → *M. nivalis*
Canis etruscus → *C. lupus*
Vulpes alopecoides → *V. vulpes*
Vulpes praecorsac → *V. corsac*

Ursus etruscus → *U. deningeri*
Ursus etruscus → *U. arctos*
Ursus arctos → *U. maritimus* (as side branch)
Equus stenonis → *E. süssenbornensis*
Equus bressanus → *E. mosbachensis*
Equus mosbachensis → *E. germanicus*
Equus stehlini → *E. hydruntinus*
Orthogonoceros verticornis → *O. cazioti*
Alces gallicus → *A. latifrons*
Alces latifrons → *A. alces*
Sciurus whitei → *S. vulgaris*
Sicista praeloriger → *S. betulina, subtilis*
Muscardinus pliocaenicus → *M. avellanarius*

Additional items might easily be found, for instance among transitions that took place outside Europe (the evolution of *Bos namadicus* into *B. primigenius*, or of *Acinonyx pardinensis* into *A. jubatus*) or if the Pliocene ancestry of various Villafranchian species were included.

Although transitions between species are known in many cases, the actual record of the transition may not be detailed enough to permit analysis in terms of evolutionary rates or else has not yet been analysed in this manner. The elephant lineages are probably better documented than any others but it would be necessary to study all the existing collections. In the glutton sequence there is some evidence that might suggest a rate increase in the C-Cromer to 2-Mindel; as regards the evolution of the cave bear line on the other hand the trend appears to have been steady, without marked rate changes.

Perhaps a stepping-up of the rate of evolution is a sign that the new species has changed its mode of life, as the glutton indeed appears to have done in the 2-Mindel.

Chapter 19

Faunal Turnover

Evolution of the fauna

THE fauna was changing in composition constantly throughout the Pleistocene: new species were introduced by local evolution or immigration, while old species vanished through extinction or by evolving into other species.

The rates at which the origination and extinction take place are not always exactly similar. For instance, out of 122 species recorded in the 1-Günz II, 27 make their first appearance, while 22 become extinct in or immediately after this stadial. The rate of origination is thus 22 per cent, while the rate of extinction is somewhat lower or 18 per cent. But in the long run the two processes, origination and extinction, will tend to balance out so that the fauna contains roughly the same number of species from age to age. The most useful measure of faunal change would then seem to be an average of the origination and extinction rates. This may be termed the rate of faunal turnover and in this case it is 20 per cent.

Analyses of the turnover rate have been made for all the stages and substages recognized in the species list (table 15); that is to say, the Villafranchian was divided into six substages beginning with the Etouaires phase and ending with the A-Tiglian, and so on.

The rate of turnover in the Villafranchian stages varied between 3 and 24 per cent, but most values were between 10 and 11 and the average was 10·8 per cent. For the five phases of the early Middle Pleistocene (1-Günz I to 2-Mindel) the corresponding average was found to be 14·1 per cent, while the figure for the four final phases (D-Holstein to 4-Würm) is 9·2 per cent. The grand mean is 11·4. The phases, whatever their temporal duration, seem thus to have been roughly equivalent with each other as regards evolutionary change, though with some concentration of activity in the early Middle Pleistocene.

The total of 15 phases recognized corresponds to a time lapse of about 3 million years, so that the average length of each phase is 200,000 years. The mean longevity of a species corresponding to an average turnover of 11·4 per cent per 200,000 years is slightly more than 3 million years.

However, this measure is an average covering both the Villafranchian and the 'glacial' Pleistocene. If the analysis is limited to the sequence from the D-Holsteinian to the 4-Würm, quite different figures are obtained. The average phase length is only about 75,000 years and the average turnover rate 9·2 per cent; this corresponds to a mean species longevity of about 1·6 million years, or only about one-half of that for the Pleistocene as a whole. This would seem to reflect a general intensification of the rate of evolution in the later Pleistocene as compared to the Villafranchian. If exact figures for the duration of the Villafranchian and the early Middle Pleistocene were available, a more detailed analysis would be possible.

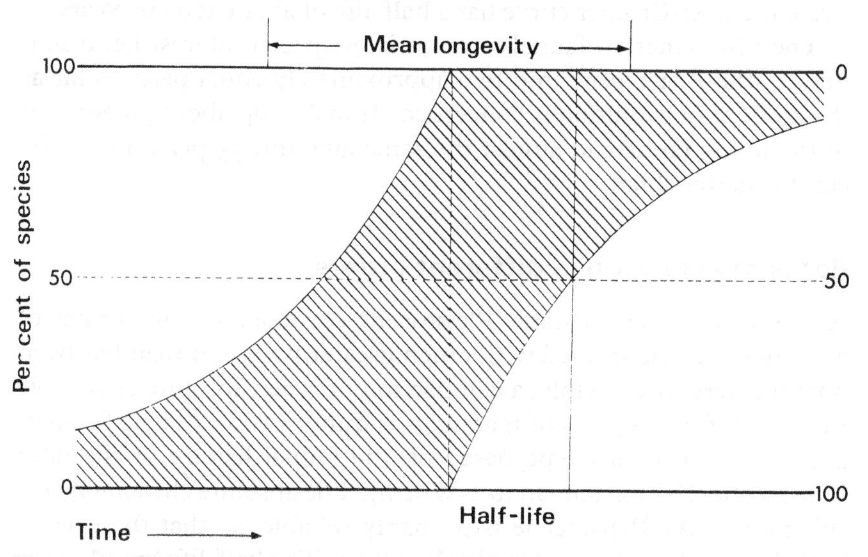

Figure 110. Diagrammatic representation of the history of a temporal stratum of species (cross-hatched) in a fauna with constant turnover rates through time. The relationship between half-life and mean longevity of species is shown. After Kurtén.

The species that are in existence at a given point in time may be said to form the faunal stratum of that time; for instance, table 6 records the history of the Recent faunal stratum. But we might also study the history of some earlier stratum, for instance the C-Cromerian. In faunas preceding or succeeding the C-Cromer in time, that stratum will make up a certain percentage which will generally dwindle with increasing temporal distance from the culmination of the stratum. It has been shown [172; 173] that the stratum will wax and wane according to the curve shown in figure 110 provided that the rate of turnover is constant. The rate of change may be expressed by the half-life, or the time in which the fauna is halved or doubled. The mean longevity of the species is approximately 2·9 times the half-life; the relationship between these parameters is shown in figure 110.

In figure 111, which shows the rise and fall of the C-Cromerian faunal stratum, the temporal position of the C-Cromerian has been determined on the basis of the date of 230,000 years for the D-Holsteinian and the assumption of constant rates of change. This puts the C-Cromer at about 500,000 BC but also necessitates the assumption that rates of change were much slower in pre-Cromer times. The pre-Cromer curve in figure 111 has a half-life of about 1·5 million years, while the post-Cromer curve has a half-life of about 400,000 years.

The C-Cromerian faunal stratum is of special interest because it divides the Pleistocene into two approximately equal parts as far as the amount of evolution is concerned. It makes up about 36 per cent of the living fauna and almost the same amount, 33 per cent, of the early Villafranchian.

Rates of evolution in different orders

As soon as we have a stable Pleistocene chronology, faunal rates of evolution may be studied with much greater precision than has been possible here. Meanwhile, a comparison between absolute or relative rates in different orders of mammals is quite feasible. Table 8 shows a comparison of this type based on the faunal history of the later Pleistocene (D-Holsteinian to 4-Würm). The absolute chronology for this part of the Pleistocene is probably reliable, so that the species half-life may be expressed in absolute time. The half-life found in the various orders may be compared with the value of 540,000 years for the fauna as a whole. The lower the figure, the higher the average rate of evolution within the order.

The highest average rate of evolution is not unexpectedly found in the Proboscidea, represented at this time only by the Elephantidae,

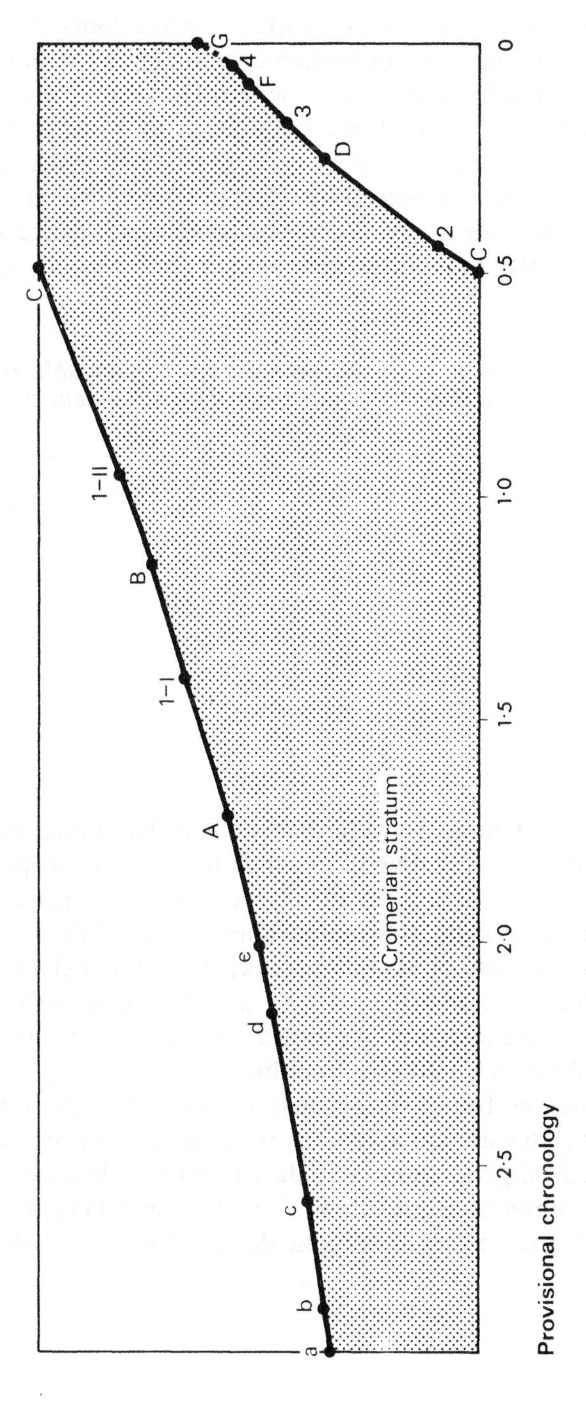

Figure 111. History of the Cromerian faunal stratum. The provisional chronology is based on the assumption that species half-life was 1·5 million years in pre-Cromerian times, and 0·4 million years in post-Cromerian. Stratigraphic levels numbered as in table 2.

which evolved rapidly in contrast with the mastodonts. Next come the Primates with high rates of human evolution and the Perissodactyla with their rapidly changing horse and rhinoceros species. The Rodentia, Artiodactyla and Insectivora form a group with almost average evolutionary rate, while rates in the Lagomorpha and Carnivora averaged a little slower. The Chiroptera take an extreme and separate position with a half-life about three times as long as the faunal average: the bats are a very conservative, slowly evolving group.

Table 8. Half-life of species in different orders of mammals, based on faunal turnover during the time span from the D-Holstein to 4-Würm inclusive.

Insectivora	490,000
Chiroptera	1,600,000
Primates	230,000
Carnivora	610,000
Proboscidea	180,000
Perissodactyla	260,000
Artiodactyla	490,000
Rodentia	490,000
Lagomorpha	600,000
Total fauna	540,000

The results are tentative and probably somewhat biased especially for the Rodentia and Insectivora, in which mostly long-range species have been included. Inclusion of all the described species would greatly modify the results, shorten the apparent half-life and indicate much greater intensity of evolution [173]. Unfortunately, the taxonomic situation in these orders is very unstable; discovery and description of new species are proceeding at such a rate that any attempt at definite evaluation would be premature.

It is possible that future studies may vindicate the opinion voiced in Kurtén [173] that evolution on the species level tends to be correlated with length of generation. The data of table 8, however, do not support that contention but rather that of Zeuner [315] that other factors are of greater importance in determining the evolutionary tempo.

Chapter 20

Animal Geography

MOST mammalian species in the Pleistocene fauna of Europe ranged well beyond the geographic boundaries of this continent. Many fossil species are only known from European deposits, but it is very hard to prove that they were actually endemic to Europe and the probability is against the assumption in most cases.

One group of species may however be regarded as endemic: the aberrant inhabitants of ancient Tyrrhenis, the Mediterranean islands. Here belong the species of *Nesiotites*, the dwarf elephants and hippopotami (if given specific status), the dwarf deer *Praemegaceros cazioti* etc., the cave goat genus *Myotragus*, the Sardinian pika *Prolagus sardus* and the aberrant rodent genera *Hypnomys*, *Leithia*, *Tyrrhenicola* and *Rhagamys*. Even if the elephants and hippos are not regarded as separate species, this gives at least eight distinct endemic forms in the Mediterranean area.

Otherwise species may only be regarded as endemic to Europe if clearly distinct vicarious forms are found in the surrounding areas. At present the cave bear (*Ursus spelaeus*) is almost the only certainly endemic Pleistocene species, but *Hipparion crusafonti* may also be endemic and some of the Villafranchian bovids, for instance, may turn out to be so. In the living European fauna there are several endemic species, most of them insectivores and rodents, but also two larger animals, *Felis pardina* and *Rupicapra rupicapra*.

Europe is part of the Palaearctic faunal province, which also includes Africa north of the Sahara and Asia north of the Himalayas. Most of the present-day European mammals range beyond Europe into some part of the remaining Palaearctic region. A small number of species have mainly African associations (*Macaca sylvana, Genetta genetta, Herpestes ichneumon, Hystrix cristata* and the locally extinct *Hippopotamus amphibius*), but most have an appreciable distribution in Asia. Fossil European species of this type are *Homo neanderthalensis*,

Hyaena brevirostris, H. hyaena, Megantereon megantereon, Homotherium sainzelli, Felis issiodorensis, Acinonyx pardinensis, Cuon majori, C. alpinus, Vulpes alopecoides, Nyctereutes megamastoides, Ursus thibetanus, Zygolophodon borsoni, Palaeoloxodon antiquus, Mammuthus trogontherii, Dicerorhinus kirchbergensis, D. hemitoechus, Coelodonta antiquitatis, Equus przewalski, E. hemionus, Sus strozzii, Megaloceros giganteus, Bubalus murrensis (?), *Bos primigenius, Trogontherium cuvieri.*

A few species range widely both in the Palaearctic and Ethiopian (African) region. Of these, *Crocuta crocuta, Felis leo* and *F. pardus* are now extinct in Europe, while *Felis silvestris, F. chaus, Capra ibex* and *Lepus capensis* remain as European members of this select company. Many fossil forms are related to both African and Asiatic species (for instance the small-clawed otters) but the precise taxonomic relationships remain to be worked out.

Finally, there are some species with a circumpolar distribution. The Palaearctic and Nearctic (North American) provinces may be grouped together, forming the Holarctic faunal province. Both *Saiga tatarica* and *Ovibos moschatus* have had a Holarctic distribution, which has been diminished since the Pleistocene. *Mammuthus primigenius* was certainly Holarctic and *Archidiskodon meridionalis* probably so; *Homotherium latidens* of Europe may possibly be the same species as *H. serum* in North America. The wisent, *Bison bonasus*, is now often regarded as conspecific with the American bison. Other living, definitely Holarctic species include glutton, ermine, least weasel, wolf, red fox, brown (and grizzly) bear, polar bear, red deer (wapiti, American 'elk'), true elk or moose, reindeer, marmot, tundra vole and Arctic lemming.

Data on geographic distribution have been brought together in table 9, giving the ranges in 119 living and 45 fossil species. The relative distribution in the fossil group is almost exactly the same as in the living, but it should be remembered that the main part of the fossil endemics consist of Mediterranean island forms.

The geographic origin of a species does not always coincide with its range in later times. The origin of several European species is known or may be inferred with high probability (table 10).

In most instances the ancestral form is also found in Europe, so that the evolution of the species appears to have taken place *in situ*, perhaps as part of an evolutionary process occurring simultaneously all over the range of the population. A second group is formed by the

Table 9. Geographic distribution of Quaternary species of mammals.

	Recent %	*Fossil* %
Europe, endemic	20	22
Europe and Palaearctic	62	58
Europe and Holarctic	12	11
Europe, Palaearctic and Ethiopian	3	7
Europe and Ethiopian	3	2
Total species considered	119	45

immigrants from Asia; future studies of Asiatic faunas will probably result in a great increase of the percentage of this group. As it is we have direct evidence for the Asiatic origin of some species (*Crocuta crocuta, Ursus arctos, U. thibetanus, Dicerorhinus kirchbergensis, Coelodonta antiquitatis, Bison priscus, Bubalus murrensis, Bos primigenius, Marmota bobak, Allactaga jaculus*) and good circumstantial evidence for many others.

Forms of African origin appear to be decidedly rarer; more or less well-documented examples include *Genetta genetta, Herpestes ichneumon, Hyaena hyaena, Canis aureus* (?), *Lycaon lycaonoides, Hippopotamus amphibius, Hesperoceras merlae* and *Syncerus iselini*.

A few species may have originated more or less directly in North America (though they probably immigrated by way of Asia). This may perhaps be suggested for *Equus stenonis* and very tentatively for *Hypolagus brachygnathus* and seems fairly certain for *Bison bonasus*.

Again, the distributions for the fossil and recent forms are fairly similar, indicating that no radical long-term changes of migration

Table 10. Geographic origin of European mammal species.

	Recent %	*Fossil* %
Europe	77	53
Asia	15	29
Africa	6	13
North America	2	5
Total species considered	48	38

routes took place in the Pleistocene, whatever the short-term effect of glaciation barriers and interglacial transgressions.

The migration between the Old World and the New has been repeatedly discussed [182; 224; 232; 249]. It is generally assumed that it followed the route across the Bering Straits. Generally, the migrations may be assumed to have taken place during glacial regressions which exposed the Bering Bridge. For this reason, appearance of a migrant in an interglacial fauna may probably be taken as evidence for migration during the preceding glaciation. All of the forms recorded as migrants during glacial phases are evidently hardy northern types, well equipped to withstand the rigours of life in this area.

The earliest migration came at the very beginning of the Villafranchian, when at least three different lineages crossed from North America to Eurasia. Two of the populations reached Europe in the shape of *Felis issiodorensis* and *Equus stenonis*, while the third (ancestral camels) colonized Asia, eastern Europe and Africa but did not reach western Europe. The distribution of the camels supports the idea that the Bering Bridge was the route for this interchange.

Also perhaps at the beginning of the Villafranchian, a form of hunting hyena reached North America to give rise to the American *Chasmaporthetes*. Whether this migration followed the same route is uncertain, for *Euryboas* has not so far been found in Asia but of course such negative evidence is not conclusive.

Perhaps at a somewhat later date in the Villafranchian some elephant at the *Archidiskodon meridionalis* level seems also to have entered North America.

Migrations in the 1-Günz are uncertain, but it is probable that steppe mammoths of *Mammuthus trogontherii* type entered the Nearctic at about this time and the same holds for the *Mimomys* voles.

In the 2-Mindel, black bears migrated to North America, where they gave rise to *Ursus americanus*. A dirk-tooth of the *Megantereon* group also migrated, giving rise to *Smilodon*. It is also possible that scimitar-toothed cats (*Homotherium*) crossed from Asia to North America at this time, for we know no homotheres in Asia in post-Mindel times. There is no record of a migration in the opposite direction.

Migration items in the 3-Riss from the Old World to the New comprise *Gulo gulo, Mustela erminea, Canis lupus, Vulpes vulpes, Cervus elaphus* and some form of *Bison*, destined to give rise to the highly varied bison fauna of the Late Pleistocene in North America. In the other direction there is a possible migrant in *Lutra lutra*, which appears in the F-Eemian in Europe.

Two species appear to have migrated from America to Eurasia in the 4-Würm: *Mustela lutreola* and *Bison bonasus*. Migrations in the opposite direction are numerous and include *Homo sapiens, Alopex lagopus, Ursus arctos, U. maritimus, Mammuthus primigenius, Alces alces, Rangifer tarandus, Saiga tatarica* and *Ovibos moschatus*.

The data are summarized in table 11. It may be noted that the number of migrating species tended to increase for each new glacial phase. The impression may well be spurious, since we do not know enough about faunal exchange during the earlier glaciations. On the other hand it is possible that the increasing adaptation of many species to a cold climate tended to make the Bering Bridge climatically acceptable to an increasing number of migrants as time went by. Finally, it should be noted that the bridge probably grew broader for each new glacial phase as the sea level has receded more deeply every time.

Table 11. Mammal migrations between Eurasia and North America in the Pleistocene.

	America to Eurasia	*Eurasia to America*	*Total*
4-Würm	2	9	11
3-Riss	1?	6	7
2-Mindel	—	3	3
1-Günz	—	2	2
Villafranchian	3	2	5
Total	6	22	28

Chapter 21

Man and the Fauna

HUMAN BEINGS have been present in Europe ever since the C-Cromer Interglacial. During the earlier glaciations, man probably was forced to retreat to the southern peninsulas. In the 4-Würm he was able to cope with the Arctic environment.

But wherever he lived, the mammals that have been surveyed here formed the dominant element in his environment. This is especially true for the large game animals. From the earliest times man in Europe was a hunter. He was also, naturally, a food collector with a richly varied diet including berries, fruits, fungi, snails, grubs and other invertebrates, lower vertebrates and so on. As early as the C-Cromer his main tool was the hand-axe, made for the skinning of game. It seems almost certain that the first tools were made to help pierce and cut the hide of a dead animal about to be eaten, whether killed by man himself or by something else.

The all-encompassing importance of the game to Ice Age man is very evident when we study the Late Pleistocene cave paintings, engravings and sculptures. Almost all of the pictures represent animals and usually animals that were used for food: deer, wild horse, bison, aurochs, rhinoceros, mammoth. The large predators are also fairly common in cave art – lion, bear, wolf: the powerful, dangerous enemies or rivals. On the other hand, the small animals, whether carnivores or small game, play a quite subordinate role. And pictures of human beings are not only scarce but also in most cases quite clumsily done in comparison with the beautifully executed animal pictures.

Different tribes had different favourite game. The selection of game was perhaps mainly dominated by availability but doubtless tribal tradition also played a role. We find, for instance, evidence of specialized mammoth hunters, who even used the bones and tusks of their prey to build palisades and huts. Other tribes were horse hunters and their middens contain the remains of thousands of wild horses. In yet other instances the reindeer was the all-important game.

Sometimes we find evidence that men at different times but at exactly the same spot may have selected quite different types of game. In the Vogelherd Cave near the township of Stetten in Wurttemberg [187] almost all the fossil bones are midden remains from the human occupation, so that we get a very good idea of what these people used for food (table 12). The Neandertalers that lived here in the 4-Würm I

Table 12. Relative distribution of the three main forms of game (horse, mammoth and reindeer) at different levels in the cave of Vogelherd (Wurttemberg). Data from Lehmann.

	Mousterian (*4-Würm 1*) %	*Aurignacian* (*Interstadial*) %	*Magdalenian* (*4-Würm II*) %
Wild Horse, *Equus germanicus*	45	0	0
Wild Horse, *E. przewalskii*	0	23	22
Mammoth, *Mammuthus primigenius*	13	24	6
Reindeer, *Rangifer tarandus*	2	12	22
Other mammals	40	40	50

were evidently specialized wild horse hunters; their main prey was the large, powerful *germanicus*-type horse. Next comes a deposit from the early time of *Homo sapiens* in Europe, containing Stone Age tools of Aurignacian type and probably dating from the later part of the 4-Würm I–II interstadial. These people also hunted wild horse but the species of that time was the small *Equus przewalskii* and it does not predominate in the same way. There is in fact an equal number of woolly mammoth, and of course the total biomass of mammoth represented greatly exceeds that of horse. There is a possibility, however, that mammoth teeth were collected as souvenirs or prizes.

Finally, in the uppermost strata, with a Magdalenian industry (end of 4-Würm), reindeer has increased to parity with horse, while mammoth is scarce.

At the end of the Ice Age the Arctic mammals once again retreated northward and a woodland fauna of interglacial type became established in central Europe. In comparison with the F-Eemian fauna, however, it was much impoverished. This is also true of the present-day Arctic fauna of northern Europe compared with that of the 4-Würm.

Species present in the Late Pleistocene of Europe but now extinct include three species of *Nesiotites*; Neandertal man; scimitar cat, Corsica otter, cave bear; straight-tusked elephant, woolly mammoth; Merck's rhinoceros, steppe rhinoceros, woolly rhinoceros and European wild ass; *Myotragus* (cave goat), Bonal's tahr, steppe wisent, woodland wisent; *Hystrix vinogradovi, Leithia melitensis,* two species of *Hypnomys, Cricetulus bursae, Rhagamys orthodon* and *Prolagus sardus*; a few of these survived well into the Postglacial and even into historical times. Furthermore, a great number of species have become extinct locally in Europe though they survive elsewhere; these are the striped and spotted hyenas, steppe cat, lion, leopard, dhole, wild horse, kulan, hippopotamus, saiga, musk ox, great jerboa, Arctic lemming and steppe pika.

Table 13. Extinction of mammals at the end of the Pleistocene and in the Postglacial, compared with modern European fauna. Relative ordinal distribution of species in both samples.

	Extinct %	*Living* %
Insectivora	8	12
Chiroptera	0	22
Primates	3	2
Carnivora	25	18
Proboscidea	6	0
Perissodactyla	17	0
Artiodactyla	17	9
Rodentia	22	35
Lagomorpha	3	3

The great preponderance of large mammals in these lists is striking. Table 13 gives the ordinal distribution of the extinct species in comparison with that of the living European fauna. It is clear that the Postglacial mass extinction took a much heavier toll, relatively speaking, of the large mammals (Carnivora, Proboscidea, Perissodactyla, Artiodactyla) than of the small (Insectivora, Chiroptera, Rodentia).

The mass death can hardly be ascribed to climatic causes alone, for there was no similar mass extinction in earlier interglacials. It seems fairly certain that modern man has played a dominant role in the wiping out of many species, although perhaps by indirect influence as much as by actual hunting. Species like the cave bear may have been seriously affected by competition for living sites (caves) with humans. Perhaps also a rapid climatic change contributed to the outcome by reducing the size of some populations to such a low number of individuals that they became especially vulnerable. However, the effect of man on his environment may be the main factor. This problem may

Table 14. Length of hindmost lower molar in aurochs (Bos primigenius) *and in domestic cattle, all from Denmark, showing temporal decrease in size. The material from the Ertebølle culture is intermediate between aurochs and domestic cattle. Data from Degerbøl.*

Length (mm.)	*Aurochs*	*Ertebølle Culture*	*Neolithic Stone Age*	*Bronze Age*	*Iron Age*	*Middle Ages*
27	—	—	—	—	—	1
28	—	—	—	—	—	1
29	—	—	—	—	—	—
30	—	—	—	—	—	—
31	—	—	—	1	2	2
32	—	—	—	3	3	6
33	—	—	—	4	11	12
34	—	—	1	11	23	11
35	—	—	2	14	24	6
36	—	—	1	12	19	2
37	—	—	7	8	16	1
38	—	—	13	3	4	2
39	—	—	12	2	—	—
40	—	2	15	3	—	—
41	—	—	12	—	—	—
42	—	1	5	—	—	—
43	—	2	3	—	—	—
44	—	1	2	—	—	—
45	—	—	1	—	—	—
46	3	1	—	—	—	—
47	6	1	—	—	—	—
48	8	—	—	—	—	—
49	7	—	—	—	—	—
50	2	—	—	—	—	—
51	3	—	—	—	—	—
52	3	—	—	—	—	—
53	1	—	—	—	—	—
54	1	—	—	—	—	—

be studied further by combined study of changes in mean size and population numbers of wild animals accompanying the evolution of human cultures, as shown above. Probably a common denominator can be found for the general trend of size decrease since the Pleistocene, and the extinction of the large mammals.

There is quite another way in which man influenced the fauna: by domesticating part of it [319]. Trends and changes in this connection may be studied by the same quantitative methods that have been outlined here. Often it can be shown that a sudden character shift represents the actual period of domestication, as in the case of the taming of cattle in Denmark where Degerbøl [54] has shown that the Ertebølle levels give us the very transition (see table 14).

Seen in a longer perspective, the Pleistocene as a whole becomes a faunal revolution. Geologists have long recognized some turning points in the history of the earth's fauna, when as it were a great edifice would topple over and new structures are built upon the ruins of the old. Such was the case as the Permo-Carboniferous Ice Age was at its height and again when the reptiles of the Cretaceous died out and the relay was picked up by the mammals. In the Pleistocene we have such another revolution at our very geological doorstep.

Appendix

Stratigraphic Range of Species

THE stratigraphic ranges of the nearly three hundred species described here have been set forth in table 15. They are recorded by the total span as established, continuously from first to last appearance, even if not actually found in deposits of all the intervening phases. In actual fact the record of many species is spotty, either because of the incompleteness of the fossil record, or because they were absent during times with unsuitable climate. Nevertheless, they must have been in existence somewhere. One type of migration has, however, been noted in the table: if a species was in existence outside Europe well before its appearance in the European fauna, or if it survived elsewhere after its local extinction in Europe, this is indicated by a separate symbol.

Six phases of the Villafranchian and most of the glacial and interglacial phases of the Middle and Late Pleistocene have been recorded separately in the table, excepting the incompletely known E-Ilford interglacial. Astian records are given only for species that survived in the Pleistocene; no full recording of the Astian fauna is intended.

Table 15. *Stratigraphic span of Pleistocene mammal species in Europe.*

	Astian	Villafranchian Phases					Middle Pleistocene							Late Pleistocene			Recent
		a	*b*	*c*	*d*	*e*	*A*	*1(i)*	*B*	*1(ii)*	*C*	*2*	*D*	*3*	*F*	*4*	*G*
INSECTIVORA																	
Erinaceus sp.	—	—	—	—	×	×	×	×	×	×	×	×	×	×	×	×	×
Sorex minutus	×	×	×	×	×	×	×	×	×	×	×	×	×	×	×	×	×
S. araneus, etc.	—	—	—	—	—	—	—	×	×	×	×	×	×	×	×	×	×
S. praealpinus	—	—	—	—	—	—	—	×	×	×	—	—	—	—	—	—	—
S. alpinus	—	—	—	—	—	—	—	—	—	—	—	?	—	—	×	×	×
S. runtonensis	×	×	×	×	×	×	×	×	×	×	×	×	—	—	—	—	—
S. kennardi	—	—	—	—	—	—	—	—	—	?	—	—	×	×	×	×	—
S. margaritodon, etc.	—	—	—	—	—	—	×	×	×	×	×	×	×	—	—	—	—
Neomys fodiens, etc.	—	—	—	—	—	—	—	—	—	—	×	×	×	×	×	×	×
Beremendia fissidens	×	×	×	×	×	×	×	×	×	×	×	×	×	—	—	—	—
Petenyia hungarica	×	×	×	×	×	×	×	×	×	×	×	—	—	—	—	—	—
Soriculus kubinyi	×	×	×	×	×	×	×	×	—	—	—	—	—	—	—	—	—
Suncus hungaricus	×	×	×	×	×	×	×	—	—	—	—	—	—	—	—	—	—
Crocidura leucodon	—	—	—	—	—	—	—	—	—	—	—	—	×	×	×	×	×
C. russula	—	—	—	—	—	—	—	—	—	—	—	—	—	—	×	×	×
C. suaveolens	—	—	—	—	—	—	—	—	—	—	—	—	×	×	×	×	×
C. kornfeldi	×	×	×	×	×	×	×	×	×	×	—	—	—	—	—	—	—
Nesiotites spp.	—	—	—	—	—	—	—	?	?	?	×	×	×	×	×	×	—
Desmana moschata	—	—	—	—	—	—	—	—	—	—	×	×	×	×	×	×	×
D. thermalis	—	—	—	—	—	—	—	—	×	×	—	—	—	—	—	—	—
Talpa fossilis	×	×	×	×	×	×	×	×	×	×	×	×	—	—	—	—	—
T. europaea	—	—	—	—	—	—	—	—	—	—	—	—	×	×	×	×	×
T. gracilis	×	×	×	×	×	×	×	×	×	×	×	×	×	—	—	—	—
T. caeca	—	—	—	—	—	—	—	—	—	—	—	—	—	—	×	×	×

T. episcopalis	—	—	—	—	—	—	—	—	×	×	×	×	×	?	?	—	—
CHIROPTERA																	
Rhinolophus euryale	—	—	—	—	×	×	×	×	×	×	×	×	×	×	×	×	×
R. delphinensis	×	×	×	×	×	×	×	×	—	—	—	—	—	—	—	—	—
R. ferrum-equinum	—	—	—	—	—	—	—	—	?	?	?	?	×	×	×	×	×
R. hipposideros	—	—	—	—	—	—	—	—	—	—	—	—	×	×	×	×	×
Myotis daubentoni	—	—	—	—	—	—	—	—	×	×	×	×	×	×	×	×	×
M. dasycneme	×	×	×	×	×	×	×	×	×	×	×	×	×	×	×	×	×
M. mystacinus	—	—	—	—	—	—	—	—	—	—	×	×	×	×	×	×	×
M. emarginatus	—	—	—	—	—	—	—	—	×	×	×	×	×	×	×	×	×
M. nattereri	—	—	—	—	—	—	—	—	—	×	×	×	×	×	×	×	×
M. bechsteini	—	—	—	—	—	—	—	—	—	×	×	×	×	×	×	×	×
M. baranensis	—	—	—	—	×	×	×	×	×	×	×	—	—	—	—	—	—
M. myotis	—	—	—	—	—	—	—	—	—	—	—	—	×	×	×	×	×
M. oxygnathus	—	—	—	—	—	—	—	—	×	×	×	×	×	×	×	×	×
Plecotus auritus	—	—	—	—	—	—	—	—	×	×	×	×	×	×	×	×	×
P. crassidens	×	×	×	×	×	×	×	×	×	—	—	—	—	—	—	—	—
Miniopterus schreibersi	×	×	×	×	×	×	×	×	×	×	×	×	×	×	×	×	×
Barbastella barbastella	—	—	—	—	—	—	—	—	—	—	—	×	×	×	×	×	×
B. leucomelas	—	—	—	—	—	—	—	—	—	—	×	×	×	O	O	O	O
Pipistrellus pipistrellus	—	—	—	—	—	—	—	—	—	—	—	—	×	×	×	×	×
Vespertilio serotinus	—	—	—	—	—	—	—	—	—	—	—	×	×	×	×	×	×
V. nilssoni	—	—	—	—	—	—	—	—	—	—	×	×	×	×	×	×	×
V. murinus	—	—	—	—	—	—	—	—	—	—	—	—	—	—	×	×	×
V. praeglacialis	—	—	—	—	—	—	—	×	×	—	—	—	—	—	—	—	—
Nyctalus noctula	—	—	—	—	—	—	—	—	—	—	—	—	×	×	×	×	×
PRIMATES																	
Macaca florentina	—	×	×	×	×	×	×	×	×	×	×	×	—	—	—	—	—
M. sylvana	—	—	—	—	—	—	—	—	—	—	—	—	×	×	×	×	×
Dolichopithecus arvernensis	—	×	×	×	×	×	×	—	—	—	—	—	—	—	—	—	—
Homo heidelbergensis	—	—	—	—	—	—	—	—	—	—	×	×	—	—	—	—	—
H. neanderthalensis	—	—	—	—	—	—	—	—	—	—	—	—	×	×	×	×	—
H. sapiens	—	—	—	—	—	—	—	—	—	—	—	—	—	—	×	×	×

Table 15—continued

	Astian	Villafranchian Phases					Middle Pleistocene								Late Pleistocene		Recent
		a	b	c	d	e	A	1(i)	B	1(ii)	C	2	D	3	F	4	G
CARNIVORA																	
Genetta genetta	—	—	—	—	—	—	—	—	—	—	—	—	—	—	○	○	×
Herpestes ichneumon	—	—	—	—	—	—	—	—	—	—	—	—	—	—	○	○	×
Hyaena perrieri	—	×	×	×	×	×	×	×	×	×	×	—	—	—	—	—	—
H. brevirostris	—	—	—	—	—	×	×	×	×	×	—	—	—	—	—	—	—
H. hyaena	—	?	?	?	?	?	○	○	○	○	○	○	×	×	×	○	○
Euryboas lunensis	—	×	×	×	×	×	—	—	—	—	—	—	—	—	—	—	—
Crocuta crocuta	—	—	—	—	—	—	—	—	—	×	×	×	×	×	×	×	○
Megantereon megantereon	—	×	×	×	×	×	—	—	—	—	—	—	—	—	—	—	—
Homotherium sainzelli	—	×	×	×	×	×	×	×	×	—	—	—	—	—	—	—	—
H. latidens	—	—	—	—	—	—	—	—	×	×	×	×	×	×	×	×	—
Felis lunensis	—	—	?	?	?	×	×	×	×	×	×	—	—	—	—	—	—
F. silvestris	—	—	—	—	—	—	—	—	—	—	—	—	×	×	×	×	×
F. manul	—	—	—	—	—	—	—	—	—	—	?	?	?	?	?	×	○
F. chaus	—	—	—	—	—	—	—	—	—	—	—	—	—	—	×	○	○
F. issiodorensis	—	×	×	×	×	×	—	—	—	—	—	—	—	—	—	—	—
F. pardina	—	—	—	—	—	—	—	×	×	×	×	×	×	×	×	×	×
F. lynx	—	—	—	—	—	—	—	—	—	—	—	—	—	—	×	×	×
F. toscana	—	×	×	×	×	×	×	×	×	×	×	×	—	—	—	—	—
F. pardoides	—	—	—	×?	—	—	—	—	—	—	—	—	—	—	—	—	—
F. leo	—	—	—	—	—	—	—	—	—	—	×	×	×	×	×	×	○
F. pardus	—	—	—	—	—	—	—	—	—	—	×	×	×	×	×	×	○
Acinonyx pardinensis	—	×	×	×	×	×	×	×	×	×	—	—	—	—	—	—	—
Gulo schlosseri	—	—	—	—	—	—	—	—	×	×	×	—	—	—	—	—	—
G. gulo	—	—	—	—	—	—	—	—	—	—	—	×	×	×	×	×	×
Martes vetus	—	—	—	—	—	—	—	×	×	×	×	?	—	—	—	—	—

M. martes	—	—	—	—	—	—	—	—	—	—	—	—	×	×	×	×	×
M. foina	—	—	—	—	—	—	—	—	—	—	—	—	—	—	?	×	×
Baranogale antiqua	×	×	×	×	×	×	×	×	—	—	—	—	—	—	—	—	—
Vormela beremendensis	×	×	×	×	×	×	×	×	×	×	—	—	—	—	—	—	—
V. peregusna	—	—	—	—	—	—	—	—	—	—	—	—	—	—	○	○	×
Enhydrictis ardea	—	×	×	×	×	×	×	×	×	×	—	—	—	—	—	—	—
Pannonictis pliocaenica	×	×	×	×	×	×	×	×	×	×	×	—	—	—	—	—	—
Mustela lutreola	—	—	—	—	—	—	—	—	—	—	—	—	—	—	—	× ?	×
M. stromeri	—	—	—	—	×	×	×	×	×	×	—	—	—	—	—	—	—
M. putorius	—	—	—	—	—	—	—	—	—	—	×	×	×	×	×	×	×
M. eversmanni	—	—	—	—	—	—	—	—	—	—	—	—	—	—	×	×	×
M. palerminea	—	—	—	—	×	×	×	×	×	×	×	—	—	—	—	—	—
M. erminea	—	—	—	—	—	—	—	—	—	—	—	?	?	× ?	× ?	×	×
M. praenivalis	—	× ?	× ?	× ?	×	×	×	×	×	×	×	—	—	—	—	—	—
M. nivalis	—	—	—	—	—	—	—	—	—	—	—	×	×	×	×	×	×
M. rixosa	—	—	—	—	—	—	—	—	—	—	—	—	—	—	—	×	×
Meles thorali	—	—	—	×	—	—	—	—	—	—	—	—	—	—	—	—	—
M. meles	—	—	—	—	○?	○?	○?	×	×	×	×	×	×	×	×	×	×
Aonyx bravardi	—	×	×	×	×	×	×	× ?	× ?	× ?	—	—	—	—	—	—	—
A. antiqua	—	—	—	—	—	—	—	—	—	—	—	—	×	×	×	×	—
Lutra simplicidens	—	—	—	—	—	—	—	—	—	×	×	—	—	—	—	—	—
L. lutra	—	—	—	—	—	—	—	—	—	—	—	—	—	—	×	×	×
Canis falconeri	—	—	—	—	—	×	—	—	—	—	—	—	—	—	—	—	—
C. arnensis	—	?	—	—	—	×	×	× ?	× ?	× ?	—	—	—	—	—	—	—
C. etruscus	—	×	×	×	×	×	?	?	—	—	—	—	—	—	—	—	—
C. lupus	—	—	—	—	—	—	—	—	?	×	×	×	×	×	×	×	×
C. aureus	—	—	—	—	—	—	—	—	—	—	—	○	○	○	○	○	×
Cuon majori	—	—	—	—	—	×	×	×	×	—	—	—	—	—	—	—	—
C. alpinus	—	—	—	—	—	—	—	—	—	×	×	×	×	×	×	×	○
Lycaon lycaonoides	—	—	—	—	—	—	—	—	×	×	—	—	—	—	—	—	—
Vulpes alopecoides	—	—	×	×	×	×	×	×	×	×	—	—	—	—	—	—	—
V. vulpes	—	—	—	—	—	—	—	—	—	—	× ?	× ?	×	×	×	×	×

Table 15—continued

	Astian	Villafranchian Phases					Middle Pleistocene							Late Pleistocene			Recent
		a	b	c	d	e	A	1(i)	B	1(ii)	C	2	D	3	F	4	G
V. praecorsac	×	×	×	×	×	×	×	×	×	×	—	—	—	—	—	—	—
V. corsac	—	—	—	—	—	—	—	—	—	—	—	O	O	O	×	×	×
Alopex lagopus	—	—	—	—	—	—	—	—	—	—	—	—	—	×	×	×	×
Nyctereutes megamastoides	—	×	×	×	×	—	—	—	—	—	—	—	—	—	—	—	—
Parailurus anglicus	×	×	—	—	—	—	—	—	—	—	—	—	—	—	—	—	—
Agriotherium insigne	×	×	—	—	—	—	—	—	—	—	—	—	—	—	—	—	—
Ursus minimus	×	×	—	—	—	—	—	—	—	—	—	—	—	—	—	—	—
U. etruscus	—	—	×	×	×	×	×	—	—	—	—	—	—	—	—	—	—
U. deningeri	—	—	—	—	—	—	—	?	×	×	×	×	—	—	—	—	—
U. spelaeus	—	—	—	—	—	—	—	—	—	—	—	—	×	×	×	×	—
U. arctos	—	—	—	—	—	—	—	—	—	—	—	O	×	×	×	×	×
U. maritimus	—	—	—	—	—	—	—	—	—	—	—	—	—	—	×	×	×
U. thibetanus	—	—	—	—	—	—	—	×	×	×	×	×	×	×	O	O	O
PROBOSCIDEA																	
Anancus arvernensis	×	×	×	×	×	×	×	×	×	×	—	—	—	—	—	—	—
Zygolophodon borsoni	×	×	—	—	—	—	—	—	—	—	—	—	—	—	—	—	—
Archidiskodon meridionalis	—	—	—	×	×	×	×	×	×	×	×	—	—	—	—	—	—
Palaeoloxodon antiquus	—	—	—	—	—	—	—	—	—	—	×	×	×	×	×	×	—
Mammuthus trogontherii	—	—	—	—	—	—	—	—	—	×	×	×	×	—	—	—	—
M. primigenius	—	—	—	—	—	—	—	—	—	—	—	—	—	×	×	×	—
PERISSODACTYLA																	
Tapirus arvernensis	×	×	×	×	×	×	×	—	—	—	—	—	—	—	—	—	—
Dicerorhinus megarhinus	×	×	—	—	—	—	—	—	—	—	—	—	—	—	—	—	—
D. etruscus	—	—	×	×	×	×	×	×	×	×	×	×	—	—	—	—	—
D. kirchbergensis	—	—	—	—	—	—	—	—	—	—	—	×	×	×	×	×	—
D. hemitoechus	—	—	—	—	—	× ?	?	?	?	?	?	?	×	×	×	O	—

Coelodonta antiquitatis	—	—	—	—	—	—	—	—	—	—	—	○	○	×	×	×	—
Hipparion crusafonti, etc.	—	×	×	—	—	—	—	—	—	—	—	—	—	—	—	—	—
Equus stenonis	—	—	—	×	×	×	×	—	—	—	—	—	—	—	—	—	—
E. süssenbornensis	—	—	—	—	—	—	—	×	×	×	—	—	—	—	—	—	—
E. bressanus	—	—	×	×	×	×	×	× ?	× ?	?	—	—	—	—	—	—	—
E. mosbachensis	—	—	—	—	—	—	—	—	?	?	×	×	—	—	—	—	—
E. germanicus	—	—	—	—	—	—	—	—	—	—	—	—	×	×	×	×	—
E. przewalskii	—	—	—	—	—	—	—	—	—	—	—	—	—	—	—	×	× ○
E. hydruntinus	—	—	—	—	× ?	×	×	×	×	×	×	×	×	×	×	×	—
E. hemionus	—	—	—	—	—	—	—	—	—	—	—	—	—	×	×	×	○
ARTIODACTYLA																	
Sus arvernensis	× ?	×	—	—	—	—	—	—	—	—	—	—	—	—	—	—	—
S. strozzii	—	—	○	○	×	×	×	—	—	—	—	—	—	—	—	—	—
S. scrofa	—	—	—	—	—	—	—	—	—	×	×	×	×	×	×	×	×
Hippopotamus amphibius	—	—	—	—	—	×	×	×	×	×	×	×	×	×	×	×	○
'Cervus' ardei	—	×	—	—	—	—	—	—	—	—	—	—	—	—	—	—	—
'C.' issiodorensis	—	×	—	—	—	—	—	—	—	—	—	—	—	—	—	—	—
'C.' cusanus	—	×	×	×	×	—	—	—	—	—	—	—	—	—	—	—	—
'C.' perrieri	—	×	×	×	×	×	×	—	—	—	—	—	—	—	—	—	—
Anoglochis ramosus	×	×	×	×	×	×	×	—	—	—	—	—	—	—	—	—	—
Euctenoceros ctenoides	—	—	—	—	×	×	×	—	—	—	—	—	—	—	—	—	—
E. senezensis, etc.	—	—	×	×	×	×	×	×	×	×	×	—	—	—	—	—	—
Eucladoceros dicranios	—	—	—	—	—	×	—	—	—	—	—	—	—	—	—	—	—
E. falconeri-sedgwicki	—	×	×	×	×	×	×	×	×	× ?	× ?	—	—	—	—	—	—
Cervus etuerarium	—	×	×	×	×	×	×	?	?	?	?	× ?	—	—	—	—	—
C. elaphus	—	—	—	—	—	—	—	—	—	×	×	×	×	×	×	×	×
Praemegaceros verticornis	—	?	—	—	—	× ?	× ?	× ?	×	×	×	×	—	—	—	—	—
P. cazioti, etc.	—	—	—	—	—	—	—	—	—	—	—	×	×	×	×	×	× —
Megaloceros savini	—	—	—	—	—	—	—	—	—	×	×	×	—	—	—	—	—
M. giganteus	—	—	—	—	—	—	—	—	—	—	—	—	×	×	×	×	× —
Dama nestii	—	—	—	—	×	×	×	×	×	×	×	—	—	—	—	—	—
D. clactoniana	—	—	—	—	—	—	—	—	—	—	?	?	×	×	—	—	—

Table 15—continued

	Astian	Villafranchian Phases					Middle Pleistocene							Late Pleistocene			Recent
		a	b	c	d	e	A	1(i)	B	1(ii)	C	2	D	3	F	4	G
D. dama	—	—	—	—	—	—	—	—	—	—	—	—	—	—	×	×	○×
Capreolus capreolus	—	—	—	—	—	—	—	—	—	×	×	×	×	×	×	×	×
Alces gallicus	—	—	×	×	×	×	—	—	—	—	—	—	—	—	—	—	—
A. latifrons	—	—	—	—	—	—	—	—	×	×	×	×	—	—	—	—	—
A. alces	—	—	—	—	—	—	—	—	—	—	—	—	—	×	×	×	×
Rangifer tarandus	—	—	—	—	—	—	—	—	—	×	×	×	×	×	×	×	×
Gazella borbonica	—	×	×	×	×	×	—	—	—	—	—	—	—	—	—	—	—
Gazellospira torticornis	—	—	×	×	×	×	—	—	—	—	—	—	—	—	—	—	—
Saiga tatarica	—	—	—	—	—	—	—	—	—	—	—	—	—	—	○	×	×○
Procamptoceras brivatense	×	×	×	×	×	×	—	—	—	—	—	—	—	—	—	—	—
Gallogoral meneghinii	—	—	×	×	×	×	—	—	—	—	—	—	—	—	—	—	—
?Nemorhaedus melonii	—	—	—	—	—	—	—	—	—	?	?	× ?	× ?	×	×	×	×—
Rupicapra rupicapra	—	—	—	—	—	—	—	—	—	—	—	—	—	—	×	×	×
Myotragus balearicus	—	—	—	—	—	—	—	—	—	—	—	—	×	×	×	×	×—
Deperetia ardea	—	×	×	×	×	—	—	—	—	—	—	—	—	—	—	—	—
Praeovibos priscus	—	—	—	—	—	—	—	—	—	×	×	×	—	—	—	—	—
Ovibos moschatus	—	—	—	—	—	—	—	—	—	×	×	×	×	×	×	×	○
Megalovis latifrons	—	—	×	×	×	—	—	—	—	—	—	—	—	—	—	—	—
Ovis spp.	—	—	—	—	×	—	—	—	—	—	×	—	—	—	×	×	×
Soergelia elisabethae	—	—	—	—	—	—	—	—	—	×	×	×	—	—	—	—	—
Hesperoceras merlae	—	—	×	—	—	—	—	—	—	—	—	—	—	—	—	—	—
Capra ibex	—	—	—	—	—	—	?	—	—	?	?	—	—	×	×	×	×
Hemitragus bonali	—	—	—	—	—	—	—	—	—	—	—	—	—	—	×	×	—
H. stehlini	×	×	×	×	×	×	×	×	×	×	—	—	—	—	—	—	—
Leptobos elatus	—	×	×	×	×	×	—	—	—	—	—	—	—	—	—	—	—
L. etruscus	—	—	—	—	×	×	×	× ?	× ?	× ?	× ?	—	—	—	—	—	—

Bison priscus	—	—	—	—	—	—	—	—	—	×	×	×	×	×	×	×	—
B. schoetensacki	—	—	—	—	—	—	—	—	×	×	×	×	×	×	×	×	—
B. bonasus	—	—	—	—	—	—	—	—	—	—	—	—	—	—	—	○	×
Bubalus murrensis	—	—	—	—	—	—	—	—	—	—	—	—	×	—	—	—	—
Syncerus iselini	—	—	—	—	—	× ?	—	—	—	—	—	—	—	—	—	—	—
Bos primigenius	—	—	—	—	—	—	—	—	—	—	—	—	×	×	×	×	×
RODENTIA																	
Sciurus whitei	—	—	—	—	?	?	?	?	?	?	×	×	×	—	—	—	—
S. vulgaris	—	—	—	—	—	—	—	—	—	—	—	—	—	—	×	×	×
Marmota marmota	—	—	—	—	—	—	—	—	—	—	—	—	—	×	×	×	×
M. bobak	—	—	—	—	—	—	—	—	—	—	—	○	○	○	×	×	×
Citellus primigenius	—	—	—	—	—	×	×	×	×	×	—	—	—	—	—	—	—
C. citellus	—	—	—	—	—	—	—	—	—	×	×	×	×	×	×	×	×
C. suslicus	—	—	—	—	—	—	—	—	—	—	—	—	—	—	—	×	×
C. major	—	—	—	—	—	—	—	—	—	—	—	—	—	—	—	×	×
Pliosciuropterus schaubi	×	×	×	×	—	—	—	—	—	—	—	—	—	—	—	—	—
Castor fiber	—	× ?	×	×	×	×	×	×	×	×	×	×	×	×	×	×	×
Trogontherium cuvieri	—	—	—	×	×	×	×	×	×	×	×	×	×	—	—	—	—
Hystrix refossa	—	×	×	×	×	×	×	—	—	—	—	—	—	—	—	—	—
H. vinogradovi	—	—	—	—	—	—	—	—	—	×	×	×	×	×	×	×	—
H. cristata	—	—	—	—	—	—	—	—	—	—	—	—	—	—	×	×	×
Sicista praeloriger	—	—	—	—	—	—	—	—	×	×	×	×	×	—	—	—	—
S. betulina-subtilis	—	—	—	—	—	—	—	—	—	—	—	—	—	—	×	×	×
Allactaga jaculus	—	—	—	—	—	—	—	—	—	—	—	—	—	—	×	×	○
Prospalax priscus	×	×	×	×	×	×	×	×	×	×	—	—	—	—	—	—	—
Spalax leucodon, etc.	—	—	—	—	—	—	—	—	—	×	×	×	×	×	×	×	×
Glirulus pusillus	×	×	×	—	—	—	—	—	—	—	—	—	—	—	—	—	—
Muscardinus pliocaenicus	×	×	×	—	—	—	—	—	—	—	—	—	—	—	—	—	—
M. avellanarius	—	—	—	—	—	—	—	×	×	×	×	×	×	×	×	×	×
M. dacicus	× ?	× ?	× ?	× ?	× ?	× ?	× ?	× ?	×	×	×	—	—	—	—	—	—
Glis glis	—	—	—	—	—	—	—	×	×	×	×	×	×	×	×	×	×
G. minor	×	×	×	—	—	—	—	—	—	—	—	—	—	—	—	—	—

Table 15—continued

	Astian	Villafranchian Phases					Middle Pleistocene							Late Pleistocene			Recent
		a	b	c	d	e	A	1(*i*)	B	1(*ii*)	C	2	D	3	F	4	G
Eliomys quercinus	—	—	—	—	—	—	—	—	—	—	—	—	×	×	×	×	×
Dryomys nitedula	—	—	—	—	—	—	—	—	—	—	—	—	×	×	×	×	×
Leithia melitensis	—	—	—	—	—	—	—	—	—	—	?	?	?	×	×	×	—
Hypnomys spp.	—	—	—	—	—	—	—	—	—	—	?	?	?	×	×	×	—
Cricetus cricetus	—	—	—	—	—	—	—	×	×	×	×	×	×	×	×	×	×
C. praeglacialis	—	—	—	—	—	—	—	×	×	×	×	×	×	—	—	—	—
C. nanus	—	—	—	—	—	×	×	×	×	×	—	—	—	—	—	—	—
Cricetulus bursae	—	—	—	—	—	×	×	×	×	×	×	×	×	×	×	—	—
C. migratorius	—	—	—	—	—	—	—	—	×	×	×	×	×	×	×	×	×
Rhinocricetus ehiki	—	—	—	—	—	×	×	×	×	×	—	—	—	—	—	—	—
Dolomys milleri	×	×	×	×	×	×	—	—	—	—	—	—	—	—	—	—	—
Pliomys episcopalis	—	—	—	—	—	—	—	×	×	×	×	×	—	—	—	—	—
P. coronensis	—	—	—	—	—	—	—	—	×	×	×	×	×	—	—	—	—
Clethrionomys glareolus	× ?	× ?	× ?	× ?	× ?	× ?	× ?	× ?	× ?	× ?	×	×	×	×	×	×	×
C. esperi	—	—	—	—	—	—	—	—	—	×	—	—	—	—	—	—	—
Mimomys stehlini	×	×	×	—	—	—	—	—	—	—	—	—	—	—	—	—	—
M. pusillus	—	×	×	×	×.	×	×	×	×	×	—	—	—	—	—	—	—
M. reidi	—	×	×	×	×	×	×	×	×	×	×	—	—	—	—	—	—
M. newtoni	—	—	—	—	×	×	×	×	×	×	×	—	—	—	—	—	—
M. pliocaenicus	×	×	×	×	×	×	×	×	×	×	×	—	—	—	—	—	—
M. intermedius	—	—	—	—	—	×	×	×	×	×	×	×	—	—	—	—	—
M. rex	—	—	—	—	—	—	—	×	×	×	—	—	—	—	—	—	—
M. cantianus	—	—	—	—	—	—	—	—	—	—	—	—	×	—	—	—	—
Arvicola greeni	—	—	—	—	—	—	—	—	—	×	×	×	×	—	—	—	—
A. mosbachensis	—	—	—	—	—	—	—	—	—	×	×	×	×	—	—	—	—
A. terrestris-amphibius	—	—	—	—	—	—	—	—	—	—	—	—	×	×	×	×	×
Lagurus pannonicus	—	—	—	—	×	×	×	×	×	×	× ?	× ?	—	—	—	—	—
L. lagurus	—	—	—	—	—	—	—	—	—	—	—	—	×	×	×	×	×

Allophaiomys sp.	—	—	—	—	—	×	×	×	×	×	×	×	—	—	—	—	—
Microtus arvalis, etc.	—	—	—	—	—	—	—	—	—	×	×	×	×	×	×	×	×
M. agrestis group	—	—	—	—	—	—	—	—	—	—	—	—	×	×	×	×	×
M. guentheri	—	—	—	—	—	—	—	—	—	—	—	—	—	—	—	○	×
M. ratticeps group	—	—	—	—	—	—	—	—	—	×	×	×	×	×	×	×	×
M. nivalis group	—	—	—	—	—	—	—	—	—	×	×	×	×	×	×	×	×
M. gregalis	—	—	—	—	—	—	—	—	—	—	—	×	×	×	×	×	○
Pitymys gregaloides, etc.	—	—	—	—	—	—	—	—	×	×	×	×	×	—	—	—	—
P. subterraneus	—	—	—	—	—	—	—	—	—	—	—	—	—	—	×	×	×
Tyrrhenicola henseli	—	—	—	—	—	—	—	—	—	—	—	× ?	× ?	×	×	×	×—
Lemmus lemmus	—	—	—	—	—	—	—	—	—	×	×	×	×	×	×	×	×
Myopus schisticolor	—	—	—	—	—	—	—	—	—	—	—	—	—	—	—	○	×
Dicrostonyx torquatus	—	—	—	—	—	—	—	—	—	× ?	—	—	—	—	—	×	○
Parapodemus coronensis	×	×	×	×	×	×	×	×	×	×	—	—	—	—	—	—	—
Apodemus sylvaticus	×	×	×	×	×	×	×	×	×	×	×	×	×	×	×	×	×
A. flavicollis group	—	—	—	—	×	×	×	×	×	×	×	×	×	×	×	×	×
A. mystacinus	—	—	—	—	—	—	—	—	—	×	×	×	×	×	×	×	×
A. agrarius	—	—	—	—	—	—	—	—	—	—	—	—	×	×	×	×	×
Micromys minutus	—	—	—	—	—	—	—	—	—	—	—	—	×	×	×	×	×
Rhagamys orthodon	—	—	—	—	—	—	—	—	—	—	—	× ?	× ?	×	×	×	×—
Mus musculus	—	—	—	—	—	—	—	—	—	—	—	—	×	×	×	×	×
LAGOMORPHA																	
Prolagus sardus	—	—	—	—	—	—	—	—	—	—	—	× ?	× ?	×	×	×	×—
Ochotona pusilla	× ?	× ?	× ?	× ?	× ?	× ?	× ?	× ?	× ?	× ?	× ?	× ?	× ?	× ?	×	×	○
Hypolagus brachygnathus	×	×	×	×	×	×	×	×	×	×	×	—	—	—	—	—	—
Oryctolagus lacosti	—	×	×	×	×	×	—	—	—	—	—	—	—	—	—	—	—
O. cuniculus	—	—	—	—	—	—	—	—	—	—	—	—	×	×	×	×	×
Lepus terraerubrae	—	—	—	—	×	×	×	×	×	×	?	?	—	—	—	—	—
L. timidus	—	—	—	—	—	—	—	—	—	—	—	—	?	?	×	×	×
L. europaeus	—	—	—	—	—	—	—	—	—	—	—	—	—	—	×	×	×
L. capensis	—	—	—	—	—	—	—	—	—	× ?	× ?	× ?	× ?	× ?	× ?	×	×

a–e, Villafranchian phases. Glaciations enumerated 1–4, interglacials A–G. Stadials separated for 1-Günz only. × denotes existence, — no record, ○ presence in extra-European deposits only.

References

1 Abel, O. & Kyrle, G. (Ed.)
(1931) Die Drachenhöhle bei Mixnitz. *Speläol. Monogr.* **7–9**: 1–953.

2 Adam, K. D.
(1953) Die Bedeutung der altpleistozänen Säugetier-Faunen Südwestdeutschlands für die Gliederung des Eiszeitalters. *Geol. Bavarica* no. 19, 357–63.

3 — (1953) *Elephas meridionalis* Nesti aus den altpleistozänen Goldshöfer Sanden bei Aalen (Württemberg). *Eiszeitalter u. Gegenwart* **3**: 84–95.

4 — (1954) Die mittelpleistozänen Faunen von Steinheim an der Murr (Württemberg). *Quaternaria* **1**: 131–44.

5 — (1959) Mittelpleistozäne Caniden aus dem Heppenloch bei Gutenberg (Württemberg). *Stuttgarter Beitr. Naturkunde* no. 27, 1–46.

6 — (1961) Die Bedeutung der pleistozänen Säugetier-Faunen Mitteleuropas für die Geschichte des Eiszeitalters. *Stuttgarten Beitr. Naturkunde* no. 78, 1–34.

7 — (1964) Die Grossgliederung des Pleistozäns in Mitteleuropa. *Stuttgarter Beitr. Naturkunde* no. 132, 1–12.

8 Andrews, C. W.
(1914) A description of the skull and skeleton of a peculiarly modified rupicaprine antelope (*Myotragus balearicus*, Bate), with a notice of a new variety, *M. balearicus* var. *major*. *Philos. Trans. Roy. Soc. London*, ser. B, **206**: 281–305.

9 Arambourg, C.
(1957) Les gros mammifères des couches tayaciennes. *Mem. Arch. Inst. Paléont. humaine* no. 29 (La Grotte de Fontéchevade), 185–229.

10 Azzaroli, A.
(1948) Revisione della fauna dei terreni fluvio-lacustri del Valdarno Superiore. 3. I cervi fossili della Toscana con particolare riguardo alle specie Villafranchiane. *Palaeontogr. Italica* **43**: 45–82.

11 — (1950) Osservazioni sulla formazione villafranchiane di Olivola in Val di Magra. *Atti Soc. Tosc. Sci. Nat.*, Mem. **57**: 3–10.

12 — (1952) L'Alce di Senèze. *Palaeontogr. Italica* **47**: 133–41.

13 — (1953) The deer of the Weybourn Crag and Forest Bed, Norfolk. *Bull. British Mus. Nat. Hist.*, (Geol.) **2**, no. 1, 1–96.

14 — (1954) Filogenesi e biologia di *Sus strozzii* e di *Sus minor*. *Palaeontogr. Italica* **58**: 41–76.

15 — (1961) Il nanismo nei cervi insulari. *Palaeontogr. Italica* **56**: 1–32.

16 Azzaroli, A. (1963) Validità della specie *Rhinoceros hemitoechus* Falconer. *Palaeontogr. Italica* **57**: 21–34.

17 — (1965) The two Villafranchian horses of the Upper Valdarno. *Palaeontogr. Italica* **59**: 1–12.

18 — (1966) Pleistocene and living horses of the Old World. *Palaeontogr. Italica* **61**: 1–15.

19 Ballesio, R.
(1963) Monographie d'un *Machairodus* du gisement villafranchien de Senèze: *Homotherium crenatidens* Fabrini. *Trav. Lab. Géol. Lyon*, new ser., **9**: 1–129.

20 Banfield, A. W. F.
(1961) A revision of the Reindeer and Caribou, genus *Rangifer*. *Bull. Natl. Mus. Canada* **177**: 1–137.

21 Bate, D. M. A.
(1919) On a new genus of extinct muscardine rodent from the Balearic Islands. *Proc. Zool. Soc. London*, 1918, 209–22.

22 — (1935) Note on the habits of *Enhydrictis galictoides*, with description of some bones of this mustelid from the Pleistocene of Sardinia. *Proc. Zool. Soc. London* 241–45.

23 — (1935) Two new mammals from the Pleistocene of Malta, with notes on the associated fauna. *Proc. Zool. Soc. London* 247–64.

24 — (1944) Pleistocene shrews from the larger western Mediterranean islands. *Ann. Mag. Nat. Hist.* **11**: 738–69.

25 — (1945) The Pleistocene mole of Sardinia. *Ann. Mag. Nat. Hist.* **12**: 448–61.

26 — (1950) The 'Licorne' of Lascaux: is it *Pantholops*? *Archaeol. News Letter* **2**: 182–84.

27 Bauer, K.
(1960) Die Säugetiere des Neusiedlersee-Gebietes (Österreich). *Bonn. Zool. Beitr.* **11**: 141–344.

28 Begouen, Comte & Koby, F. E.
(1951) Le crâne de Glouton de la caverne des Trois-Frères (Ariège). *Bull. Soc. Préhist. Ariège* **5**: 1–20.

29 Berckhemer, F.
(1927) *Buffelus murrensis* n. sp. ein diluvialer Büffelschädel von Steinheim a. d. Murr. *Jahresh. Ver. Naturk. Württemberg* **83**: 146–58.

30 — (1940) Über die Riesenhirschfunde von Steinheim an der Murr. *Jahrh. Ver. vaterl. Naturk. Württemberg* **96**: 63–88.

31 Bohlken, H.
(1961) Haustiere und zoologische Systematik. *Zeitschr. Tierzücht. Züchtungsbiol.* **76**: 107–13.

32 — (1962) Probleme der Merkmalsbewertung am Säugetierschädel, dargestellt am Beispiel des *Bos primigenius* Bojanus 1827. *Morphol. Jahrb.* **103**: 509–661.

33 Bolomey, A.
(1965) Contribution à la connaissance de la morphologie de *Pliotragus ardeus*. *Rev. Roumaine de Biol., sér. Zool.* **10**: 315–23.

34 Bonifay, M. F. & Bonifay, E.
(1963) Un gisement à faune épi-villafranchienne à Saint-Estève-Janson (Bouches-du-Rhône). *C.R. Acad. Sci. Paris* **256**: 1136–38.

35 BONIFAY, M. F. & BONIFAY, E. (1965) Age du gisement de mammifère fossiles de Lunel-Viel (Hérault). *C.R. Acad. Sci. Paris* **260**: 3441–44.

36 BOSWELL, P. G. H. (1931) The stratigraphy of the glacial deposits of East Anglia in relation to earl Man. *Proc. Geol. Assoc. London* **42**: 87–111.

37 BOULE, M. (1893) Description de l'*Hyaena brevirostris* du Pliocène de Sainzelles près L Puy. *Ann. Sci. Nat.*, Zool. **15**: 85–97.

38 (1902) La caverne à ossements de Montmaurin (Haute-Garonne). *L'Anthropologie* **13**: 305–19.

39 BOURDIER, F. (1963) *Le bassin du Rhône au Quaternaire*, vols. I–II. Paris, Centre Natl. Rech Sci.

40 BOURDIER, F. & LACASSAGNE, H. (1963) Précisions nouvelles sur la stratigraphie et la faune du gisement villefranchien de Saint-Prest (Eure-et-Loir). *Bull. Soc. Géol. France* **5**: 446–53.

41 BRAIN, C. K. (1956) Some aspects of the Ape-Man period in the Transvaal. *Bull. Sout African Mus. Assoc.* **6**: 171–76.

42 BRANDT, J. (1878) Mitteilungen über die Gattung *Elasmotherium*, besonders den Schädelbau derselben. *Mem. Acad. Sci. Petersbourg* **26**: 1–36.

43 BROECKER, W. S. (1966) Absolute dating and the astronomical theory of glaciation. *Science* **151** 299–304.

44 BRUNNER, G. (1956) Nachtrag zur Kleinen Teufelshöhle bei Pottenstein (Oberfranken). *N Jahrb. Geol. Paläont., Monatsh.*, 75–100.

45 — (1957) Die Breitenberghöhle bei Gössweinstein/Ofr. Eine Mindel-Riss und eine postglaziale Mediterran-Fauna. *N. Jahrb. Geol. Paläont., Monatsh* 352–78, 385–88.

46 BUTZER, K. (1964) *Environment and archaeology*. Chicago, Aldine.

47 CHAUVIRÉ, C. (1962) Les gisements fossilifères quaternaires de Châtillon-Saint-Jea (Drôme). Thèse no. 62, *Fac. Sci. Univ. Lyon* 1–216.

48 COLBERT, E. H. & HOOIJER, D. A. (1953) Pleistocene mammals from the limestone fissures of Szechwan, China *Bull. Amer. Mus. Nat. Hist.* **102**: 1–134.

49 COUTURIER, M. (1938) *Le Chamois*. Grenoble.

50 CRUSAFONT PAIRÓ, M. (1961) *El cuaternario español y su fauna de mamíferos. Ensayo de sintesis*. Speleon vol. 13, 181–99.

51 — (1965) Zur Obergrenze des Villafranchiums in Spanien. *Ber. Geol. Ge DDR* **10**: 19–48.

52 DEGERBØL, M.
(1933) Danmarks Pattedyr i Fortiden i Sammenligning med recente Former, I. (With an English summary.) *Vidensk. Meddel. Dansk. Naturhist. Foren.* **96**: 357–641.

53 — (1961) On a find of a Preboreal domestic dog (*Canis familiaris* L.) from Starr Carr, Yorkshire, with remarks on other Mesolithic dogs. *Proc. Prehist. Soc.* **27**: 35–55.

54 — (1963) Prehistoric cattle in Denmark and adjacent areas. *Occ. Papers Roy. Anthrop. Inst.* no. 18, 68–79.

55 DEGERBØL, M. & IVERSEN, J.
(1945) The Bison of Denmark. A zoological and geological investigation of the finds in Danish Pleistocene deposits. *Danmarks Geol. Unders.* II. Raekke, no. 73, 1–62.

56 DEHM, R.
(1962) Altpleistocäne Säuger von Schernfeld bei Eichstätt in Bayern. *Mitt. Bayer. Staatssamml.* Paläont. hist. Geol. **2**: 17–61.

57 DEL CAMPANA, D.
(1913) I cani pliocenici di Toscana. *Palaeontogr. Italica* **19**: 189–254.

58 DEPÉRET, C.
(1885) Considérations générales sur les vertébrés pliocènes de l'Europe. *Ann. Soc. Géol. France* **17**: 231–72.

59 — (1889) Sur le *Dolichopithecus ruscinensis*, nouveau singe fossile du pliocène du Roussillon. *C.R. Acad. Sci. Paris* **109**: 982–83.

60 — (1929) *Dolichopithecus arvernensis*. Nouveau singe du pliocène supérieur de Senèze (Hte-Loire). *Trav. Lab. Géol. Fac. Sci. Lyon* **15**, mem. 12.

61 DEPÉRET, C. L. M. & ROMAN, F.
(1923) Les éléphants pliocènes. I. *Elephas planifrons* Falc., des sables de Chagne et faunes de mammifères d'âge villafranchien – saint-prestien. *Ann. Univ. Lyon* **1**, no. 43: 1–87.

62 DERANIYAGALA, P. E. P.
(1951) Some hippopotamuses of Ceylon, Western Asia, East Africa and England. *Spolia Zeylanica* **26**: 125–32.

63 DIETRICH, W. O.
(1938) Zur Kenntnis der oberpliozänen echten Hirsche. *Zeitschr. deutsch. geol. Ges.* **90**: 261–67.

64 — (1953) Neue Funde des etruskischen Nashorns in Deutschland und die Frage der Villafranchium-Faunen. *Geologie* **2**: 417–30.

65 — (1958) Übergangsformen des Südelefanten (*Elephas meridionalis* Nesti) im Altpleistozän Thüringens. *Geologie* **7**: 797–807.

66 — (1959) *Hemionus* Pallas im Pleistozän von Berlin. *Vertebrata Palasiatica* **3**: 13–22.

67 DONNER, J. J. & KURTÉN, B.
(1958) The floral and faunal succession of Cueva del Toll, Spain. *Eiszeitalter u. Gegenwart* **9**: 72–82.

68 EBERL, B.
(1930) *Die Eiszeitenfolge am nördlichen Alpenvorlande.* Augsburg.

69 EHRENBERG, K.
(1938–1940) (Ed.). Die Fuchs- oder Teufelslucken bei Eggenburg, Niederdonau. *Abh. Zool. Bot. Ges. Wien* **17**: 1–301.

70 — (1942) Berichte über Ausgrabungen in der Salzofenhöhle im Toten Gebirge. II. Untersuchungen über umfassendere Skelettfunde als Beitrag zur Frage der Form- und Grössenverschiedenheiten zwischen Braunbär und Höhlenbär. *Palaeobiologica* **7**: 531–666.

71 — (1962) Über Lebensweise und Lebensraum des Höhlenbären. Eine paläobiologische Studie. *Verh. Zool. Bot. Ges. Wien* **101–102**: 18–31.

72 — (1964) Ein Jungbärenskelett und andere Höhlenbärenreste aus der Bärenhöle im Hartlesgraben bei Hieflau (Steiermark). *Ann. Naturhist. Mus. Wien* **67**: 189–252.

73 ELLERMAN, J. R. & MORRISON-SCOTT, T. C. S.
(1951) *Checklist of Palaearctic and Indian mammals, 1758 to 1946*. London, British Museum (Natural History).

74 EMILIANI, C.
(1958) Ancient temperatures. *Scient. American* 2–11.

75 — (1961) Cenozoic climatic changes as indicated by the stratigraphy and chronology of deep-sea cores of *Globigerina*-ooze facies. *Ann. New York Acad. Sci.* **95**: 521–36.

76 ERDBRINK, D. P.
(1953) *A review of fossil and recent bears of the Old World*. 2 vols. Deventer, Jan de Lange.

77 ERICSSON, D. B., EWING, M. & WOLLIN, G.
(1964) The Pleistocene Epoch in deep-sea sediments. *Science* **146**: no. 3645; 723–32.

78 EVERNDEN, J. F. & JAMES, G. T.
(1964) Potassium-argon dates and the Tertiary floras of North America. *American Jour. Sci.* **262**: 945–74.

79 EVERNDEN, J. F., SAVAGE, D. E., CURTIS, G. H. & JAMES, G. T.
(1964) Potassium-argon dates and the Cenozoic mammalian chronology of North America. *American Jour. Sci.* **262**: 145–98.

80 EWER, R. F.
(1964) Large Carnivora of Bed II. In: L. S. B. Leakey, *Olduvai Gorge 1951–61*; **1**: 21–22. Cambridge Univ. Press.

81 FABRINI, E.
(1896) La Lince del Plioceno italiano. *Palaeontogr. Italica* **2**: 1–24.

82 FEJFAR, O.
(1961) Review of Quaternary vertebrata in Czechoslovakia. *Prace Instytutu Geol. Warszawa* **34**: 109–18.

83 — (1964) The Lower Villafranchian vertebrates from Hajnáčka near Filákovo in Southern Slovakia. *Rozpravy, Ustredního ú. geol. Prague* no. 30, 1–115.

84 FUCHS, T.
(1880) Referat über die pliocäne Säugetierfauna Ungarns. *N. Jahrb. Mineral* **2**: 388.

85 GARUTT, V. E.
(1964) *Das Mammut* (Mammuthus primigenius *Blumenb.*). Neue Brehm-Bücherei, Wittenberg-Lutherstadt, Ziemsen.

86 GIGNOUX, M.
(1913) Les formations marines pliocènes et quaternaires de l'Italie du Sud et de la Sicile. *Ann. Univ. Lyon* **1**: no. 36.

87 GROMOVA, V.
(1949) Histoire des chevaux (genre *Equus*) de l'Ancien Monde. *Trav. Inst. Paléont. Acad. Sci. URSS* **7**: 1–373. (French translation by Piedresson de St. Auboin, *Ann. Cent. Etud. Docum. Paléont.* no. 13, Paris 1955.)

88 GROSS, H.
(1965) Die geochronologischen Befunde der Bären- oder Tischoferhöhle bei Kufstein am Inn. *Quartär* **15–16**: 133–41.

89 GUENTHER, E. W.
(1964) Säugetierreste aus eiszeitlichen Ablagerungen von Schleswig-Holstein. *Lauenburgische Heimat* no. 45, 48–52.

90 GUÉRIN, C.
(1965) *Gallogoral* (nov. gen.) *meneghinii* (Rütimeyer, 1878), un rupicapriné du Villafranchien d'Europe occidentale. *Doc. Lab. Géol. Lyon* **11**: 1–353.

91 HARLÉ, E.
(1910) La *Hyaena intermedia* et les ossemens humatiles des cavernes de Lunel-Viel. *Bull. Soc. Géol. France* ser. 4, **10**: 34–50.

92 HAUG, E.
(1911) *Traité de géologie*, **2**: Les périodes géologiques, no. 3, 1397–2024.

93 HEINTZ, A.
(1958) On the pollen analysis of the stomach contents of the Beresovka mammoth. (In Norwegian, with an English summary.) *Blyttia*, Oslo, **16**: 122–42.

94 HEINTZ, A. & GARUTT, V. E.
(1965) Determination of the absolute age of the fossil remains of Mammoth and Woolly Rhinoceros from the permafrost in Siberia by the help of radio carbon (C_{14}). *Norsk. Geol. Tidsskr.* **45**: 73–79.

95 HELBING, H.
(1934) *Felis* (*Catolynx*) *chaus* Güld. aus dem Travertin von Untertürkheim bei Stuttgart. *Ecl. Geol. Helvetiae* **27**: 443–57.

96 — (1935) *Cyrnaonyx antiqua* (Blainv.), ein Lutrine aus dem europaeischen Pleistocaen. *Ecl. Geol. Helvetiae* **28**: 563–77.

97 HELLER, F.
(1930) Eine Forest-Bed-Fauna aus der Sackdillinger Höhle (Oberpfalz). *N. Jahrb. Mineral.* Beil.-Band **83**, Abt. B, 281–82.

98 — (1936) Eine Forest-Bed-Fauna aus der Schwäbischen Alb. *Sitzber. Heidelberg Akad. Wiss.* 2 Abh., 1–29.

99 — (1955) Zur Diluvialfauna des Fuchsloches bei Siegmannsbrunn, Ldkr. Pegnitz. *Geol. Bl. NO-Bayern* **5**: 49–70.

100 — (1956) Thomas Grebners bisher unveröffentlichte 'Descriptio antri subterranei prope Galgenreuth' aus dem Jahre 1748. *Geol. Bl. NO-Bayern* **6**: 32–40.

101 HELLER, F. (1957) Die fossilen Gattungen *Mimomys* F. Maj., *Cosomys* Wil. und *Ogmodontomys* Hibb. (Rodentia, Microtinae) in ihren systematischen Beziehungen. *Acta Zool. Cracoviensia* **2**: 219–37.

102 — (1958) Eine neue altquartäre Wirbeltierfauna von Erpfingen (Schwäbische Alb). *N. Jahrb. Geol. Paläont.*, Abh. **107**: 1–102.

103 HILDEBRAND, M.
(1961) Further studies on locomotion of the Cheetah. *Jour. Mammalogy* **42**: 84–91.

104 HINTON, M. A. C.
(1911) The British fossil shrews. *Geol. Mag.* **5**: 529–39.

105 — (1926) *Monograph of the voles and lemmings (Microtinae) living and extinct*, **1**: London, Brit. Mus. (Nat. hist.).

106 HOOIJER, D. A.
(1945) A fossil gazelle (*Gazella schreuderae* nov. spec.) from the Netherlands. *Zool. Meded.* **25**: 55–64.

107 — (1950) The fossil Hippopotamidae of Asia, with notes on the recent species. *Zool. Verhand. Leiden*, no. 8, 1–124.

108 — (1958) An early Pleistocene mammalian fauna from Bethlehem. *Bull. British Mus. Nat. Hist.* (Geol.) **3**: 267–92.

109 — (1959) *Trogontherium cuvieri* Fischer from the Neede Clay (Mindel-Riss Interglacial) of the Netherlands. *Zool. Meded.* **36**: 275–80.

110 HOPKINS, D. M.
(1959) Cenozoic history of the Bering land bridge. *Science*, **129**: no. 3362, 1519–28.

111 HOPWOOD, A. T.
(1935) Fossil elephants and Man. *Proc. Geol. Assoc.* **46**: 46–60.

112 HOWELL, F. C.
(1960) European and Northwest African Middle Pleistocene hominids. *Current Anthropology* **1**: 195–232.

113 HUGHES, T. M.
(1911) Excursions to Cambridge and Barrington. *Proc. Geol. Assoc.* **22**: 268–78.

114 IMBRIE, J.
(1957) The species problem with fossil animals. *Publ. Amer. Assoc. Adv. Sci.* no. 50, 125–53.

115 JÁNOSSY, D.
(1955) Die Vogel- und Säugetierreste der spätpleistozänen Schichten der Höhle von Istállóskö. *Acta Arch.* **5**: 149–81.

116 — (1961) Die Entwicklung der Kleinsäugerfauna Europas im Pleistozän (Insectivora, Rodentia, Lagomorpha). *Zeitschr. Säugetierk.* **26**: 1–64.

117 — (1962) Vorläufige Mitteilung über die mittelpleistozäne Vertebratenfauna der Tarkö-Felsnische (NO-Ungarn, Bükk-Gebirge). *Ann. Hist. Nat. Mus. Natl. Hung. Min. Paleont.*, **54**: 155–74.

118 — (1963) Die altpleistozäne Wirbeltierfauna von Kövesvárad bei Répáshuta (Bükk-Gebirge). *Ann. Hist. Nat. Mus. Natl. Hung.* **55**: 109–40.

119 — (1963–1964) Letztinterglaziale Vertebraten-Fauna aus der Kálmán Lambrecht-Höhle (Bükk-Gebirge, Nordost-Ungarn). 2 pts. *Acta Zool. Acad. Sci. Hung.* **9**: 293–331, **10**, 139–97.

120 KAHLKE, H. D.
(1958) Die Cervidenreste aus den altpleistozänen Tonen von Voigtstedt bei Sangerhausen. *Abh. Deutsch. Akad. Wiss.* 1956, no. 9, 1–50.

121 — (1958) Die jungpleistozänen Säugetierfaunen aus dem Travertingebiet von Taubach-Weimar-Ehringsdorf. Vorbericht. *Alt-Thüringen* **3**: 97–130.

122 — (1959) *Die Cervidenreste aus den altpleistozänen Ilmkiesen von Süssenborn bei Weimar.* 3 vols. (1956–9). Berlin, Akademie-Verlag.

123 — (1961) Revision der Säugetierfaunen der klassischen deutschen Pleistozän-Fundstellen von Süssenborn, Mosbach und Taubach. *Geologie* **10**: 493–532.

124 — (1963) *Rangifer* aus den Sanden von Mosbach. *Paläont. Zeitschr.* **37**: 277–282.

125 — (1964) Early Middle Pleistocene (Mindel/Elster) *Praeovibos* and *Ovibos*. *Comment. Biol. Soc. Sci. Fennica* **26**: 1–17.

126 — (1965) Die Cerviden-Reste aus den Tonen von Voigtstedt in Thüringen. *Paläont. Abh.*, ser. A **2**: 379–426.

127 — (1965a) Die Rhinocerotiden-Reste aus den Tonen von Voigtstedt in Thüringen. *Paläont. Abh.*, ser. A **2**: 451–519.

128 KOBY, F. E.
(1941) Contribution à l'étude de *Felis spelaea* Goldf. *Verh. Naturf. Ges. Basel* **52**: 168–88.

129 — (1951) Le Putois d'Eversmann fossile en Suisse et en France. *Ecl. Geol. Helvetiae* **44**: 394–98.

130 — (1953) Les paléolithiques ont-ils chassé l'ours des cavernes? *Actes Soc. Jurass. Emul.* 1–48.

131 — (1956) Une représentation de Tahr (*Hemitragus*) à Cougnac? *Bull. Soc. Préhist. France* **53**: 103–107.

132 — (1959) Contribution au diagnostic ostéologique différentiel de *Lepus timidus* Linné et *L. europaeus* Pallas. *Verh. Naturf. Ges. Basel* **70**: 19–44.

133 — (1960) Contribution à la connaissance des lièvres fossiles, principalement de ceus de la dernière glaciation. *Verh. Naturf. Ges. Basel* **71**: 149–73.

134 KOBY, F. & SCHAEFER, H.
(1960) Der Höhlenbär. *Veröff. Naturhist. Mus. Basel* no. 2, 1–24.

135 KOENIGSWALD, G. H. R. VON (Ed.)
(1958) *Hundert Jahre Neanderthaler. Neanderthal Centenary 1856–1956.* Köln & Graz, Böhlau.

136 KORMOS, T.
(1930) *Desmana thermalis* n. sp., eine neue Bisamspitzmaus aus Ungarn. *Ann. Mus. Natl. Hung.* **27**: 1–19.

137 — (1930) Diagnosen neuer Säugetiere aus der oberpliozänen Fauna des Somlyóberges bei Püspökfürdö. *Ann. Mus. Natl. Hung.* **27**: 287–328.

138 — (1931) *Pannonictis pliocaenicus* n. g., n. sp., a new giant mustelid from the late Pliocene of Hungary. *Ann. Inst. Reg. Hung.* (Geol.) **29**: 1–14.

139 — (1932) Neue Wühlmäuse aus dem Oberpliocän von Püspökfürdö. *N. Jahrb. Mineral.* Beil.-Band **69**: Abt. B, 323–46.

140 — (1932) Die Füchse des ungarischen Oberpliocäns. *Folia Zool. Hydrobiol.* **4**: 167–88.

141 KORMOS, T. (1934) Neue und wenig bekannte Musteliden aus dem ungarischen Oberpliozän. *Folia Zool. Hydrobiol.* **5**: 129–58.

142 — (1934) Neue Insectenfresser, Fledermäuse und Nager aus dem Oberpliocän der Villányer Gegend. *Fözlöny* **64**: 296–321.

143 — (1935) Die perlzähnige Spitzmaus (*Sorex margaritodom* Korm.) und das Anpassungsproblem. *Állattani Közlemények* **32**: 61–79.

144 — (1935) Beiträge zur Kenntnis der Gartung *Parailurus*. *Mitteil. Jahrb. Kgl. Ungar. Geol. Anst.* **30**: no. 2, 1–38.

145 — (1937) Zur Geschichte und Geologie der oberpliocänen Knochenbreccien des Villányer Gebirges. *Math. Nat. Anz. Ungar. Akad. Wiss.* **56**: 1061–1100.

146 — (1937) Zur Frage der Abstammung und Herkunft der quartären Säugetierfauna Europas. *Festschr. Embrik Strand.* Riga. **3**: 287–328.

147 — (1937) Revision der Kleinsäuger von Hundsheim in Niederösterreich. *Földtani Közlöny* **67**: 1–15; 157–71.

148 — (1938) *Mimomys newtoni* F. Major und *Lagurus pannonicus* Korm., zwei gleichzeitige verwandte Wühlmäuse von verschiedener phylogenetischen Entwicklung. *Math. Nat. Anz. Ungar. Akad. Wiss.* **57**: 353–78.

149 KOWALSKI, K.
(1956) Insectivores, bats and rodents from the early Pleistocene bone breccia of Podlesice near Kroczyce (Poland). *Acta Palaeont. Polonica* **1**: 331–94.

150 — (1958) An early Pleistocene fauna of small mammals from the Kadzielnia Hill in Kielce (Poland). *Acta Palaeont. Polonica* **3**: 1–47.

151 — (1958) Altpleistozäne Kleinsäugerfauna von Podumci in Norddalmatien. *Palaeont. Jugoslavica* no. 2, 1–30.

152 — (1959) *Katalog ssaków plejstocenu polski.* (A catalogue of the Pleistocene mammals of Poland.) Polska Akad. Nauk, Wroclaw.

153 — (1960) Pliocene insectivores and rodents from Rebielice Królewskie (Poland). *Acta Zool. Cracoviensia* **5**: 155–94.

154 — (1960) An early Pleistocene fauna of small mammals from Kamyk (Poland). *Folia Quaternaria* no. 1, 1–24.

155 — (1960) *Prospalax priscus* (Nehring) (Spalacidae, Rodentia) from the Pliocene of Poland. *Anthropos*, suppl. 1960, 109–14.

156 — (1962) Fauna of bats from the Pliocene of Weze in Poland. *Acta Zool. Cracoviensia* **7**: 39–51.

157 — (1963) The Pliocene and Pleistocene Gliridae (Mammalia, Rodentia) from Poland. *Acta Zool. Cracoviensia* **8**: 533–67.

158 — (1964) Palaeoecology of mammals from the Pliocene and early Pleistocene of Poland. *Acta Theriologica* **8**: 73–88.

159 — (1967) *Lagurus lagurus* (Pallas, 1773) and *Cricetus cricetus* (Linnaeus, 1758) (Rodentia, Mammalia) in the Pleistocene of England. *Acta Zool. Cracoviensia* **12**: 111–22.

160 KRETZOI, M.
(1938) Die Raubtiere von Gombaszög. *Ann. Mus. Natl. Hung.* **31**: 88–157.

161 — (1941) Die unterpleistozäne Säugetierfauna von Betfia bei Nagyvárad. *Földtani Közlöny* **71**: 308–35.

162 — (1956) Die altpleistozänen Wirbeltierfaunen des Villányer Gebirges. *Geol. Hungarica*, ser. Palaeont. no. 27, 1–264.

163 KRETZOI, M. (1961) Stratigraphie und Chronologie. *Prace Instytutu Geol. Warszawa* **34**: 313–29.

164 KROTT, P.
(1959) *Der Vielfrass* (Gulo gulo L. 1758). *Zur Kenntnis seiner Naturgeschichte und seiner Bedeutung für den Menschen.* Monogr. Wildsäugetiere, **13**. Jena, Fischer.

165 KURTÉN, B.
(1955) Sex dimorphism and size trends in the Cave Bear, *Ursus spelaeus* Rosenmüller and Heinroth. *Acta Zool. Fennica* no. 90, 1–48.

166 —(1956) The status and affinities of *Hyaena sinensis* Owen and *Hyaena ultima* Matsumoto. *American Mus. Novitates* no. 1764, 1–48.

167 —(1957) Mammal migrations, Cenozoic stratigraphy, and the age of Peking Man and the australopithecines. *Jour. Paleont.* **31**: 215–27.

168 —(1958) Life and death of the Pleistocene Cave Bear. A study in paleoecology. *Acta Zool. Fennica* no. 95, 1–59.

169 —(1958) The bears and hyenas of the interglacials. *Quaternaria* **4**: 69–81.

170 —(1958) A differentiation index, and a new measure of evolutionary rates. *Evolution* **12**: 146–57.

171 —(1959) On the bears of the Holsteinian Interglacial. *Stockholm Contr. Geol.* **2**: 73–102.

172 —(1959) On the longevity of mammalian species in the Tertiary. *Comment. Biol. Soc. Sci. Fennica* **21**: no. 4, 1–14.

173 —(1960) Chronology and faunal evolution of the earlier European glaciations. *Comment. Biol. Soc. Sci. Fennica* **21**: no. 5, 1–62.

174 —(1960) Rates of evolution in fossil mammals. *Cold Spring Harbor Symp. Quant. Biol.* **24**: 205–15.

175 —(1963) Villafranchian faunal evolution. *Comment. Biol. Soc. Sci. Fennica* **24**: no. 3, 1–18.

176 —(1963) Return of a lost structure in the evolution of the felid dentition. *Comment. Biol. Soc. Sci. Fennica* **24**: no. 4, 1–12.

177 —(1963) Notes on some Pleistocene mammal migrations from the Palaearctic to the Nearctic. *Eiszeitalter u. Gegenwart* **14**: 96–103.

178 —(1964) The evolution of the Polar Bear, *Ursus maritimus* Phipps. *Acta Zool. Fennica* no. 108, 1–26.

179 —(1965) The Carnivora of the Palestine caves. *Acta Zool. Fennica* no. 107, 1–74.

180 —(1965) The Pleistocene Felidae of Florida. *Bull. Florida State Mus.* **9**: no. 6, 215–73.

181 —(1965) On the evolution of the European Wild Cat, *Felis silvestris* Schreber. *Acta Zool. Fennica* no. 111, 1–26.

182 —(1966) Pleistocene mammals and the Bering bridge. *Comment. Biol. Soc. Sci. Fennica* **29**: no. 8, 1–7.

183 KUSS, S. E.
(1957) Altpleistozäne Reste des *Hippopotamus antiquus* Desmarest. *Jahresh. Geol. Landesamt Baden-Württemberg* **2**: 299–331.

184 KÜTHE, K. H.
(1932) *Sus scrofa mosbachensis. Notizbl. Verh. Erd. u. Hess. Geol. Landesanst. Darmstadt* **5**: 117–24.

185 LA BAUME, W.
(1947) Diluviale Schädel vom Ur (*Bos primigenius* Bojanus) aus Toscana. *Ecl. Geol. Helvetiae* **40**: no. 2, 299–308.

186 LEHMANN, U.
(1949) Die Ur im Diluvium Deutschlands und seine Verbreitung. *N. Jahrb. Mineral.* Abh., **90**: Abt. B, 163–266.

187 — (1954) Die Fauna des 'Vogelherds' bei Stetten ob Lontal (Württemberg). *N. Jahrb. Mineral.* **99**: 33–146.

188 — (1965) Der Eiszeitwisent (*Bison priscus* Boj.). *Erläuterung zu Schulwandbild Nr. 1, Tiere der Vorzeit.* 15 pp. Dr. te Neues.

189 — (1957) Weitere Fossilfunde aus dem ältesten Pleistozän der Erpfinger Höhle (Schwäbische Alb). *Mitteil. Geol. Staatsinst. Hamburg* no. 26, 60–99.

190 LIBBY, W. F.
(1955) *Radiocarbon dating.* 2nd ed. Chicago, Univ. Chicago Press.

191 LONA, F.
(1950) Contributi alla storia della vegetazione e del clima nella Val Padana. Analisi pollinica del giacimento villafranchiano di Leffe (Bergamo). *Atti Soc. Ital. Sci. Nat. Milano* **89**: 123–80.

192 LUNDHOLM, B.
(1949) Abstammung und Domestikation des Hauspferdes. *Zool. Bidr. Upsala* **27**: 1–287.

193 MAJOR, C. J. FORSYTH
(1905) Rodents from the Pleistocene of the Western Mediterranean region. *Geol. Magazine* 462–67, 501–506.

194 MATTHEW, W. D.
(1925) The value of palaeontology. *Natural History* **25**: 166–68.

195 MAYR, E.
(1963) *Animal species and evolution.* Cambridge, Mass.

196 MEADE, G. E.
(1961) The saber-toothed cat *Dinobastis serus. Bull. Texas Memor. Mus.* no. 2, 25–60.

197 MILANKOVICH, M.
(1920) *Théorie mathématique des phénomènes thermiques produits par la radiation solaire.* Paris.

198 MOTTL, M.
(1933) Zur Morphologie der Höhlenbärenschädel aus der Igric-Höhle. *Jahrb. Kgl. Ungar. Geol. Anst.* **29**.

199 — (1937) Einige Bemerkungen über '*Mustela robusta* Newton' (Kormos) bzw. '*M. Eversmanni soergeli* Ehik' aus dem ungarischen Pleistozän. *Földtani Közlöny* **67**: 37–45.

200 — (1958) Die fossilen Murmeltierreste in Europa mit besonderer Berücksichtigung Österreichs. *Jahrb. Österr. Arbeitskr. Wildtierforsch.* Graz, 91–102.

201 — (1964) Bärenphylogenese in Südost-Österreich. *Mitteil. Mus. Bergbau* etc., Landesmus. 'Joanneum', Graz no. 26, 1–56.

202 MOVIUS, H.
(1949) Villafranchian stratigraphy in southern and southwestern Europe. *Jour. Geology* **57**: 380–412.

203 MUSIL, R.
(1961) Die Höhle 'Švéduv Stul', ein typischer Höhlenhyänenhorst. *Anthropos* 13 (new ser., no. 5): 97–260.

204 — (1965) Die Bärenhöhle Pod hradem. Die Entwicklung der Höhlenbären im letzten Glazial. *Anthropos* 1–92.

205 — (1965) Aus der Geschichte der Stránská skála. *Acta Mus. Moraviae* **50**: 75–106.

206 NAPOLI-ALLIATA, E. DI.
(1947) Sull' existenza del Calabriano e del Siciliano, rivelata dei microfossili, nell sottosuolo della Pianura Lodigiana (Milano). *Riv. Ital. Palaeont.* **53**: 19–24.

207 OAKLEY, K. P.
(1955) Analytical methods of dating bones. *The Advancement of Science* **11**: no. 45, 3–8.

208 — (1964) *Frameworks for dating fossil Man.* London, Weidenfeld & Nicolson.

209 — (1965) Discovery of part of skull of *Homo erectus* with Buda industry at Vértesszöllös, north-west Hungary. *Proc. Geol. Soc. London*, no. 1630, 31–34.

210 OAKLEY, K. P. & LEAKEY, M.
(1937) Report on excavations at Jaywick Sands, Essex. *Proc. Prehist. Soc.* **3**: 217–60.

211 OSBORN, H. F.
(1936, 1942) *Proboscidea*, vols. I–II. New York, American Museum Press.

212 OVEY, C. D. (Ed.)
(1964) *The Swanscombe Skull. A survey on research on a Pleistocene site.* London, Roy. Anthrop. Inst.

213 OWEN, R.
(1846) *A history of British fossil mammals and birds.* London, John Van Voorst.

214 PEI, WEN-CHUNG
(1934) On the Carnivora from Locality 1 of Choukoutien. *Palaeont. Sinica* ser. C, **8**: 1–216.

215 PENCK, A. & BRÜCKNER, E.
(1909) *Die Alpen im Eiszeitalter.* 3 vols. Leipzig.

216 PERKINS, D.
(1964) The fauna from the prehistoric levels of Shanidar Cave and Zawi Chemi Shanidar. *Rept. VI INQUA Congr., Warsaw 1961*, **2**: 565–72.

217 PILGRIM, G. E. & SCHAUB, S.
(1939) Die schraubenhörnige Antilope des europäischen Oberpliocaens und ihre systematische Stellung. *Abh. Schweiz. Palaeont. Ges.* **58**: 1–30.

218 POMEL, A.
(1843) Note sur une espèce fossile du genre Loutre, dont les ossements ont été recueillis dans les alluvions volcaniques de l'Auvergne. *Bull. Soc. Géol. France* **14**: 168–71.

219 RAKOVEC, I.
(1958) (The beavers of the lacustrine age from the Ljubljana Moor and from

other Holocene find spots in Slovenia.) With English summary. *Razprav. Acad. Sci. Art. Slovenica*, Cl. IV, Hist. Nat., **4**: 211–67.

220 RAMSAY, W.
(1924) The probable solution of the climate problem in geology. *Geol. Magazine* **61**: 152–63.

221 RAUSCH, R.
(1953) On the status of some Arctic mammals. *Arctic* **6**: 91–148.

222 REED, C. A.
(1961) Osteological evidence for prehistoric domestication in southwestern Asia. *Zeitschr. Tierzüchtungsbiol.* **76**: 31–38.

223 REICHENAU, W. VON.
(1906) Beiträge zur näheren Kenntnis der Carnivoren aus den Sanden von Mauer und Mosbach. *Abh. Grhzl. Hess. Geol. Landesamt, Darmstadt* **4**: 202-285.

224 REPENNING, C. A.
(1967) Palearctic-Nearctic mammalian dispersal in the late Cenozoic. In D. M. Hopkins (Ed.): *The Bering land bridge.* Stanford Univ. Press, 288–311

225 REYNOLDS, S. H.
(1929) *Monograph of British Pleistocene Mammalia*, vol. 3, no. 3, Giant Deer 1–62. London, Palaeontogr. Soc.

226 RISTORI, G.
(1886) Contributo alla flora fossile del Valdarno superiore. *Atti Soc. Toscana Sci. Nat. Mem.* 7: 141–89.

227 ROMER, A. S.
(1928) Pleistocene mammals of Algeria. Fauna of the paleolithic station of Mechta-el-Arbi. *Bull. Logan Mus.* **1**: 80–163. Beloit, Wisc.

228 ROSHOLT, J. N., EMILIANI, C., GEISS, J., KOCZY, F. F. & WANGERSKY, P. J
(1961) Absolute dating of deep-sea cores by the Pa^{231}/Th^{230} method. *Jour. of Geology* **69**: 162–85.

229 RÜGER, L.
(1931) Ein Lebensbild von Mauer. *Badische Geol. Abh.* **3**: 121–36.

230 RYZIEWICZ, Z.
(1955) Systematic place of the fossil Musk-Ox from the Eurasian Diluvium. *Trav. Soc. Sci. Lett. Wroclaw*, ser. B, no. 49, 1–74.

231 SAINT-PÉRIER, R. DE
(1922) Nouvelles recherches dans le caverne de Montmaurin (Haute-Garonne). *L'Anthropologie* **32**: 193–202.

232 SAVAGE, D. E.
(1958) Evidence from fossil land mammals on the origin and affinities of the Western Nearctic fauna. *Amer. Assoc. Adv. Sci.* (Zoogeography), 97–129.

233 SCHAUB, S.
(1923) Neue und wenig bekannte Cavicornier von Senèze. *Ecl. Geol. Helvetiae* **18**: 281–95.

234 — (1925) Ueber die Osteologie von *Machairodus cultridens* Cuvier. *Ecl. Geol. Helvetiae* **19**: 255–66.

235 — (1928) Die Antilopen des toskanischen Oberpliocäns. *Ecl. Geol. Helvetiae* **21**: 259–66.

236 SCHAUB, S. (1930) Quartäre und jungteriäre Hamster. *Abh. Schweiz. Palaeont. Ges.* **49**: 1–50.

237 — (1932) Die Ruminantier des ungarischen Praeglacials. *Ecl. Geol. Helvetiae* **25**: 319–30.

238 — (1938) Tertiäre und quartäre Murinae. *Abh. Schweiz. Palaeont. Ges.* **59**: 1–40.

239 — (1941) Die kleine Hirschart aus dem Oberpliocaen von Senèze (Haute-Loire). *Ecl. Geol. Helvetiae* **34**: 264–71.

240 — (1941) Ein neues Hyaenidengenus von der Montagne de Perrier. *Ecl. Geol. Helvetiae* **34**: 279–86.

241 — (1941) Demonstration der Fauna des Ravin des Etouaires an der Montagne de Perrier. *Ecl. Geol. Helvetiae* **34**: no. 19, 1 p.

242 — (1944) Die oberpliocaene Säugetierfauna von Senèze (Haute-Loire) und ihre verbreitungsgeschichtliche Stellung. *Ecl. Geol. Helvetiae* **36**: 260–89.

243 — (1949) Revision de quelques carnassiers villafranchiens du niveau des Etouaires (Montagne de Perrier, Puy-de-Dôme). *Ecl. Geol. Helvetiae* **42**: 492–506.

244 — (1951) *Soergelia* n. gen., ein Caprine aus dem thüringischen Altpleistocaen. *Ecl. Geol. Helvetiae* **44**: 375–81.

245 SCHLESINGER, G.
(1922) Die Mastodonten der Budapester Sammlungen. *Geol. Hungarica* **2**: 1–284.

246 SCHMID, E.
(1940) *Variationsstatistische Untersuchungen am Gebiss pleistozäner und rezenter Leoparden und anderer Feliden.* Inaugural-Dissertation, Freiburg i. Br.

247 SCHREUDER, A.
(1945) The Tegelen fauna, with a description of new remains of its rare components (*Leptobos, Archidiskodon meridionalis, Macaca, Sus strozzii*). *Arch. Néerl. Zool.* **7**: 153–204.

248 SICKENBERG, O.
(1965) *Dama clactoniana* (Falc.) in der Mittelterrasse der Rhume – Leine bei Edesheim (Lankreis Northeim). *Geol. Jahrb.* **83**: 353–96.

249 SIMPSON, G. G.
(1947) Holarctic mammalian faunas and continental relationships during the Cenozoic. *Bull. Geol. Soc. America* **58**: 613–88.

250 — (1953) *The major features of evolution.* New York, Columbia Univ. Press.

251 — (1961) *Principles of animal taxonomy.* New York, Columbia Univ. Press.

252 SIMPSON, G. G., ROE, A. & LEWONTIN, R. C.
(1960) *Quantitative Zoology.* New York, Burlingame, Harcourt, Brace.

253 SKINNER, M. F. & KAISEN, O. C.
(1947) The fossil Bison of Alaska and preliminary revision of the genus *Bison. Bull. Amer. Mus. Nat. Hist.* **89**: 123–56.

254 SOERGEL, W.
(1915) Die Stammesgeschichte der Elefanten. *Centralbl. Min. Geol. Pal.* 179–188; 208–15; 245–53; 278–84.

255 — (1929) Das Alter der Sauerwasserkalke von Cannstatt. *Jber. Mitteil. Oberrhein. Geol. Ver.* 93–153.

256 SOERGEL, W. (1930) Die Bedeutung variationsstatistischer Untersuchunger für die Säugetier-Paläontologie. *N. Jahrb. Mineral.* Beil.-Band **63**: Abt. B 349–450.

257 — (1940) *Die Massenvorkommen des Höhlenbären.* Jena, G. Fischer.

258 SOKAL, R. R. & SNEATH, P. H. A.
(1963) *The principles of numerical taxonomy.* San Francisco & London, W. H Freeman.

259 STACH, J.
(1930) The second Woolly Rhinoceros from the diluvial strata of Starunia. In J. Nowak *et al.*, The second Woolly Rhinoceros (*Coelodonta antiquitatis* Blum.) from Starunia, Poland. *Bull. Acad. Polonaise Sci.* suppl. 1–47.

260 — (1959) On some Mustelinae from the Pliocene bone breccia of Weze *Acta Palaeont. Polonica* **4**: 101–16.

261 — (1962) On two Carnivores from the Pliocene breccia of Weze. *Acta Palaeont Polonica* **6**: 321–29.

262 STAESCHE, K.
(1941) Nashörner der Gattung *Dicerorhinus* aus dem Diluvium Württembergs. *Abh. Reichsst. Bodenfrosch., Berlin* no. 200, 1–148.

263 STEHLIK, A.
(1934) Fossilni ssavci ze Stránské skály u Brna. *Acta Soc. Sci. Nat. Moravia* **9**

264 STEHLIN, H. G.
(1930) Die Säugetierfauna von Leffe (Prov. Bergamo). *Ecl. Geol. Helvetiae* **23**: 648–81.

265 — (1933) In: Dubois, Auguste & Stehlin, H. G.: La grotte de Cotencher station moustérienne. *Mem. Soc. Paléont. Suisse* **52–53**: 1–292.

266 — (1933) Über die fossilen Asiniden Europas. *Ecl. Geol. Helvetiae* **26**: 229–32

267 — (1934) *Bubalus iselini* n. spec. aus dem obern Pliocaen von Val d'Arno. *Ecl Geol. Helvetiae* **27**: 407–12.

268 STIRTON, R. A. & CHRISTIAN, W. G.
(1940) A member of the Hyaenidae from the upper Pliocene of Texas. *Jour. Mammalogy* **21**: 445–48.

269 SULIMSKI, A.
(1959) Pliocene insectivores from Weze. *Acta Palaeont. Polonica* **4**: 119–74.

270 — (1964) Pliocene Lagomorpha and Rodentia from Weze 1 (Poland). *Acta Palaeont. Polonica* **9**: 149–261.

271 SUTCLIFFE, A. J.
(1960) Joint Mitnor Cave, Buckfastleigh. *Trans. Torquay Nat. Hist. Soc.* **13**: 1–28.

272 SUTCLIFFE, A. J. & ZEUNER, F. E.
(1962) Excavations in the Torbryan caves, Devonshire. I. Tornewton Cave. *Proc. Devon Archaeol. Explor. Soc.* **5**: 127–45.

273 SYCH, L.
(1964) Fossil Leporidae from the Pliocene and Pleistocene of Poland. *Rept. VI INQUA Congr., Warsaw 1961* **2**: 591–94.

274 SYLVESTER-BRADLEY, P. C. (Ed.)
(1956) The species concept in palaeontology. *Publ. Syst. Assoc.*, no. 2, A symposium, 1–145.

275 TACKENBERG, K. (Ed.)
(1956) *Der Neandertaler und seine Umwelt.* Bonn, Rudolf Habelt.

276 THENIUS, E.
(1954) Die Caniden (Mammalia) aus dem Altquartär von Hundsheim (Niederösterreich) nebst Bemerkungen zur Stammesgeschichte der Gattung *Cuon. N. Jahrb. Geol. Pal., Abh.* **99**: 230–86.

277 —(1954) Zur Abstammung der Rotwölfe. *Österr. Zool. Zeitschr.* **5**: 377–87.

278 —(1955) Die Verknöcherung der Nasenscheidewand bei Rhinocerotiden und ihr systematischer Wert. *Schweiz. Palaeont. Abh.* **71**: 3–18.

279 — (1956) Neue Wirbeltierfunde aus dem Ältestpleistozän von Niederösterreich. *Jahrb. Geol. Bundesanst.* **99**: 259–71.

280 — (1956) Zur Kenntnis der fossilen Braunbären (Ursidae, Mammal.). *Sitzber. Österr. Akad. Wiss.*, Math.-nat. Kl., Abt. I, **165**: 153–72.

281 — (1957) Zur Kenntnis jungpleistozäner Feliden Mitteleuropas. *Säugetierkundl. Mitteil.* **5**: 1–4.

282 — (1958) Über einen Kleinbären aus dem Pleistozän von Slowenien nebst Bemerkungen zur Phylogenese der plio-pleistozänen Kleinbären. *Razprave Acad. Slov.* **4**: 633–46.

283 —(1959) Ursidenphylogenese und Biostratigraphie. *Zeitschr. Säugetierkunde* **24**: 78–84.

284 —(1962) *Capra 'prisca'* Sickenberg und ihre Bedeutung für die Abstammung der Hausziegen. *Zeitschr. Tierzüchtung u. Züchtungsbiol.* **76**: 321–25.

285 — (1962) Die Grossäugetiere des Pleistozäns von Mitteleuropa. *Zeitschr. Säugetierk.* **27**: 65–83.

286 — (1965) Die Carnivoren-Reste aus dem Altpleistozän von Voigtstedt bei Sangerhausen in Thüringen. *Paläont. Abh.*, ser. A, **2**: 537–64.

287 — (1965a) Ein Primaten-Rest aus dem Altpleistozän von Voigtstedt in Thüringen. *Palaeont. Abh.* **2**: 681–86.

288 —(1965b) Über das Vorkommen von Streifenhyänen (Carnivora, Mammalia) im Pleistozän Niederösterreichs. *Ann. Nat. hist. Mus. Wien* **68**: 263–68.

289 TOBIEN, H.
(1935) Über die pleistozänen und postpleistozänen Prolagusformen Korsikas und Sardiniens. *Ber. Naturf. Ges. Freiburg i. Br.* **34**: 253–344.

290 — (1957) *Cuon* Hodg. und *Gulo* Frisch (Carnivora, Mammalia) aus den altpleistozänen Sanden von Mosbach bei Wiesbaden. *Acta Zool. Cracoviensia* **2**: 433–52.

291 TOEPFER, V.
(1933) Die glazialen und präglazialen Schotterterrassen im mittleren Saaletal. *Ber. Naturf. Ges. Freiburg i. Br.* **32**: 1–110.

292 —(1957) Die Mammutfunde von Pfännerhall im Geiseltal. *Veröff. Landesmus. Vorgeschichte Halle*, no. 16, 12–14.

293 —(1963) *Tierwelt des Eiszeitalters.* Leipzig, Geest & Portig.

294 TOPÁL, G.
(1963) Chiroptera of the lower Pleistocene locality of Kövesvárad at Répáshuta (Northern Hungary). *Ann. Hist. Nat. Mus. Natl. Hung.* **55**.

295 UMBGROVE, J. H. F.
(1942) *The pulse of the earth.* The Hague, M. Nijhoff.

296 VAN DEN BRINK, F. H.
(1958) *Zoogdierengids.* (Swedish translation used in the present study: *All Europas däggdjur*, translation by Lars Silén. Stockholm, Albert Bonnier.)

297 VAN DER VLERK, I. M. & FLORSCHÜTZ, F.
(1950) *Nederland in het Ijstidvak.* Utrecht, W. de Haan.

298 VILLALTA COMELLA, J. F. DE
(1952) Contribución al conocimiento de la fauna de mamíferos fósiles de Plioceno de Villarroya (Logroño). *Bol. Inst. Geol. Min. España* **64**: 1–201.

299 VILLALTA COMELLA, J. F. DE & CRUSAFONT PAIRÓ, M.
(1955) Un nuevo ovicaprino en la fauna villafranquiense de Villaroya (Logroño). *Actes INQUA IV Congr.* 5 pp.

300 VIRET, J.
(1950) *Meles thorali* n. sp. du loess villafranchien de Saint-Vallier (Drôme) *Ecl. Geol. Helvetiae* **43**: 274–87.

301 — (1950) Sur l'identité générique des mustélidés fossíles désignées sous le noms de *Pannonictis pilgrimi* et d'*Enhydrictis galictoides*. *C.R. Soc. Géol France* no. 9, 165–66.

302 — (1954) Le loess à bancs durcis de Saint-Vallier (Drôme) et sa faune de mammifères villafranchiens. *Nouv. Arch. Mus. Hist. Nat. Lyon* **4**: 1–200.

303 VOELCKER, J.
(1930) Beiträge z. Oberrhein. Fossilkatalog No. 3. *Felis issiodorensis* Croizet von Mauer a. d. Elsenz. *Sitzber. Heidelberg Akad. Wiss.*, Math.-nat. Kl., 12 Abh

304 WARREN, S.
(1923) The *Elephas-antiquus* bed of Clacton-on-Sea. *Quart. Jour. Geol. Soc London* **79**: 606–34.

305 WEST, R.
(1958) The Pleistocene epoch in East Anglia. *Jour. Glaciol.* **3**: 211–16.

306 — (1961) The glacial and interglacial deposits of Norfolk. *Trans. Norfolk & Norwich Naturalists' Soc.* **19**: 365–75.

307 — (1963) Problems of the British Quaternary. *Proc. Geol. Assoc.* **74**: 147–86.

308 — (1964) Inter-relations of ecology and Quaternary palaeobotany. *British Ecol. Soc. Jubilee Symp., Jour. Ecology* **52** (suppl.): 47–57.

309 WEST, R. & MCBURNEY, C. M. B.
(1954) The Quaternary deposits at Hoxne, Suffolk, and their archaeology. *Proc Prehist. Soc.* **20**: 131–54.

310 WEST, R. G. & WILSON, D. G.
(1966) Cromer Forest Bed series. *Nature* **209**: 497–98.

311 WOJTUSIAK, K.
(1953) Szczatki lwa jaskiniowego (*Felis spelaea* Goldf.) z jaskini 'Wierzchowskiej Górnej'. *Acta Geol. Polonica* **3**: 573–92.

312 WOLFF, B.
(1938–41) *Fauna fossilis cavernarum*, I–III. *Fossilium Catalogus, Animalia*, 82 89, 92, 1–288 & 1–320.

313 ZAPPE, H.
(1948) Die altplistozänen Bären von Hundsheim in Niederösterreich. *Jahrb. Geol. Bundesanst.* for 1946, 95–164.

314 — (1954) Beiträge zur Erklärung der Entstehung von Knochenlagerstätten in Karstspalten und Höhlen. *Geologie* no. 12, 1–59.

315 ZEUNER, F. E.
(1931) Die Insektenfauna des Böttinger Marmors. *Fortschr. Geol. Pal.*, ser. 9 **28**: 1–160.

316 — (1935) Die Beziehungen zwischen Schädelform und Lebensweise bei den rezenten und fossilen Nashörnern. *Ber. Naturf. Ges. Freiburg i. Br.* **34**: 21–80.

317 — (1937) A comparison of the Pleistocene of East Anglia with that of Germany. *Proc. Prehist. Soc.* **3**: 136–57.

318 — (1959) *The Pleistocene period. Its climate, chronology and faunal successions.* 2nd ed. London, Hutchinson.

319 — (1963) *A history of domesticated animals.* London, Hutchinson.

Index

Numbers printed in italics refer to the figures.

Abbeville, 24
Acinonyx sp., 89, 90 (*see also* Cheetah)
Agriotherium insigne Gervais, *see* Bear, Hyena
Ailuropoda melanoleuca Milne-Edwards, *see* Panda, Giant
Ailurus fulgens F. Cuvier, *see* Panda
Alces sp., *see* Elk
Allactaga sp., *see* Jerboa, Great
Allophaiomys pliocaenicus Kormos, *see* Lemming, Kormos's Steppe
Alopex sp., *see* Fox
Alpine Shrew (*Sorex alpinus* Schinz), 43
  anterior teeth, *16*
Alps, 4
  glacial terraces, 15
Amphibious mammals,
  relative numbers in Villafranchian, *6*
Anancus arvernensis Croizet & Jobert, *see* Mastodont, Auvergne
Anoglochus sp., 159 (*see also* Deer)
Antelope,
  Saiga (*Saiga tatarica* Linné), 173–4
  Chiru (*Patholops hodgsoni* Abel), 174
    head, *76*
  Chamois (*Procamptoceras brivatense* Schaub), 174
  Ardé (*Deperetia ardea* Depéret), 178
Antlers,
  Ramosus Deer, *65*
  Senèze Deer, *66*
  Bush-antlered Deer, *67*
Aonyx sp., 105 (*see also* Otter)
'Ape', Gibraltar, 58–9
Apodemus sp., 223 (*see also* Mouse)
Archidiskodon sp., 13, 133 (*see also* Elephant)
Arctic species,
  origination of, 253–4
  first appearances in Pleistocene, Table 5
Arno valley, *4*
Artefacts, 29, 60
  early European, 59
  pebble-tool industry, 60
Artiodactyla, 153–90
  stratigraphic range of species, 281–3
Arvicola sp., 216 (*see also* Vole)
Ass,
  European Wild (*Equus hydruntinus* Regalia), 151
    teeth, *62F*
  Asiatic Wild (*Equus hemionus* Pallas), 151–2
Astian, caves, 16
Astian faunas, 9
  Ivanovce fissure, 16
  Weze fissure, 17
Astian floras, 9
Astian stage, *see* Late Pliocene stage
Auroch (*Bos primigenius* Bojanus), 30, 188–90, *84*
Australopithecinae, 59

Bacton, 23
Badger, 36, 251, 254
  Thoral's (*Meles thorali* Viret), 103–4
    skull and mandible, *42*
  Thoral's (*Meles meles* Linné), 104–5
Banahilk, 190
Baranogale sp., *see* Polecat
Barbastelle,
  (*Barbastella barbastella* Schreber), 56
  Asiatic (*Barbastella leucomelas* Cretzschmar), 56
Barrington, 30
Bat (*see also* Barbastelle, Noctule)
  Blasius' Horseshoe (*Rhinolophus blasii* Peters), 51
  Mehely's Horseshoe (*Rhinolophus mehelyi* Matschie), 51
  Mediterranean Horseshoe (*Rhinolophus euryale* Blasius), 51, 254
  Greater Horseshoe (*Rhinolophus ferrumequinum* Schreber), 51–2
    teeth, *20*

Bat (*contd.*)
Lesser Horseshoe (*Rhinolophus hipposideros* Bechstein), 52
Water (*Myotis daubentoni* Leisler), 52
Pond (*Myotis dasycneme* Boie), 52–3, 254
Whiskered (*Myotis mystacinus* Leisler), 53
teeth, *20*
Geoffroy's (*Myotis emarginatus* Geoffroy), 53
Natterer's (*Myotis nattereri* Kuhl), 53–4
Bechstein's (*Myotis bechsteini* Leisler), 54
Large Mouse-eared (*Myotis Myotis* Borkhausen), 54
teeth, *20*
Lesser Mouse-eared (*Myotis oxygnathus* Monticelli), 54
Long-eared (*Plecotus auritus* Linné), 54
mandibles, *21*
Grey Long-eared (*Plecotus austriacus* Fischer), 55
Long-winged (*Miniopterus schreibersi* Natterer), 55–6, 254
mandible, *21*
Common (*Pipistrellus pipistrellus* Schreber), 56
Serotine (*Vespertilio serotinus* Schreber), 56
Northern (*Vespertilio nilssoni* Keyserling & Blasins), 57
Parti-coloured (*Vespertilio murinus* Linné), 57
European Free-tailed (*Tadarida teniotis* Rafinesque), 57
Bear,
Deninger's (*Ursus deningeri* Reichenau), 120–2
mandibular dentition, *48C*
Cave (*Ursus spelaeus* Rosenmuller & Heinroth), 122–7, 265
skull, *49*
restoration, *50*
geographic distribution, *51*
width of lower canines, *102*
Brown (*Ursus arctos* Linné), 127–8
size oscillation, 245, 248, *104*
mandibular dentition, *48A, B*
skull and mandible, *49*
engraving, *52*
Hyena (*Agriotherium insigne* Gervais), 119
Auvergne (*Ursus minimus* Devéze & Bouillet), 119–20
Etruscan (*Ursus etruscus* Cuvier), 120
Polar (*Ursus maritimus* Phipps), 128–9, 254
Asiatic Black (*Ursus thibetanus* Cuvier), 129, 268
Bear caves, 33, 36, *14*
Bear stratum,
Tornewton cave, 33
Beaver,
(*Castor fiber* Linné), 10, 197–8, 254
skull and mandible, *87*
European Giant (*Trogontherium cuvieri* Fischer; *Conodontes boisvilletti* Laugel), 198–9
skull and mandible, *87*
Beresovka, 137
Bergmann's Rule, 27, 211
and mammoth, 138
Beremendia fissidens Petenyi, *see* Shrew, Beremend
Bering Bridge, 268, 269
Bering Straits,
migrations, 268
Bison, 268
in Villafranchian, 8
Bison sp., 186, 187 (*see also* Wisent)
Blancan, 21
Bobcat, 251, 252
Bore holes,
Leffe, *5*
Bos sp., 189 (*see also* Auroch)
Boswell's sequence, 22
Bovidae, 171–90
Bovids, cattle-like, in Villafranchian, 8
Brassó, 25
Brentford, 30
Bresse Plain, 12
Brown's Orchard, Acton, 30
Bubalus murrensis Berckhemer, *see* Buffalo, Murr
Buffalo,
Murr (*Bubalus murrensis* Berckhemer), 187
Stehlin's (*Syncerus iselini* Stehlin), 187, 235
horn cores, *83*

Calcareous tufas, 31
Canidae, 108–17
Canis sp., 109, 113 (*see also* Dog, Jackal, Wolf)
Cannstatt, 31
Capra sp., 181, 182 (*see also* Ibex)
Capreolus sp., 166 (*see also* Deer)
Carnivora, 62–129
stratigraphic range of species, 278–80
Carnivores, size oscillation, *105*, *106*
Castor sp., 197 (*see also* Beaver)
Castoridae, 197–9
Castoroides, 197
Cat (*see also* Tiger, Leopard, Dirk-tooth, Lynx, Panther, Cheetah)
Greater Scimitar (*Homotherium sainzelli* Aymard), 75–6, 268
restoration, *29*
Lesser Scimitar (*Homotherium* (*Dinobastis*) *latidens* Owen), 76–7

Cat (*contd.*)
Martelli's Wild (*Felis lunensis* Martelli), 77–8
Steppe (*Felis manul* Pallas), 78, 79–80
European Wild (*Felis silvestris* Schreber), 78–9, 251
mandible, *30*
Jungle (*Felis chaus* Gueldenstaidt), 80
Cave deposits,
Villafranchian, 8, 15
Late Pleistocene, 31–6
Würm, 35
Cave orientation, Hungary, 16
Cave paintings, 270
Cave sediments, Cueva del Toll, *15*
Caves, bear, 33, 36, *14*
hyena, 36
Cenozoic Era, 5
Cercopithecidae, 58–9
Cervidae, 157–71
Cervus sp., 163 (*see also* Deer)
Chagny, 10, 11, 12, 13, 16
Chalk Sea, 3
Chalon, 12
Chamois (*Rupicapra rupicapra* Linné), 175–6, 265
skull, *77*
Châtillon-Saint-Jean, 30
Choukoutien, 165
Cheetah, Giant (*Acinonyx pardinensis* Croizet & Jobert), 88–90
restoration, *35*
Chelles Terrace, 31
Chilhac, 13, 16
Chillesford Crag, 22
Chiroptera, 51–7
evolution rate, 264
stratigraphic range of species, 277
Chronologies, alternative absolute, 255, 257, Table 7
Citellus sp., *see* Souslik
Civets, *see* Genet, Ichneumon
Clacton, 29
Clethrionomys sp., 213 (*see also* Vole)
Climate,
Cretaceous to Tertiary, 3, 4
Villafranchian, 8
Astian, 9
Climatic change, Pliocene period, 4
Climatic fluctuations,
Ice Age, 4
Pleistocene, 5–6
Leffe, 13, 16
Cromer, 25
Climatic indicators, flora and fauna as, 6
Climatic phases, dating, Table 1
Coelodonta antiquitatis Blumenbach, *see* Rhinoceros, Woolly
'Cold' deposits, 7
Conodontes boisvilletti Laugel, *see* Beaver, European Giant
Coralline Crag, 22
Cordillera, 4
Corsac,
Primitive (*Vulpes praecorsac* Kormos; *Vulpes corsac* Linné), 116
Crags, East Anglian, 22
Creodonts, 72
Cretaceous period, 3
Cricetidae, 209–12
size fluctuation, 27, 9
Cricetulus sp., 211, 240 (*see also* Hamster)
Crocidura sp., 48 (*see also* Shrew)
Crocuta sp., *see* Hyena
Cromer, 23
Cromer fauna, 24
Cromer interglacial, date, Table 1
Cromerian, 24
faunal stratum, 262, *111*
Cromerian interglacial, 19
Csarnóta, 16
Cueva del Toll, 35
sediment profile, *15*
Cuon sp., 112, 113 (*see also* Dhole)

Dama sp., *see* Deer
Danube complex, 15
Dating, 19–21
from sedimentation rates, 20
radioactive, 19–21
Deep-sea sediments,
dating, 21
Deer, 157–8
Etouaires Deer ('*Cervus*' *ardei* Croizet & Jobert; '*Cervus*' *issiodorensis* Croizet & Jobert; '*Cervus*' *cladocerus* Pomel; '*Cervus*' *cusanus* Croizet & Jobert; '*Cervus*' *perrieri* Croizet & Jobert), 158
Ramosus (*Anoglochis ramosus* Croizet & Jobert), 158–9
frontlet with antlers, *65*
Tegelen (*Euctenoceros ctenoides* Nesti), 159
Senèze (*Euctenoceros senezensis* Déperet), 159
skull and antlers, *66*
Philis (*Cervus etuerarium* Croizet & Jobert), 160–1
restoration, *68*
Bush-antlered (*Eucladoceros dicranios* Nesti), 159–60
skull and antlers, *67*
Red (*Cervus elaphus* Linné), 33, 161–3, 268
restoration, *69*
Verticornis (*Praemegaceros verticornis* Dawkins), 163–4
restoration, *70*

Deer (*contd.*)
Savin's Giant (*Megaloceros savini* Dawkins), 164
restoration, *70*
Giant (*Megaloceros giganteus* Blumenbach), 30, 164–5, *13*
restoration, *70*
Nesti's Fallow (*Dama nestii* Major), 165
Clacton Fallow (*Dama clactoniana* Falconer), 29, 166
Fallow (*Dama dama* Linné), 166
Roe (*Capreolus capreolus* Linné), 166–7
Deperetia ardea Depéret, *see* Antelope, Ardé
Deposits,
Villafranchian, 8, 15, *5*
Middle Pleistocene, *5*
Desmana moschata Pallas, *see* Desman, European
Desman,
Pyrenean (*Desmana pyrenaica* Geoffroy), 48
European (*Desmana moschata* Pallas), 48
skull and mandible, *19*
Desmania thermalis Kormos, 49
Desmania tegelensis Schreuder, 49
Dhole,
Primitive (*Cuon alpinus* Pallas), 112–14
skull and mandible, *46*
(*Cuon majori* Del Camoana), 111–12
Sardinian (*Cuon alpinus sardous* Studiati) mandible, *47*
Dicerorhinus sp., *see* Rhinoceros
Dicrostonyx sp., 221 (*see also* Vole)
Diluvium, *see* Pleistocene Epoch
Dinobastis latidens Owen, *see* Cat, Lesser Scimitar
Dinosaurs, age of, 3
Dipodidae, 202–3
Dirk-tooth (*Megantereon megantereon* Croizet & Jobert), 73–5, 268
skull, *27*
mandible, *27*
restoration, *28*
Dog, 108–9 (*see also* Wolf, Jackal, Dhole, Fox)
Arno (*Canis arnensis* Del Campana), 109
European Hunting (*Lycaon lycaonoides* Kretzoi), 114
Great Raccoon, *see* Raccoon-Dog
Dolichodoryceros süssenborensis Kahlke, *see* Deer, Savin's Giant
Dolichopithecus arvernensis Depéret, *see* Monkey, Auvergne
Dolomys sp., *see* Vole
Domestication, effect on the fauna, 274
Donau complex, 15
Dormouse,
Heller's (*Glirulus pusillus* Heller), 205–6
Pliocene (*Muscardinus pliocaenicus* Kowalski; *Muscardinus avellanarius* Linné), 206
teeth, *92C*
Dacian (*Muscardinus dacicus* Kormos), 206
Fat (*Glis glis* Linné), 206–8
mandible, *92A*
teeth, *92E*
Garden (*Eliomys quercinus* Linné), 208
mandible, *93C*
Forest (*Dryomys nitedula* Pallas), 208
mandible, *92B*
teeth, *92D*
Maltese (*Leithia melitensis* Leith Adams), 209
mandible, *93A*
Balearic (*Hypnomys mahonensis* Bate; *Hypnomys morphaeus* Bate), 209
mandible, *93B*
Drachenhöhle, 36
Dryomys nitedula Pallas, *see* Dormouse, Forest
Dryopithecus, 58
Durfort, 15
Dwarfing and population decline, 252

Early Pleistocene stage, 5
climatic oscillations in, 5–6
East Anglian Crags, 11, 22
East Mersea, 30
Eastern Torrs Quarry Cave, 33
Eemian fauna, 29–30
Joint Mitnor Cave, 33
restoration, *13*
Eastern Torrs Quarry Cave, 33
Tornewton Cave, 33
Eemian interglacial, 19
date, Table 1
Eemian Terrace, 29–30
Ehringsdorf, 31
Elasmotherium sibiricum Fischer, *see* 'Unicorn', Giant
Elephant, 8, *13A*
Straight-tusked (*Paleoloxodon antiquus* Falconer & Cautley), 29, 30, 31, 134–5, *13*
teeth, *54B*
restoration, *55*
Southern (*Archidiskodon meridionalis* Nesti), 132–4
teeth, *54A*
Dwarf, 265
Elephantidae, 132–8
Eliomys quercinus Linné, *see* Dormouse, Garden
Elk, Irish, *see* Deer, Giant

Elk, 254
Gallic (*Alces gallicus* Azzaroli), 167–8
restoration, *71*
Broad-fronted (*Alces latifrons* Johnson), 168
restoration, *71*
Broad-fronted (*Alces alces* Linné), 168–9
restoration, *71*
geographic distribution, *72*
Elk stratum, Tornewton cave, 33
Elster glaciation, 19
Endemic forms, Mediterranean, 265
Enhydrictis ardea Bravard, *see* Polecat, Ardé
Eocene Epoch, 5
Eohippus, 145
Epimachairodus crenatidens Fabrini, *see* Cat, Greater Scimitar
Episcopia, 25
Epochs, division of Periods into, 5
Equidae, 145, 147–52
Equus, sp., 150, 151 (*see also* Horse, Ass, Onager)
Eras, division of geological time into, 5
Erinaceidae, 42
Erinaceus sp., 42 (*see also* Hedgehog)
Ermine, *see* Stoat
Erpfingen, 11
cave deposits, 15
Eskers, 6
Etouaire Ravine, 9
Etouaires, 10, 16
Eucladoceros sp., 160 (*see also* Deer)
Euctenoceros sp., 159 (*see also* Deer)
Europe, climate in Tertiary period, 4
Euryboas, 268
Evolution,
new species, 257–9
fauna, 260–2
rate in different orders, 262, 264, Table 8
Extinct species of Pleistocene, 272
Extinction, rate of, 260

Fauna,
Astian, 9, 16
Villafranchian, 9, 16
chronological sequences, 8–9
forest, 9–10
at Leffe, 15
Pliocene, 17
Forest Bed, 23–4
Cromerian, 25–6
Faunal provinces, 265, 266
Faunal stratum, 261–2
Cromerian, 262, *111*
Faunal succession, later Middle Pleistocene, 28
'Faunal waves', 9
Felidae, 72–90
Felis sp., 79, 83, 84–5, 88 (*see also* Cat, Lynx, Lion, Leopard, Panther)
Ferret, *see* Polecat, Steppe
Flora,
Astian, 9
Villafranchian, 10
Fontéchevade, 61
Forest Bed Series, 23
Forests,
in Villafranchian, 10
and pluvials, 10
Fossiliferous gravels,
Sainzelles, *8*
Fox, 36
Alopecoid (*Vulpes alopecoides* Major), 114–15
teeth, *44*
Red (*Vulpes vulpes* Linné), 115, 252, 268
teeth, *44*
Steppe, *see* Corsac
Arctic (*Alopex lagopus* Linné), 116–17, 254
teeth, *44*
Friesenhahn Cave, 76

Gailenreuth Cave, 36
section, *14*
Gallogoral meneghinii Rütimeyer, *see* Goral, European
Game, variations between tribes, 270–1
Gazelle, 10
Bourbon (*Gazella borbonica* Depéret), 171–2
skull, *74*
Gazelle-antelope,
European (*Gazellospira torticornis* Aymard), 172–3
skull, *75*
Gazellospira torticornis Aymard, *see* Gazelle-antelope, European
Genet (*Genetta genetta* Linné), 62
Geographic distribution of species, Table 9, 267
Geographic origin, European mammal species, Table 10, 267
Gibraltar 'Ape', 58–9
Gibraltar macaque, 31
Glacial deposits, characteristic fauna, 6
Glacial terraces, Alps, 15
Glaciations, 6, 13, 15
Alpine,
Mindel, 15, 18
Günz, 18
Riss, 18
Würm, 18
Scandinavian,
Elster, 19
Saale, 19
Warthe-Weichsel, 19

Glaciations (*contd.*)
North American, 19
Glaciers, continental, 3
Gliridae, 205–9
Glirulus pusillus Heller, *see* Dormouse, Heller's
Glis sp., 206, 207 (*see also* Dormouse)
Glutton,
Schlosser's (*Gulo schlosseri* Kormos), 90–2
teeth, *36*
Glutton or Wolverine (*Gulo gulo* Linné), 36, 92–3, 254, 268
engraving, *37*
Glutton stratum, Tornewton cave, 33
Goat,
Cave (*Myotragus balearicus* Bate), 176–7, 265
skull, *78*
Soergel's (*Soergelia elisabethae* Schaub), 180
Merla's (*Hesperoceras merlae* Villalta & Crusafont), 180, 235
Gomphotheriidae, 131–2
Goral,
European (*Gallogoral meneghinii* Rutimeyer), 174–5
Sardinian (*Nemorhaedus melonii* Dehaut), 175
Grafenrain pit, profile of, *7*
Grasslands, in Early Pliocene, 4
Gray's Thurrock, 29, 31
Great interglacial, *see* Holstein interglacial, 29
Grison, 97–8
Gulo sp., *see* Glutton
Günz, date, Table 1
Günz glaciation, 18

Half-life, 261, 262, 264, Table 8
Hamster, 209
Common (*Cricetus cricetus* Linné), 210
size fluctuations, 245
Giant (*Cricetus cricetus runtonensis* Hinton), 209
mandible, *94C*
Preglacial (*Cricetus praeglacialis* Schaub), 211
Dwarf (*Cricetus nanus* Schaub), 211
mandible, *94A*
Schaub's Dwarf (*Cricetulus bursae* Schaub), 211
mandible, *94B*
Migratory (*Cricetulus migratorius* Pallas), 212
Ehik's Dwarf (*Rhinocricetus ehiki* Schaub), 212
Hare,
Varying (*Lepus timidus* Linné), 229–30
skull and mandible, *99*
Brown (*Lepus europaeus* Pallas), 230
skull and mandible, *99*
Cape or Tolai (*Lepus capensis* Linné; *Lepus tolai* Pallas), 230–1
Hedgehog, European (*Erinaceus europaeus* Linné), 42, 254
Heidelberg Man, 24
Hemitragus sp., 183 (*see also* Tahr)
Heppenloch cave, 31
Herpestes ichneumon Linné, *see* Ichneumon
Hesperoceras merlae Villalta & Crusafont, *see* Goat, Merla's
Himalayas, 4
Hipparion, 10
Crusafont's (*Hipparion crusafonti* Villalta), 147–8
teeth, *62A*
Hipparion crusafonti Villalta, *see* Hipparion, Crusafont's
Hippopotamidae, 156–7
Hippopotamus (*Hippopotamus amphibius* Linné), 24, 29, 33, 156–7, 254
Dwarf, 265
(*Hippopotamus antiquus* Desmarest), 156
(*Hippopotamus melitensis* Major), 157
(*Hippopotamus minor* Desmarest), 157
(*Hippopotamus pentlandi* Meyer), 157
Hog, Eurasian Wild (*Sus scrofa* Linné), 155
Holarctic faunal province, 266
Holocene, date, Table 1
Holocene Epoch, 5
Holocene interglacial, 19
Holstein fauna, 29
Holstein terrace, 29
Holsteinian interglacial, 19, 28, 31, Table 1
Hominidae, 59–61
Homo sp., *see* Man
Homotheriini, 73
Homotherium sp., 76 (*see also* Cat)
Horse, 10, 29, 30, 145, 147 (*see also* Hipparion, Ass)
Zebrine (*Equus stenonis* Cocchi; *Equus süssenbornensis* Wüst), 148–9, 268
teeth, *62B*, *62C*
Caballine (*Equus bressanus* Viret; *Equus robustus* Pomel; *Equus mosbachensis* Reichenau; *Equus germanicus* Nehring; *Equus przewalski* Poliakoff), 149–51
wall painting, *63*
teeth, *62D*, *62E*
One-toed, 8, 10
Horseshoe Bat, *see* Bat
Hoxne, 29
Human remains, 30, 31, 33
Hundsheim fissure, 25
Hyaena sp., 65, 71 (*see also* Hyena)
Hyaenidae, 63–72

Hyena,
Indian (*Crocuta sivalensis* Falconer & Cautley), 71
Cave (*Crocuta crocuta spelaea* Goldfuss), 71–2, 30
ivory sculpture of, *26*
Perrier (*Hyaena perrieri* Croizet & Jobert), 9, 63–4
Brown (*Hyaena brunnea* Thunberg), 64–5
Hunting (*Euryboas lunensis* Del Campana), 68–9
mandible and tibia, *24*
Spotted or Cave (*Crocuta crocuta* Erxleben), 65, 66, 69–72, 251, 252
tibia, *24*
skull and mandible, *25*
Short-faced (*Hyaena brevirostris* Aymard), 65–6, 240, 242
restoration, *22*
Striped (*Hyaena hyaena* Linné), 66–8, 251, 254
skull and mandible, *23*
Hyena caves, 36
Hyena stratum, Tornewton cave, 33
Hypnomys sp., *see* Dormouse
Hypolagus brachygnathus Kormos, *see* Rabbit, Beremend
Hystricidae, 199–201
Hystrix sp., 200 (*see also* Porcupine)

Ibex (*Capra ibex* Linné), 181
cave painting, *80*
Ice Age,
climatic oscillation, 4
causes, 5
Ichneumon (*Herpestes ichneumon* Linné), 62
Igric Cave, 36
Ilford fauna, 29
Ilford interglacial, 19, 28
date, Table 1
Ilford Terrace, 29
Insectivora, 42–50
stratigraphic range of species, 276–7
Interglacials,
Tiglian, 15, 19
Waalian, 19
Cromerian, 19
Holsteinian, 19
Ilford, 19
Eemian, 19
Holocene, 19
East Anglian, 23
Interglacial phases, 5
Interglacial shore line, 7
Interstadials, 6, 18
Ivanovce fissure, faunas, 17

Jackal,
Golden (*Canis aureus* Linné), 111
teeth, *44*
Jaguar, 251
Jarmo, clay figurines from, 111
Jerboa,
Great (*Allactaga jaculus* Pallas), 202–3
geographic distribution, *90*
teeth, *89C, D*
Joint Mitnor Cave, 33
reconstruction, *13*

Kadzielnia, 11, 17
Kamyk, 25
Kanjera skulls, 61
Kent's Cavern, 69
Kirkdale Cave, 69
Kulan, *see* Ass, Asiatic Wild

La Niche, 31
Lagomorpha, 226–31
stratigraphic range of species, 285
Lagurus sp., *see* Lemming
Lake sediments, Hajnácka, 17
Lascaux frieze, 163
Late Pleistocene, cave deposits, 31–6
Late Pleistocene stage, 5
climatic oscillations, 5–6
Late Pliocene stage, 9
Leffe, 11, 13, 15
deposits at, *5*
glaciations at, 13, 15
Leithia melitensis Leith, *see* Dormouse, Maltese
Lemming,
Pannonian Steppe (*Lagurus pannonicus* Kormos), 216
Steppe (*Lagurus lagurus* Pallas), 216–17
Kormos's Steppe (*Allophaiomys pliocaenicus* Kormos), 217
Norway (*Lemmus lemmus* Linné), 220, 253
teeth, *96E*
Wood (*Myopus schistocolor* Lilljeborg), 220
Arctic or Varying (*Dicrostomyx torquatus* Pallas), 220–1
teeth, *96F*
Lemmus lemmus Linné, *see* Lemming, Norway
Leopard, Clouded (*Felis nebulosa* Griffith), 72–3
Leopard, 251
Villafranchian (*Felis schaubi* Viret), 85
skull, *32*
Felis pardus Linné, 87–8
Snow (*Felis uncia* Schreber), 88
Leporidae, 227–31
Leptobos sp., 183, 184 (*see also* Ox)
Lepus sp., 229, 230 (*see also* Hare, Rabbit)
Linnaeus, 40

Lion,
Tuscany (*Felis toscana* Schaub), 83–5
Cave (*Felis leo spelaea* Linné), 36, 85–7, 240
skull and mandible, *33*
engraving, *34*
Loess, 7
mammal-bearing Villafranchian, *3*
Lower Pleistocene stage, *see* Early Pleistocene stage
Lunel-Viel cave, 31
Lutra sp., 105, 107 (*see also* Otter)
Lutreola lutreola Linné, *see* Mink, European
Lycaon sp., 114 (*see also* Dog)
Lynx,
Issoire (*Felis issiodorensis* Croizet & Jobert), 80, 268
mandible, *31*
Northern (*Felis lynx* Linné), 80
mandible, *31*
Rexroad, 82
Pardel (*Felis pardina* Oken), 82–3, 265

Macaca prisca Gervais, 58
Macaque, Florentine (*Macaca florentina* Cocchi), 58
Machairodus, 76
Magra valley, 13
Malusteni, 9
Mammals as climatic indicators, 6
Mammoth, 8, 30, 33
Steppe (*Mammuthus trogontherii* Pohlig), 29, 30, 31, 135–6, 253, 268
teeth, *54C*
Woolly (*Mammuthus primigenius* Blümenbach), 29, 30, 136–8, 254
teeth, *54D*
engraving in ivory, *56*
geographical distribution, *57*
size variation, 244, *103*
Mammutidae, 132
Man,
Heidelberg (*Homo heidelbergensis* Schoetensack), 59, 60
Neandertal (*Homo neanderthalensis* King), 60–1
Modern (*Homo sapiens* Linné), 61
Fontéchevade, 61
Man and extinction of species, 273
Mandible, 64, 87, 98, 99, *16*, *17*, *18*, *19*, *21*, *23*, *24*, *25*, *27*, *30*, *31*, *33*, *38*, *39*, *40*, *41*, *42*, *43*, *46*, *47*, *49*, *61*, *92A*, *B*, *93A*, *B*, *C*, *94A*, *B*, *C*, *95B*, *C*, *97B*, *D*
Mandibular dentition,
Brown Bear, *48A*
Deninger's Bear, *48C*
Mariupol, 11
Marine deposits, Villafranchian, 10
Marmot,
Alpine (*Marmota marmota* Linné), 194
teeth, *85D–F*
geographic distribution, *86*
Grey (*Marmota caligota* Eschscholz), 194
Bobak (*Marmota bobak* Muller), 194, 196
geographic distribution, *86*
Marmota primigenia Kaup, 194
Marsupials, meat-eating, 72
Marten,
Primitive (*Martes vetus* Kretzoi), 93
Pine (*Martes martes* Linné), 93–4
skull and mandibles, *38*
Beech (*Martes foina* Erxleben), 95
teeth, *38*
Martes sp., 93 (*see also* Marten, Sable)
Mastodont, 9
Pig-toothed, 131–2
Auvergne (*Anancus arvernensis* Croizet & Jobert), 131–2
restoration, 131
True, 132
Borson's (*Zygolophodon borsoni* Hays), 132
Mauer, gravel pit at, 24
Mediterranean, endemic forms, 265
Megantereon sp., 74 (*see also* Dirk-tooth)
Megaloceros sp., 165 (*see also* Deer)
Megalovis latifrons Schaub, *see* Sheep, Giant
Meles sp., *see* Badger
Metacervocerus pardinensis Croizet & Jobert, 162
Microtidae, 212–21
Micromys minutus, *see* Mouse, Harvest
Microtus sp., 217, 218 (*see also* Vole)
Middle Pleistocene stage, 5
climatic oscillations in, 5–6
Migrations, 268–9, Table 11, 269
Milan, Villafranchian floras near, 10
Milankovitch, radiation curve of, 20
Mimomys cantianus Hinton, 215
Mimomys newtoni Major, 24, 25, 214
Mimomys pliocaenicus Major, 214
mandible, *95A*
teeth, *96A*
Mimomys pusillus Mehely, 214
Mimomys reidi Hinton, 214
Mimomys stehlini Kormos, 24, 25, 214
Mindel glacial, date, Table 1
Mindel glaciation, 18, 27
Miniopterus schreibersi Natterer, *see* Bat, Long-winged
Mink, European (*Mustela lutreola* Linné), 98
Miocene Epoch, 5
Mole,
Roman (*Talpa romana* Thomas), 50
Blind (*Talpa caeca* Savi), 50
Episcopal (*Talpa episcopalis* Kormos), 50

Mole (*contd.*)
European (*Talpa europea* Linné), 49
skull and mandible, *19*
Mole Rat, *see* Rat
Molossidae, 57
Monkey (*see also* Macaque, Ape)
Auvergne (*Dolichopithecus arvernensis* Depéret), 59
Montmaurin cave, 31
Montpellier, 9
Moose, *see* Elk
Morainic deposits, 6
Morphological boundaries, 238, *100*
Mortières terrace, 31
Mosbach, 27
Mosbach faunas, 24
Mouflon (*Ovis musimon* Schreber), 180
Mountain chains, in Tertiary period, 3, 4
Mouse,
Northern Birch (*Sicista betulina* Pallas), 202
Southern Birch (*Sicista subtilis* Pallas), 202
teeth, *89A*, *89B*
Schaub's Field (*Parapodemus coronensis* Schaub), 221–2
Common Field (*Apodemus sylvaticus* Linné), 222–3, 254
Yellow-necked Field (*Apodemus flavicollis* Melchior), 223, 254
Broad-toothed Field (*Apodemus mystacinus* Danford & Alson), 223
Striped Field (*Apodemus agrarius* Pallas), 223
skull, *97A*
mandible, *97B*
Harvest (*Micromys minutus*), 223–4
Hensel's Field (*Rhagamys orthodon* Hensel), 224
skull and mandible, *98*
House (*Mus musculus* Linné), 224–5
skull, *97C*
mandible, *97D*
Mouse-Hare, *see* Pika, Steppe
Muridae, 221–5
Mus musculus Linné, *see* Mouse, House
Muscardinus sp., *see* Dormouse
Mustela sp., 100, 101, 102 (*see also* Polecat, Stoat, Mink, Weasel)
Mustelidae, 90–107
Myopus schistocolor Lilljeborg, *see* Lemming, Wood
Myotis sp., *see* Bat
Myotragus sp., 177 (*see also* Goat)

Nagyharsányhegy, 16
Nemorhaedus melonii Dehaut, *see* Goral, Sardinian
Neomys sp., 45 (*see also* Shrew)
Nesiotites sp., 48, 265 (*see also* Shrew)
Nesiotites corsicanus Bate, 48
Nesiotites hidalgo Bate, 48
Nesolutra euena Bate, 106
Newer Red Crag, 22
Niah Great Cave, 61
Noctule,
Common (*Nyctalus noctula* Schreber), 57
Lesser (*Nyctalus leisleri* Kuhl), 57
Nomenclature, 39–41
Norwich Crag, 22
Nyctalus noctula Schreber, *see* Noctule, Common
Nyctereutes sp., 117 (*see also* Raccoon-Dog)

Ochotonidae, 226–7
Ochotona pusilla Pallas, *see* Pika, Steppe
Older Red Crag, 22
Olduvai, 21
Oligocene Epoch, 5
Olivola, 13, 16
Oltet, 11
Onager (*Equus onager* Pallas), 152
Origination, rate of, 260
Oryctolagus sp., *see* Rabbit
Otter, 10
Bravard's (*Aonyx bravardi* Pomel), 105
Corsican (*Aonyx antiqua* Blainville), 106–7
mandible, *43*
Hundsheim (*Lutra simplicidens* Thenius), 107
Lutra lutra Linné, 107, 268
teeth, *43*
Ovibos moschatus Linné, *see* Ox, Musk
Ovis sp., 180 (*see also* Mouflon)
Owls, and cave deposits, 31
Ox,
Giant Musk (*Praeovibos priscus* Staudinger), 178
Musk (*Ovibus moschatus* Linné), 178, 253
figurine, *79*
Perrier (*Leptobos elatus* Pomel), 183–4
Etruscan (*Leptobos etruscus* Falconer), 184, 188, 189
skull, *81*

Palaearctic faunal province, 265
Palaeoloxodon sp., 135 (*see also* Elephant)
Paleocene Epoch, 5
Panda,
English (*Parailurus anglicus* Boyd Dawkins; *Ailurus fulgens* F. Cuvier), 118
Giant (*Ailuropoda melanoleuca* Milne-Edwards), 118
Pannonictis sp., *see* Polecat
Panther, Owen's (*Felix pardoides* Owen), 85, 88

Parailurus anglicus Boyd Dawkins, *see* Panda, English
Parapodemus coronensis Schaub, *see* Mouse, Schaub's Field
Pardines, 16
Patholops hodgsoni Abel, *see* Antelope, Chiru
Pearsonian coefficient of variation, skeletal measurements of cave lion, Table 2
Pebble-tool industry, 60
Pellets, 16
Periglacial zone, 7
Periods, division of Eras into, 5
Perissodactyla, 139–52
  evolution rate, 264
  stratigraphic range of species, 280–1
Permian Ice Age, 4
Perrier (Mt), 11
  steppe fauna, 10
Petauria helleri Dehm, 197
Petenyia hungarica Kormos, 46
  mandibles, *17*
Pig, 153–4 (*see also* Hog)
  Strozzi's (*Sus strozzi* Meneghini), 154–5
    skull and mandible, *64*
Pika,
  Sardinian (*Prolagus sardus* Wagner), 226–7
  Steppe (*Ochotona pusilla* Pallas), 227, 254
Pipistrellus sp., 56 (*see also* Bat)
Pithecanthropus erectus Dubois, 60
Pitymys sp., 219 (*see also* Vole)
Plecotus sp., 54–5 (*see also* Bat)
Pleistocene,
  climatic change, 4
  caves, 16
  date, Table 1
Pleistocene, Late, 29
Pleistocene, Later Middle, 28
  faunal succession in, 28
Pleistocene, Middle, 27
Pleistocene Epoch, 5
Pliolagus sp., *see* Rabbit
Pliomys episcopolis Mehely, *see* Vole, Episcopal
Pliosciuropterus schaubi Sulimski, 197
Pluvials, in Villafranchian, 10
Podlesice, 17
Polecat,
  Baranya (*Baranogale antiqua* Pomel), 95–6
    skull and mandible, *39*
  Beremend (*Vormela beremendensis* Pétenyi), 96–7
  Marbled (*Vormela peregusna* Gueldenstaedt), 97
  Ardé (*Enhydrictis ardea* Bravard), 97–8
  Pannonian (*Pannonictis pliocaenica* Kormos), 98
  Stromer's (*Mustela stromeri* Kormos), 98–9
    mandible, *40*
  Mustela putorius Linné, 99–100
    mandible, *40*
  Steppe (*Mustela eversmanni* Lesson), 100–1
    skull and mandible, *40*
Pollen analysis, 6, 13, 23
Population trends, 249–52
Populations, evolving, diagrammatic representation, *100*
Porcupine, 199
  geographic distribution, *88*
  Crested (*Hystrix cristata* Linné), 200–1
Post-glacial Epoch, *see* Holocene Epoch
Postglacial, date, Table 1
Potassium-argon dating method, 20, 21
Praemegaceros sp., 164 (*see also* Deer)
Praeovibos priscus Staudinger, *see* Ox, Giant Musk
Primates, 58–61
  rate of evolution, 264
  stratigraphic range of species, 277
Procamptoceras brivatense Schaub, *see* Antelope, Chamois
Procyonidae, 118
Prolagus sp., 226 (*see also* Pika)
Prospalax priscus Nehring, *see* Rat, Primitive Mole
Protactinium/ionium dating method, 21
Pteromys volans Linné, *see* Squirrel, Flying
Putorius sp., *see* Polecat

Quaternary Period, 5
Quaternary, species as unit of correlation in, 237

Rabbit,
  Beremend (*Hypolagus brachygnathus* Kormos; *Pliolagus beremendensis* Kormos; *Pliolagus tothi* Kretzoi), 82, 228
  Arno (*Oryctolagus lacosti* Pomel; *Lepus etruscus* Bosco; *Lepus valdarnensis* Weithofer), 82, 228
  Arno (*Oryctolagus cuniculus* Linné), 228–9
Raccoon-Dog, Great (*Nyctereutes megamastoides* Pomel), 9, 117
Radiation curve of Milankovitch, 20
Radioactive dating, 19–20
Radiocarbon dating, 20
Ramapithecus, 59
Rangifer tarandus Linné, *see* Reindeer
'Raised beach', 7
Rat, Primitive Mole (*Prospalax priscus* Nehring), 204–5
    teeth, *89E*
    geographic distribution, *90*
  Lesser Mole (*Spalax leucodon* Nardmann), 205
  Black (*Rattus rattus* Linné), 225
  Brown (*Rattus norvegicus* Berkenhout), 225

Ratte, 12
Rebielice, 11
Red Crags, *see* Newer Red Crag, Older Red Crag
Regression, 5, 7
Reindeer (*Rangifer tarandus* Linné), 25, 33, 170–1, 253
  engraving, *73*
Reptiles, giant, 3
Rhagamys orthodon Hensel, *see* Mouse Hensel's Field
Rhinoceros, 9
  Christol's (*Dicerorhinus megarhinus* Christol), 140
  Etruscan (*Dicerorhinus etruscus* Falconer), 141–2
    restoration, *58*
  Merck's (*Dicerorhinus kirchbergensis* Jäger), 29, 31, 142–3
  Steppe (*Dicerorhinus hemitoechus* Falconer), 29, 33, 143
  Woolly (*Coelodonta antiquitatis* Blumenbach), 30, 33, 143–4, 254
    restoration, *59*
    geographical distribution, *60*
    skull and mandible, *61*
Rhinocerotidae, 140–4
Rhinocricetus ehiki Schaub, *see* Hamster, Ehik's Dwarf
Rhinolophidae (Horseshoe Bats), 51–2
Rhinolophus sp., 51, 52 (*see also* Bat)
Rodentia, 191–225
  stratigraphic range of species, 283–5
Rhone Valley, mammal-bearing Villafranchian loess, *3*
Riss, date, 20
Riss fauna, Tornewton cave, 33
Riss glaciation, 18, 28
Riss interstadial, fossiliferous sites, 31
Roccaneyra, 16, 20 n. 1
Roussillon, 9
Rupicapra rupicapra Linné, *see* Chamois

Saale glacial terraces, 31
Saale glaciation, 19
Sable, *Martes zibellina* Linné, 94
Sabre-tooth, 251, 252
Saiga tatarica Linné, *see* Antelope, Saiga
Saint-Cosme, Villafranchian sands and clays at, 12
Saint-Prest, 24
Saint-Vallier, 10, 11, 13, 16
Sainzelles, 24, 25
Sambar, *see* Deer, Philis
Savannas, in early Pliocene, 4
Schernfeld fissure, 11, 25
Sciuridae, 192–7
Sciurus sp., 192 (*see also* Squirrel)
Sea level, 3, 7
Sediments, geological, as climatic indicators, 6
Selsey beach, 30
Senèze, 11, 13, 16, 24
Serotine, *see* Bat, Serotine
Sheep, Giant (*Megalovis latifrons* Schaub), 179–80
Shore line, interglacial, 7
'Short-wave' climatic oscillation, 4–5
Shrew, Common (*Sorex araneus* Linné), 43
    anterior teeth, *16*
  Masked (*Sorex caecutiens* Laxmann), 43
  Pygmy (*Sorex minutus* Linné), 43, 254
  Kennard's (*Sorex kennardi* Hinton), 44–5
  Runton (*Sorex runtonensis* Hinton), 44–5
  Pearl-toothed (*Sorex margaritodon* Kormos), 45
  Water (*Neomys fodiens*), 45
  Beremend (*Beremendia fissidens* Petényi), 45–6
  Etruscan (*Suncus etruscus* Savi), 46
  Bicolour White-toothed (*Crocidura leucodon* Hermann), 46
  Common White-toothed (*Crocidura russula* Hermann), 47
  Tyrrhenian, skull and mandible, *18*
  Lesser White-toothed (*Crocidura suaveolens* Pallas), 48
Sicista sp., 202 (*see also* Mouse)
Size fluctuation, Middle Pleistocene species, 27, 9
Size decrease and species extinction, 274
Skeletal measurements, Pearsonian coefficient of variation, Table 2
Skull, *18*, *19*, *21*, *23*, *25*, *27*, *32*, *33*, *38*, *39*, *40*, *41*, *42*, *46*, *49*, *61*, *64*, *66*, *67*, *74*, *75*, *77*, *78*, *81*, *87*, *97A*, *C*, *98*, *99*
Skulls,
  Swanscombe, 29
  Kanjera, 61
  Niah Great Cave, 61
Smilodon neogaeus Lund, 74
Smilodontini, 73
Soergelia elisabethae Schaub, *see* Goat, Soergel's
Solifluction deposits, 7
Sorex sp., 43–6, *16*, *17* (*see also* Shrew)
Soricidae, 43–8
Soriculus kubinyi Kormos, 46
Souslik,
  Kormos's (*Citellus primigenius* Kormos), 196
  European (*Citellus citellus* Linné), 196
  Spotted (*Citellus suslicus* Gueldenstaedt), 196
  Red-cheeked (*Citellus major* Pallas; *Citellus rufescens* Keyserling & Blasius), 196–7
Spalacidae, 203–5

Spalax sp., 205 (*see also* Rat)
Species,
 nomenclature, 39–41
 unit for correlation in Quaternary, 237
 definition, 237
 division, 238, 240
  and tooth length, 240, *101*
 permissible variation, 240, 242
 evolutionary changes within, 244
 competing, 250–1
 origination of Arctic, 253–4
 age of living, 254–5, 257, *109*
 rate of increase of, 255, *109*
 evolution of new, 257–9
 geographic distribution, Table 9, 267
 geographic origin of European mammal, Table 10, 267
 stratigraphic range, 275–85
Squirrel,
 White's (*Sciurus whitei* Hinton), 192
 Red (*Sciurus vulgaris* Linné), 192–3
  teeth, *85A–C*
 Flying (*Pteromys volans* Linné), 197
Stadials, 18
Steinheim, 29
 quantitative faunal distribution, *11*
Steppe fauna, 10, 15, 24, 26
 Villafranchian, *11*
Steppe-living mammals,
 relative numbers in Villafranchian, *6*
Steppes, in early Pliocene, 4
Stoat,
 Primitive (*Mustela palerminea* Petényi), 27, 101, *9*
  mandible, *41*
 Mustela erminea Linné, 101–2
  skull and mandible, 101–2, 268, *41*
Stránská Skála cave,
 faunal sequence at, 25
Stratigraphic range of species, 275–85
Subspecies, nomenclature, 40
Suidae, 153–5
Suncus sp., 46 (*see also* Shrew)
Sus sp., 154, 155 (*see also* Pig)
Süssenborn, 25, 27
Swalecliff, 30
Swanscombe, 29, *10*
Syncerus iselini Stehlin, *see* Buffalo, Stehlin's
Systema naturae, 40

Tadarida teniotis Rafinesque, *see* Bat, European Free-tailed
Taganrog, 11
Tahr,
 European (*Hemitragus bonali* Harle & Stehlin), 183
  cave painting, *80*
Talpa sp., 49, 50 (*see also* Mole)
Talpidae, 48–50
Tapir, 9
 Auvergne (*Tapirus arvernensis* Croizet & Jobert), 139
Tapiridae, 139–40
Tapirus arvernensis Croizet & Jobert, *see* Tapir, Auvergne
Tarpan, 150
Taubach, 31
Teeth, *16*, *20A–F*, *36*, *38*, *43*, *44*, *54A–D*, *62A–F*, *85A–C*, *89A–E*, *92C–D*, *96A–F*
Tegelen, 11, 15, 24
Tegelen interglacial, 15, 19
 date, Table 1
Temperature changes, in geological time, 3
Temperature oscillation and size, 245 (*see also* Bergmann's Rule)
Terraces, 29
Terrace relationship, in Lower Thames Valley, *10*
Tertiary Period, 3, 5
Tethys Sea, 4
Teufelslucken, 36, 69
Thames Valley, sequence of terrace relationship, *10*
Therailurus, *76*
Tibia,
 Hunting Hyena, *24*
 Spotted Hyena, *24*
Tiglian interglacial, *see* Tegelen interglacial
Tills, 6–7
Tischofer Cave, relative numbers of different species, Table 3
Tools, 270
Tooth length, as basis of species division, 240, *101*
Tornewton cave, 31, 33, 69, *12*
Torre, 21
Trafalgar Square, 29
Transgression, 5, 7
Travertines, 31
Trogontherium cuvieri Fischer, *see* Beaver, European Giant
Turnover rate, 260–1
Tyrrhenicola henseli Forsyth Major, *see* Vole, Tyrrhenian

'Unicorn', Giant (*Elasmotheriun sibiricum* Fischer), 144–5
 skull and mandible, *61*
Upnor, 30
Upper Pleistocene stage, *see* Late Pleistocene stage
Uranium-lead dating, 20
Ursidae, 118–29
Ursus sp., 121, 129 (*see also* Bear)

Val d'Arno basin, *4*
Valverde de Calatrava, 11
Variation,
  within populations, 240, 242
  coefficient of, 242
Vernacular names, 40
Vertesszöllös, 60
Vespertilio sp., 57 (*see also* Bat)
Vespertilionidae, 52–7
Vialette, 9, 10, 11, 16
Vicarious forms, 242
Villafranca d'Asti, 9, 10, 11, 16
Villafranchian (*see also* Early Pleistocene)
  climate, 10
  deposits at Leffe, *5*
  fauna, 5, 6, 10, 16, 17, 24
    geographic distribution, *1*
  flora, 10
  mammal-bearing loess, *3*
  pluvials in, 10
    variations in numbers of mammal types, 6
  marine deposits, 10
  caves, 16
  alternation of forest and steppe faunas, 15
  date, Table 1
Villány, 11, 16, 24
Villanyian faunal stage, 16
Villaroya, 10, 11, 16
Vincelles, 12
Viverridae, 62
Vogelherd Cave, 271
  distribution of game in, Table 12, 271
Volcanic eruption, Hajnácka, 17
Volcanoes, Senèze, 13
Vole (*see also* Arvicola, Mimomys),
  Nehring's Snow (*Dolomys milleri* Nehring), 212
  Martino's Snow (*Dolomys bogdanovi* Martino), 212–13
  Episcopal (*Pliomys episcopalis* Méhely), 213
  Bank (*Clethrionomys glareolus* Schreber), 213, 254
  Mimomys, 214–15, 268
  Water (*Arvicola terrestris* Linné; *Arvicola amphibius* Linné, 24, 215–16
    mandible, *95B*
    teeth, *96B*
  Common (*Microtus arvalis* Pallas), 217
  Field (*Microtus agrestis* Linné), 217–18
    mandible, *95C*
    teeth, *96C*
  Mediterranean (*Microtus guentheri* Danford & Alston), 218
  Root or Tundra (*Microtus ratticeps* Keyserling & Blasius; *Microtus oeconomus* Pallas), 218, 253
  Snow (*Microtus nivalis* Martins), 218–19
  Gregarious (*Microtus gregalis* Pallas; *Microtus anglicus* Hinton), 219
    teeth, *96D*
  Gregarious Pine (*Pitymys gregaloides* Hinton), 219
  Pine (*Pitymys subterraneus* De Selys Longchamps), 219–22
  Tyrrhenian (*Tyrrhenicola henseli* Forsyth Major), 221
'Vole spectra', 249, *107*
Vormela sp., *see* Polecat
Vulpes sp., 115, 116 (*see also* Fox, Corsac)

Waalian fauna, 24
Waalian interglacial, 19, 25
  date, Table 1
'Warm' deposits, 7
Warsaw, 29
Warthe-Weichsel glaciation, 19
Weasel,
  Primitive (*Mustela praenivalis* Kormos), 102
    mandible, *41*
  Primitive (*Mustela nivalis* Linné), 103
  Least (*Mustela rixosa* Bangs), 103
Weybourne Crag, 22
Weze fissure, faunas, 17
Wiezchowska Cave, 36, 86
Windloch, 25
Wisent, 268
  Steppe (*Bisen priscus* Bojanus), 185–6
    skull, *81*
  restoration, *82*
  Woodland (*Bison schoetensacki* Freudenberg), 186
  Woodland (*Bison bonasus* Linné), 186–7
Wolf,
  Canis lupus Linné, 36, 109, 251, 268
    teeth, *44*
    engraving, *45*
    skull and mandible, *46*
  Falconer's Dire (*Canis falconeri* Major), 109
  Etruscan (*Canis etruscus* Major), 109
Wolf Jackal, 251
Wolverine or Glutton (*Gulo gulo* Linné), 92–3
Woodland mammals, relative numbers in Villafranchian, *6*
Würm,
  date, Table 1
  fauna, 30, 33
  glaciation, 18
  Terrace, 30

Zapodidae, 201–2
Zebra, 148–9 (*see also* Horse, Zebrine)
Zygolophodon borsoni Hays, *see* Mastodont, Borson's